PLASTICS ENGINEERING

PLASTICS ENGINEERING

Third Edition

R.J. Crawford, BSc, PhD, DSc, FEng, FIMechE, FIM

*Department of Mechanical, Aeronautical
and Manufacturing Engineering
The Queen's University of Belfast*

John Oliver Giaquinto

ELSEVIER
BUTTERWORTH
HEINEMANN

AMSTERDAM • BOSTON • HEIDELBERG • LONDON • NEW YORK • OXFORD
PARIS • SAN DIEGO • SAN FRANCISCO • SINGAPORE • SYDNEY • TOKYO

Elsevier Butterworth-Heinemann
Linacre House, Jordan Hill, Oxford OX2 8DP
200 Wheeler Road, Burlington, MA 01803

First published 1981
Second edition 1987
Reprinted with corrections 1990, 1992
Third edition 1998
Reprinted 1999, 2001, 2002, 2004

British Library Cataloguing in Publication Data
Crawford, R.J. (Roy J.)
  Plastics engineering. 3rd ed.
  1. Plastics
  I. Title
  668.4

ISBN 0 7506 3764 1

Library of Congress Cataloguing in Publication Data
Crawford, R.J.
  Plastics engineering/R.J. Crawford – 3rd ed.
    p.   cm.
  Includes index.
  ISBN 0 7506 3764 1 (pbk).
  1 Plastics.   I Title
  TP1120 C74   97–36604
  668.4 – dc21   CIP

For information on all Butterworth-Heinemann publications
visit our website at www.bh.com

Typeset by Laser Words, Chennai, India
Printed and bound by Antony Rowe Ltd, Eastbourne

Contents

Preface to the Third Edition

Plastics continue to be exciting materials to use and a dynamic area in which to work. Every year new application areas are being developed to utilise more fully the unique properties of this class of materials. In addition, new processing technologies are emerging to exploit the versatility of plastics and to take advantage of their ease of manufacture into all types of end products. It is very important that students and those already working in the industry are kept fully informed about these new developments. In this new edition an attempt has been made to bring existing subject material up to date and many new sections have been added to cover the innovations introduced over the past decade. The number of Worked Examples has been increased and there are many more Set Questions at the end of each Chapter. As in the previous editions, a full set of solutions to the Set Questions is provided at the end of the book.

In this new edition, some re-structuring of the content has taken place. The subject material on Fracture that previously formed Chapter 3 has been brought forward to Chapter 2. This chapter now provides a more unified approach to the deformation and fracture behaviour of non-reinforced plastics. Chapter 3 is new and deals with all aspects of the mechanical behaviour of composites in much more detail than the previous editions. Composites are an extremely important class of material for modern design engineers and they must form an integral part of undergraduate and postgraduate teaching. There are many excellent textbooks devoted to this subject but it was felt that an introduction to the analysis of laminates would be a valuable addition to this text. It is hoped that the many worked examples in this new chapter will help the student, and the practising engineer, to gain a better understanding of this apparently complex subject area. Chapters 4 and 5 are essentially as before but they have been extensively updated. A more unified approach to the analysis of processing has also been adopted.

As other authors will know, the preparation of a textbook is a demanding, challenging and time-consuming occupation. I have been very fortunate to receive many encouraging comments on the previous editions and this has given me the enthusiasm to continue developing the subject material in the book. I am very grateful to all of those who have taken the trouble to contact me in the past and I continue to welcome comments and advice as to how the book could be improved in the future.

R.J. Crawford
September 1997

Preface to the Second Edition

In this book no prior knowledge of plastics is assumed. The text introduces the reader to plastics as engineering materials and leads on to the design procedures which are currently in use. Since the publication of the first edition the subject has developed in some areas, particularly processing and so this second edition contains the new and up-to-date information. Other modifications have also been made to improve the presentation of the contents. In particular, Chapter 1 has been completely re-written as an introduction to the general behaviour characteristics of plastics. The introduction to the structure of plastics which formed the basis of Chapter 1 in the first edition has been condensed into an Appendix in the new edition. Chapter 2 deals with the deformation behaviour of plastics. It has been expanded from the first edition to include additional analysis on intermittent loading and fibre composites. Chapter 3 deals with the fracture behaviour of plastics and here the importance of fracture mechanics has been given greater emphasis.

Chapter 4 describes in general terms the processing methods which can be used for plastics. All the recent developments in this area have been included and wherever possible the quantitative aspects are stressed. In most cases a simple Newtonian model of each of the processes is developed so that the approach taken to the analysis of plastics processing is not concealed by mathematical complexity.

Chapter 5 deals with the aspects of the flow behaviour of polymer melts which are relevant to the processing methods. The models are developed for both Newtonian and Non-Newtonian (Power Law) fluids so that the results can be directly compared.

Many more worked examples have been included in this second edition and there are additional problems at the end of each chapter. These are seen as an important aspect of the book because in solving these the reader is encouraged to develop the subject beyond the level covered in the text. To assist the reader a full set of solutions to the problems is provided at the back of the book.

R.J. Crawford
January 1987

Preface to the First Edition

This book presents in a single volume the basic essentials of the properties and processing behaviour of plastics. The approach taken and terminology used has been deliberately chosen to conform with the conventional engineering approach to the properties and behaviour of materials. It was considered that a book on the engineering aspects of plastics was necessary because there is currently a drive to attract engineers into the plastics industry and although engineers and designers are turning with more confidence to plastics there is still an underlying fear that plastics are difficult materials to work with. Their performance characteristics fall off as temperature increases and they are brittle at low temperatures. Their mechanical properties are time dependent and in the molten state they are non-Newtonian fluids. All this presents a gloomy picture and unfortunately most texts tend to analyse plastics using a level of chemistry and mathematical complexity which is beyond most engineers and designers. The purpose of this text is to remove some of the fears, by dealing with plastics in much the same way as traditional materials. The major part of this is to illustrate how quantitative design of plastic components can be carried out using simple techniques and how apparently complex moulding operations can be analysed without difficulty.

Many of the techniques illustrated have been deliberately simplified and so they will only give approximate solutions but generally the degree of accuracy can be estimated and for most practical purposes it will probably be acceptable. Once the engineer/designer has realised that there are proven design procedures for plastics which are not beyond their capabilities then these materials will be more readily accepted for consideration alongside established materials such as woods and metals. On these terms plastics can expect to be used in many new applications because their potential is limited only by the ingenuity of the user.

This book is intended primarily for students in the various fields of engineering but it is felt that students in other disciplines will welcome and benefit from the engineering approach. Since the book has been written as a general introduction to the quantitative aspects of the properties and processing of plastics, the depth of coverage is not as great as may be found in other texts on the physics, chemistry and stress analysis of viscoelastic materials. this has been done deliberately because it is felt that once the material described here has been studied and understood the reader will be in a better position to decide if he requires the more detailed viscoelastic analysis provided by the advanced texts.

In this book no prior knowledge of plastics is assumed. Chapter 1 provides a brief introduction to the structure of plastics and it provides an insight to the way in which their unique structure affects their performance. There is a resume of the main types of plastics which are available. Chapter 2 deals with the mechanical properties of unreinforced and reinforced plastics under the general heading of deformation. The time dependent behaviour of the materials is introduced and simple design procedures are illustrated. Chapter 3 continues the discussion on properties but concentrates on fracture as caused by creep, fatigue and impact. The concepts of fracture mechanics are also introduced for reinforced and unreinforced plastics.

Chapter 4 describes in general terms the processing methods which can be used for plastics and wherever possible the quantitative aspects are stressed. In most cases a simple Newtonian model of each of the processes is developed so that the approach taken to the analysis of plastics processing is not concealed by mathematical complexity. Chapter 5 deals with the aspects of the flow behaviour of polymer melts which are relevant to the processing methods. The models are developed for both Newtonian and Non-Newtonian (Power Law) fluids so that the results can be directly compared.

Throughout the book there are worked examples to illustrate the use of the theory and at the end of each chapter there are problems to be solved by the reader. These are seen as an important part of the book because in solving the problems the reader is encouraged to develop the subject material beyond the level covered in the text. Answers are given for all the questions.

R.J. Crawford

CHAPTER 1 – General Properties of Plastics

1.1 Introduction

It would be difficult to imagine our modern world without plastics. Today they are an integral part of everyone's lifestyle with applications varying from commonplace domestic articles to sophisticated scientific and medical instruments. Nowadays designers and engineers readily turn to plastics because they offer combinations of properties not available in any other materials. Plastics offer advantages such as lightness, resilience, resistance to corrosion, colour fastness, transparency, ease of processing, etc., and although they have their limitations, their exploitation is limited only by the ingenuity of the designer.

The term *plastic* refers to a family of materials which includes nylon, polyethylene and PTFE just as zinc, aluminium and steel fall within the family of *metals*. This is an important point because just as it is accepted that zinc has quite different properties from steel, similarly nylon has quite different properties from PTFE. Few designers would simply specify *metal* as the material for a particular component so it would be equally unsatisfactory just to recommend *plastic*. This analogy can be taken still further because in the same way that there are different grades of steel there are also different grades of, say, polypropylene. In both cases the good designer will recognise this and select the most appropriate material and grade on the basis of processability, toughness, chemical resistance, etc.

It is usual to think that plastics are a relatively recent development but in fact, as part of the larger family called *polymers*, they are a basic ingredient of animal and plant life. Polymers are different from metals in the sense that their structure consists of very long chain-like molecules. Natural materials such as silk, shellac, bitumen, rubber and cellulose have this type of structure. However, it was not until the 19th century that attempts were made to develop a synthetic

1

polymeric material and the first success was based on cellulose. This was a material called *Parkesine*, after its inventor Alexander Parkes, and although it was not a commercial success it was a start and it led to the development of *Celluloid*. This material was an important breakthrough because it became established as a good replacement for natural materials which were in short supply – for example, ivory for billiard balls.

During the early 20th century there was considerable interest in these new synthetic materials. Phenol-formaldehyde (*Bakelite*) was introduced in 1909, and at about the time of the Second World War materials such as nylon, polyethylene and acrylic (*Perspex*) appeared on the scene. Unfortunately many of the early applications for plastics earned them a reputation as being cheap substitutes. It has taken them a long time to overcome this image but nowadays the special properties of plastics are being appreciated, which is establishing them as important materials in their own right. The ever increasing use of plastics in all kinds of applications means that it is essential for designers and engineers to become familiar with the range of plastics available and the types of performance characteristics to be expected so that these can be used to the best advantage.

This chapter is written as a general introduction to design with plastics. It outlines the range of plastics available, describes the type of behaviour which they exhibit and illustrates the design process involved in selecting the best plastic for a particular application.

1.2 Polymeric Materials

Synthetic large molecules are made by joining together thousands of small molecular units known as **monomers**. The process of joining the molecules is called **polymerisation** and the number of these units in the long molecule is known as the **degree of polymerisation**. The names of many polymers consist of the name of the monomer with the suffix **poly-**. For example, the polymers polypropylene and polystryene are produced from propylene and styrene respectively. Names, and symbols for common polymers are given in Appendix F.

It is an unfortunate fact that many students and indeed design engineers are reluctant to get involved with plastics because they have an image of complicated materials with structures described by complex chemical formulae. In fact it is not necessary to have a detailed knowledge of the structure of plastics in order to make good use of them. Perfectly acceptable designs are achieved provided one is familiar with their performance characteristics in relation to the proposed service conditions. An awareness of the structure of plastics can assist in understanding why they exhibit a time-dependent response to an applied force, why acrylic is transparent and stiff whereas polyethylene is opaque and flexible, etc., but it is not necessary for one to be an expert in polymer chemistry in order to use plastics. Those who wish to to have a general introduction to the structure of plastics may refer to Appendix A. ←

The words **polymers** and **plastics** are often taken as synonymous but in fact there is a distinction. The polymer is the pure material which results from the process of polymerisation and is usually taken as the family name for materials which have long chain-like molecules (and this includes rubbers). Pure polymers are seldom used on their own and it is when additives are present that the term plastic is applied. Polymers contain additives for a number of reasons. The following list outlines the purpose of the main additives used in plastics.

Antistatic Agents. Most polymers, because they are poor conductors of current, build up a charge of static electricity. Antistatic agents attract moisture from the air to the plastic surface, improving its surface conductivity and reducing the likelihood of a spark or a discharge.

Coupling Agents. Coupling agents are added to improve the bonding of the plastic to inorganic filler materials, such as glass fibres. A variety of silanes and titanates are used for this purpose.

Fillers. Some fillers, such as short fibres or flakes of inorganic materials, improve the mechanical properties of a plastic. Others, called *extenders*, permit a large volume of a plastic to be produced with relatively little actual resin. Calcium carbonate, silica and clay are frequently used extenders.

Flame Retardants. Most polymers, because they are organic materials, are flammable. Additives that contain chlorine, bromine, phosphorous or metallic salts reduce the likelihood that combustion will occur or spread.

Lubricants. Lubricants such as wax or calcium stearate reduce the viscosity of the molten plastic and improve forming characteristics.

Pigments. Pigments are used to produce colours in plastics.

Plasticisers. Plasticisers are low molecular weight materials which alter the properties and forming characteristics of the plastic. An important example is the production of flexible grades of polyvinyl chloride by the use of plasticisers.

Reinforcement. The strength and stiffness of polymers are improved by adding fibres of glass, carbon, etc.

Stabilisers. Stabilisers prevent deterioration of the polymer due to environmental factors. Antioxidants are added to ABS, polyethylene and polystyrene. Heat stabilisers are required in processing polyvinyl chloride. Stabilisers also prevent deterioration due to ultra-violet radiation.

There are two important classes of plastics.

(a) Thermoplastic Materials

In a thermoplastic material the very long chain-like molecules are held together by relatively weak Van der Waals forces. A useful image of the structure is a mass of randomly distributed long strands of sticky wool. When the material is heated the intermolecular forces are weakened so that it becomes soft and flexible and eventually, at high temperatures, it is a viscous melt.

When the material is allowed to cool it solidifies again. This cycle of softening by heat and solidifying on cooling can be repeated more or less indefinitely and is a major advantage in that it is the basis of most processing methods for these materials. It does have its drawbacks, however, because it means that the properties of thermoplastics are heat sensitive. A useful analogy which is often used to describe these materials is that, like candle wax, they can be repeatedly softened by heat and will solidify when cooled.

Examples of thermoplastics are polyethylene, polyvinyl chloride, polystyrene, nylon, cellulose acetate, acetal, polycarbonate, polymethyl methacrylate and polypropylene.

An important subdivision within the thermoplastic group of materials is related to whether they have a **crystalline** (ordered) or an **amorphous** (random) structure. In practice, of course, it is not possible for a moulded plastic to have a completely crystalline structure due to the complex physical nature of the molecular chains (see Appendix A). Some plastics, such as polyethylene and nylon, can achieve a high degree of crystallinity but they are probably more accurately described as *partially crystalline* or *semi-crystalline*. Other plastics such as acrylic and polystyrene are always amorphous. The presence of crystallinity in those plastics capable of crystallising is very dependent on their thermal history and hence on the processing conditions used to produce the moulded article. In turn, the mechanical properties of the moulding are very sensitive to whether or not the plastic possesses crystallinity.

In general, plastics have a higher density when they crystallise due to the closer packing of the molecules. Typical characteristics of crystalline and amorphous plastics are shown below.

Amorphous	Crystalline
• *Broad softening range* – thermal agitation of the molecules breaks down the weak secondary bonds. The rate at which this occurs throughout the formless structure varies producing broad temperature range for softening.	• *Sharp melting point* – the regular close-packed structure results in most of the secondary bonds being broken down at the same time.
• *Usually transparent* – the looser structure transmits light so the material appears transparent.	• *Usually opaque* – the difference in refractive indices between the two phases (amorphous and crystalline) causes interference so the material appears translucent or opaque.
• *Low shrinkage* – all thermoplastics are processed in the amorphous state. On solidification, the random	• *High shrinkage* – as the material solidifies from the

Amorphous *(continued)*

arrangement of molecules produces little volume change and hence low shrinkage.
- *Low chemical resistance* – the more open random structure enables chemicals to penetrate deep into the material and to destroy many of the secondary bonds.
- *Poor fatigue and wear resistance* – the random structure contributes little to fatigue or wear properties.

Crystalline *(continued)*

amorphous state the polymers take up a closely packed, highly aligned structure. This produces a significant volume change manifested as high shrinkage.
- *High chemical resistance* – the tightly packed structure prevents chemical attack deep within the material.
- *Good fatigue and wear resistance* – the uniform structure is responsible for good fatigue and wear properties.

Examples of amorphous and crystalline thermoplastics

Amorphous

Polyvinyl Chloride (PVC)
Polystyrene (PS)
Polycarbonate (PC)
Acrylic (PMMA)
Acrylonitrile-butadiene-styrene (ABS)
Polyphenylene (PPO)

Crystalline

– Polyethylene (PE)
Polypropylene (PP)
– Polyamide (PA)
Acetal (POM)
– Polyester (PETP, PBTP)
Fluorocarbons (PTFE, PFA, FEP and ETFE)

(b) Thermosetting Plastics

A thermosetting plastic is produced by a chemical reaction which has two stages. The first stage results in the formation of long chain-like molecules similar to those present in thermoplastics, but still capable of further reaction. The second stage of the reaction (**cross-linking** of chains) takes place during moulding, usually under the application of heat and pressure. The resultant moulding will be rigid when cooled but a close network structure has been set up within the material. During the second stage the long molecular chains have been interlinked by strong bonds so that the material cannot be softened again by the application of heat. If excess heat is applied to these materials they will char and degrade. This type of behaviour is analogous to boiling an egg. Once the egg has cooled and is hard, it cannot be softened again by the application of heat.

Since the cross-linking of molecules is by strong chemical bonds, thermosetting materials are characteristically quite rigid materials and their mechanical properties are not heat sensitive. Examples of thermosets are phenol formaldehyde, melamine formaldehyde, urea formaldehyde, epoxies and some polyesters.

1.3 Plastics Available to the Designer

Plastics, more than any other design material, offer such a wide spectrum of properties that they must be given serious consideration in most component designs. However, this does not mean that there is sure to be a plastic with the correct combination of properties for every application. It simply means that the designer must have an awareness of the properties of the range of plastics available and keep an open mind. One of the most common faults in design is to be guided by pre-conceived notions. For example, an initial commitment to plastics based on an irrational approach is itself a serious design fault. A good design always involves a judicious selection of a material from the whole range available, including non-plastics. Generally, in fact, it is only against a background of what other materials have to offer that the full advantages of plastics can be realised.

In the following sections most of the common plastics will be described briefly to give an idea of their range of properties and applications. However, before going on to this it is worthwhile considering briefly several of the special categories into which plastics are divided.

1.3.1 Engineering Plastics

Many thermoplastics are now accepted as engineering materials and some are distinguished by the loose description **engineering plastics**. The term probably originated as a classification distinguishing those that could be substituted satisfactorily for metals such as aluminium in small devices and structures from those with inadequate mechanical properties. This demarcation is clearly artificial because the properties on which it is based are very sensitive to the ambient temperature, so that a thermoplastic might be a satisfactory substitute for a metal at a particular temperature and an unsatisfactory substitute at a different one.

A useful definition of an engineering material is that it is able to support loads more or less indefinitely. By such a criterion thermoplastics are at a disadvantage compared with metals because they have low time-dependent moduli and inferior strengths except in rather special circumstances. However, these rather important disadvantages are off-set by advantages such as low density, resistance to many of the liquids that corrode metals and above all, easy processability. Thus, where plastics compete successfully with other materials in engineering applications it is usually because of a favourable balance of properties rather than because of an outstanding superiority in some particular respect, although the relative ease with which they can be formed into complex shapes tends to be a particularly dominant factor. In addition to conferring the possibility of low production costs, this ease of processing permits imaginative designs that often enable plastics to be used as a superior alternative to metals rather than merely as a tolerated substitute.

Currently the materials generally regarded as making up the **engineering plastics** group are Nylon, acetal, polycarbonate, modified polyphenylene oxide (PPO), thermoplastic polyesters, polysulphone and polyphenylene sulphide. The newer grades of polypropylene also possess good basic *engineering* performance and this would add a further 0.5 m tonnes. And then there is unplasticised polyvinyl chloride (uPVC) which is widely used in industrial pipework and even polyethylene, when used as an artificial hip joint for example, can come into the reckoning. Hence it is probably unwise to exclude any plastic from consideration as an engineering material even though there is a sub-group specifically entitled for this area of application.

In recent years a whole new generation of high performance engineering plastics have become commercially available. These offer properties far superior to anything available so far, particularly in regard to high temperature performance, and they open the door to completely new types of application for plastics.

The main classes of these new materials are

(i) Polyarylethers and Polyarylthioethers

polyarylethersulphones (PES)
polyphenylene sulphide (PPS)
polyethernitrile (PEN)
polyetherketones (PEK and PEEK)

(ii) Polyimides and Polybenzimidazole

polyetherimide (PEI)
thermoplastic polyimide (PI)
polyamideimide (PAI)

(iii) Fluoropolymers

fluorinated ethylene propylene (FEP)
perfluoroalkoxy (PFA)

A number of these materials offer service temperatures in excess of 200°C and fibre-filled grades can be used above 300°C.

1.3.2 Thermosets

In recent years there has been some concern in the thermosetting material industry that usage of these materials is on the decline. Certainly the total market for thermoset compounds has decreased in Western Europe. This has happened for a number of reasons. One is the image that thermosets tend to have as old-fashioned materials with outdated, slow production methods. Other reasons include the arrival of high temperature engineering plastics and miniaturisation in the electronics industry. However, thermosets are now fighting back and have a very much improved image as colourful, easy-flow moulding materials with a superb range of properties.

Phenolic moulding materials, together with the subsequently developed easy-flowing, granular thermosetting materials based on urea, melamine, unsaturated polyester (UP) and epoxide resins, today provide the backbone of numerous technical applications on account of their non-melting, high thermal and chemical resistance, stiffness, surface hardness, dimensional stability and low flammability. In many cases, the combination of properties offered by thermosets cannot be matched by competing engineering thermoplastics such as polyamides, polycarbonates, PPO, PET, PBT or acetal, nor by the considerably more expensive products such as polysulphone, polyethersulphone and PEEK.

1.3.3 Composites

One of the key factors which make plastics attractive for engineering applications is the possibility of property enhancement through fibre reinforcement. Composites produced in this way have enabled plastics to become acceptable in, for example, the demanding aerospace and automobile industries. Currently in the USA these industries utilise over 100,000 tonnes of reinforced plastics out of a total consumption of over one million tonnes.

Both thermoplastics and thermosets can reap the benefit of fibre reinforcement although they have developed in separate market sectors. This situation has arisen due to fundamental differences in the nature of the two classes of materials, both in terms of properties and processing characteristics.

Thermosetting systems, hampered on the one hand by brittleness of the crosslinked matrix, have turned to the use of long, indeed often continuous, fibre reinforcement but have on the other hand been able to use the low viscosity state at impregnation to promote maximum utilization of fibre properties. Such materials have found wide application in large area, relatively low productivity, moulding. On the other hand, the thermoplastic approach with the advantage of toughness, but unable to grasp the benefit of increased fibre length, has concentrated on the short fibre, high productivity moulding industry. It is now apparent that these two approaches are seeking routes to move into each other's territory. On the one hand the traditionally long-fibre based thermoset products are accepting a reduction in properties through reduced fibre length, in order to move into high productivity injection moulding, while thermoplastics, seeking even further advances in properties, by increasing fibre length, have moved into long-fibre injection moulding compounds and finally into truly structural plastics with continuous, aligned fibre thermoplastic composites such as the advanced polymer composite (APC) developed by ICI and the stampable glass mat reinforced thermoplastics (GMT) developed in the USA.

Glass fibres are the principal form of reinforcement used for plastics because they offer a good combination of strength, stiffness and price. Improved strengths and stiffnesses can be achieved with other fibres such as aramid (**Kevlar**) or carbon fibres but these are expensive. The latest developments also include the use of hybrid systems to get a good balance of properties at an acceptable price. For example, the impact properties of carbon-fibre composites can be improved by the addition of glass fibres and the stiffness of gfrp can be increased by the addition of carbon fibres.

Another recent development is the availability of reinforced plastics in a form very convenient for moulding. One example is polyester dough and sheet moulding compounds (DMC and SMC respectively). DMC, as the name suggests, has a dough-like consistency and consists of short glass fibres (15–20%) and fillers (up to 40%) in a polyester resin. The specific gravity is in the range 1.7–2.1. SMC consists of a polyester resin impregnated with glass fibres (20–30%). It is supplied as a sheet would into a roll with a protective polythene film on each side of the sheet. The specific gravity is similar to that of DMC and both materials are usually formed using heat and pressure in a closed mould (see Chapter 4 on moulding of gfrp).

1.3.4 Structural Foam

The concept of structural foams offers an unusual but exciting opportunity for designers. Many plastics can be foamed by the introduction of a blowing agent so that when moulded the material structure consists of a cellular rigid foam core with a solid tough skin. This type of structure is of course very efficient in material terms and offers an excellent strength-to-weight ratio.

The foam effect is achieved by the dispersion of inert gas throughout the molten resin directly before moulding. Introduction of the gas is usually carried out by pre-blending the resin with a chemical blowing agent which releases gas when heated, or by direct injection of the gas (usually nitrogen). When the compressed gas/resin mixture is rapidly injected into the mould cavity, the gas expands explosively and forces the material into all parts of the mould. An internal cellular structure is thus formed within a solid skin.

Polycarbonate, polypropylene and modified PPO are popular materials for structural foam moulding. One of the main application areas is housings for business equipment and domestic appliances because the number of component parts can be kept to the absolute minimum due to integral moulding of wall panels, support brackets, etc. Other components include vehicle body panels and furniture.

Structural foam mouldings may also include fibres to enhance further the mechanical properties of the material. Typical performance data for foamed polypropylene relative to other materials is given in Table 1.1.

1.3.5 Elastomers

Conventional rubbers are members of the polymer family in that they consist of long chain-like molecules. These chains are coiled and twisted in a random

Table 1.1

Comparison of structural foams based on various grades of polypropylene with some traditional materials

	Unfilled copolymer		40% talc-filled homopolymer		30% coupled glass-reinforced		Chip-board	Pine	Alumi-nium	Mild steel
	Solid	Foam	Solid	Foam	Solid	Foam				
Flexural modulus MN/m^2	1.4	1.2	4.4	2.5	6.7	3.5	2.3	7.9	70	207
Specific gravity	0.905	0.72	1.24	1.00	1.12	0.90	0.650	0.641	2.7	7.83
Relative thickness at equivalent rigidity	1	1.05	0.68	0.81	0.59	0.74	0.85	0.56	0.27	0.19
Relative weight at equivalent rigidity	1	0.84	0.94	0.90	0.74	0.73	0.61	0.40	0.81	1.65

manner and have sufficient flexibility to allow the material to undergo very large deformations. In the *green* state the rubber would not be able to recover fully from large deformations because the molecules would have undergone irreversible sliding past one another. In order to prevent this sliding, the molecules are anchored together by a curing (**vulcanisation**) process. Thus the molecules are cross-linked in a way similar to that which occurs in thermosets. This linking does not detract from the random disposition of the molecules nor their coiled and twisted nature so that when the rubber is deformed the molecules stretch and unwind but do not slide. Thus when the applied force is removed the rubber will snap back to its original shape.

Vulcanised rubbers possess a range of very desirable properties such as resilience, resistance to oils, greases and ozone, flexibility at low temperatures and resistance to many acids and bases. However, they require careful (slow) processing and they consume considerable amounts of energy to facilitate moulding and vulcanisation. These disadvantages led to the development of **thermoplastic rubbers (elastomers)**. These are materials which exhibit the desirable physical characteristics of rubber but with the ease of processing of thermoplastics.

At present there are five types of thermoplastic rubber (TPR). Three of these, the polyurethane, the styrenic and the polyester are termed segmented block copolymers in that they consist of thermoplastic molecules grafted to the rubbery molecules. At room temperature it is the thermoplastic molecules which clump together to anchor the rubbery molecules. When heat is applied the thermoplastic molecules are capable of movement so that the material may be shaped using conventional thermoplastic moulding equipment.

The olefinic type of TPR is the latest development and is different in that it consists of fine rubber particles in a thermoplastic matrix as shown in Fig. 1.1.

Fig. 1.1 Typical structure of olefinic TPR

The matrix is usually polypropylene and it is this which melts during processing to permit shaping of the material. The rubber filler particles then contribute the flexibility and resilience to the material. The other type of TPR is the polyamide and the properties of all five types are summarised in Table 1.4.

1.3.6 Polymer Alloys

The development of new polymer alloys has caused a lot of excitement in recent years but in fact the concept has been around for a long time. Indeed one of the major commercial successes of today, ABS, is in fact an alloy of acrylonitrile, butadiene and styrene. The principle of alloying plastics is similar to that of alloying metals – to achieve in one material the advantages possessed by several others. The recent increased interest and activity in the field of polymer alloys has occurred as a result of several new factors. One is the development of more sophisticated techniques for combining plastics which were previously considered to be incompatible. Another is the keen competition for a share of new market areas such as automobile bumpers, body panels etc. These applications call for combinations of properties not previously available in a single plastic and it has been found that it is less expensive to combine existing plastics than to develop a new monomer on which to base the new plastic.

In designing an alloy, polymer chemists choose candidate resins according to the properties, cost, and/or processing characteristics required in the end product. Next, compatibility of the constituents is studied, tested, and either optimised or accommodated.

Certain polymers have come to be considered standard building blocks of the polyblends. For example, impact strength may be improved by using polycarbonate, ABS and polyurethanes. Heat resistance is improved by using polyphenylene oxide, polysulphone, PVC, polyester (PET and PBT) and acrylic. Barrier properties are improved by using plastics such as ethylene vinyl alchol (EVA). Some modern plastic alloys and their main characteristics are given in Table 1.2.

Table 1.2

Typical plastic alloys

Alloy	Features
PVC/acrylic	Tough with good flame and chemical resistance
PVC/ABS	Easily processed with good impact and flame resistance
Polycarbonate/ABS	Hard with high heat distortion temperature and good notch impact strength
ABS/Polysulphone	Less expensive than unmodified polysulphone
Polyphenylene oxide/HIPS	Improved processability, reduced cost
SAN/olefin	Good weatherability
Nylon/elastomer	Improved notched impact strength
Modified amorphous nylon	Easily processed with excellent surface finish and toughness
Polycarbonate/PBT	Tough engineering plastic

1.3.7 Liquid Crystal Polymers

Liquid crystal polymers (LCP) are a recent arrival on the plastics materials scene. They have outstanding dimensional stability, high strength, stiffness, toughness and chemical resistance all combined with ease of processing. LCPs are based on thermoplastic aromatic polyesters and they have a highly ordered structure even in the molten state. When these materials are subjected to stress the molecular chains slide over one another but the ordered structure is retained. It is the retention of the highly crystalline structure which imparts the exceptional properties to LCPs.

Typical Characteristics of Some Important Plastics

(a) Semi-crystalline plastics

Low density polyethylene (LDPE). This is one of the most widely used plastics. It is characterised by a density in the range 918–935 kg/m^3 and is very tough and flexible. Its major application is in packaging film although its outstanding dielectric properties means it is also widely used as an electrical insulator. Other applications include domestic ware, tubing, squeeze bottles and cold water tanks.

Linear Low Density Polyethylene (LLDPE). This new type of polyethylene was introduced in 1977. LLDPE is produced by a low pressure process and it has a regular structure with short chain branches. Depending on the cooling rate from the melt, the material forms a structure in which the molecules are linked together. Hence for any given density, LLDPE is stiffer than LDPE and exhibits a higher yield strength and greater ductility. Although the difference melt processing characteristics of LLDPE take a little getting used to, this new material has taken over traditional LDPE markets.

High Density Polyethylene (HDPE). This material has a density in the range 935–965 kg/m^3 and is more crystalline than LDPE. It is also slightly more expensive but as it is much stronger and stiffer it finds numerous applications in such things as dustbins, bottle crates, general purpose fluid containers and pipes.

One of the most exciting recent developments in this sector has been the introduction to the marketplace of metallocene-based polyethylenes. Metallocenes have been recognised as suitable catalysts for the manufacture of polyethylenes since the 1950s. However, it is only recently that their use has been perfected. Their big advantage is that they are single site catalysts so that the polymer molecules which are produced tend to be all the same – a fact which offers an array of superior properties. Traditional catalysts for polyethylene (Ziegler Natta catalysts) are multi-sited so that they produce polymers with short, medium and long molecules. In the new metallocene grades of polyethylene, the absence of low molecular weight species results in low extractables, a narrow melting range and free-flowing material even at low densities. The absence of high molecular weight species contributes excellent melting point control, clarity and improved flexibility/toughness at low temperatures.

Metallocene-based polyethylene does not offer the lower production costs associated with LLDPE. Hence there will be a price premium for the new materials but this is felt to be justified in view of their improved property profile.

Cross-linked Polyethylene (XLPE). Some thermoplastic materials such as polyethylene can have their structure altered so that the molecular chains become cross-linked and the material then behaves like a thermoset. In the case of polyethylene, a range of cross-linking methods are available. These include the use of radiation, peroxides and silanes. In some cases the cross-linking can occur during moulding whereas in other cases the end-product shape is created before the cross-linking is initiated. The action of cross-linking has a number of beneficial effects including improved stress crack resistance, improved creep resistance, better chemical resistance, improved toughness and better general thermo-mechanical stability.

Polypropylene (PP). Polypropylene is an extremely versatile plastic and is available in many grades and also as a copolymer (ethylene/propylene). It has the lowest density of all thermoplastics (in the order of 900 kg/m^3 and this combined with strength, stiffness and excellent fatigue and chemical resistance make it attractive in many situations. These include crates, small machine parts, car components (fans, fascia panels etc), chair shells, cabinets for TV, tool handles, etc. Its excellent fatigue resistance is utilised in the moulding of integral hinges (e.g. accelerator pedals and forceps/tweezers). Polypropylene is also available in fibre form (for ropes, carpet backing) and as a film (for packaging).

Polyamides (nylon). There are several different types of nylon (e.g. nylon 6, nylon 66, nylon 11) but as a family their characteristics of strength, stiffness and toughness have earned them a reputation as *engineering plastics*. Table 1.3 compares the relative merits of light metal alloys and nylon.

Table 1.3

Comparison between die casting alloys and nylons

Points for comparison	Die casting alloys	Nylon
Cost of raw material/tonne	Low	High
Cost of mould	High	Can be lower – no higher
Speed of component production	Slower than injection moulding of nylon	Lower component production costs
Accuracy of component	Good	Good
Post moulding operations	Finishing – painting. Paint chips off easily	Finishing – not required – painting not required. Compounded colour retention permanent.
Surface hardness	Low – scratches easily	Much higher. Scratch resistant.
Rigidity	Good to brittleness	Glass reinforced grades as good or better
Elongation	Low	GR grades comparable unfilled grades excellent
Toughness (flexibility)	Low	GR grades comparable unfilled grades excellent
Impact	Low	All grades good
Notch sensitivity	Low	Low
Youngs modulus (E)	Consistent	Varies with load
General mechanical properties	Similar to GR grades of 66 nylon	Higher compressive strength
Heat conductivity	High	Low
Electrical insulation	Low	High
Weight	High	Low
Component assembly	Snap fits difficult	Very good

Typical applications for nylon include small gears, bearings, bushes, sprockets, housings for power tools, terminal blocks and slide rollers. An important design consideration is that nylon absorbs moisture which can affect its properties and dimensional stability. Glass reinforcement reduces this problem and produces an extremely strong, impact resistant material. Another major application of nylon is in fibres which are notoriously strong. The density of nylon is about 1100 kg/m^3.

Acetals. The superior properties of acetal in terms of its strength, stiffness and toughness have also earned it a place as an engineering plastic. It is more dense than nylon but in many respects their properties are similar and they can be used for the same types of light engineering application. A factor which may favour acetal in some cases is its relatively low water absorption. The material is available as both a homopolymer and a copolymer. The former is slightly stronger and stiffer whereas the copolymer has improved high temperature performance. This latter feature makes this material very attractive for hot water plumbing applications and as the body for electric kettles.

Polytetrafluoroethylene (PTFE). The major advantages of this material are its excellent chemical resistance and its extremely low coefficient of friction. Not surprisingly its major area of application is in bearings particularly if the environment is aggressive. It is also widely used in areas such as insulating tapes, gaskets, pumps, diaphragms and of course non-stick coatings on cooking utensils.

Thermoplastic Polyesters. These linear polyesters are highly crystalline and exhibit toughness, strength, abrasion resistance, low friction, chemical resistance and low moisture absorption. Polyethylene terephthalate (PET) has been available for many years but mainly as a fibre (e.g. Terylene). As a moulding material it was less attractive due to processing difficulties but these were overcome with the introduction of polybutylene terephthalate (PBT). Applications include gears, bearings, housings, impellers, pulleys, switch parts, bumper extensions, etc. and of course PET is now renowned for its success as a replacement for glass in beverage bottles. PBT does not have such a high performance specification as PET but it is more readily moulded.

Polyetheretherketone. This material, which is more commonly known as PEEK, is one of the new generation plastics which offer the possibility of high service temperatures. It is crystalline in nature which accounts in part for its high resistance to attack from acids, alkalis and organic solvents. It is easily processed and may be used continuously at 200°C where it offers good abrasion resistance, low flammability, toughness, strength and good fatigue resistance. Its density is 1300 kg/m^3. Applications include wire coatings, electrical connections, fans, impellers, fibres, etc.

(b) Amorphous plastics

Polyvinyl Chloride (PVC). This material is the most widely used of the amorphous plastics. It is available in two forms – plasticised or unplasticised. Both types are characterised by good weathering resistance, excellent electrical insulation properties, good surface properties and they are self extinguishing. Plasticised PVC is flexible and finds applications in wire covering, floor tiles, toy balls, gloves and rainwear. Unplasticised PVC (uPVC) is hard, tough, strong material which is widely used in the building industry. For example, pipes, gutters, window frames and wall claddings are all made in this material. The familiar credit cards are also made from uPVC.

Polymethyl Methacrylate (PMMA). This material has exceptional optical clarity and resistance to outdoor exposure. It is resistant to alkalis, detergents, oils and dilute acids but is attacked by most solvents. Its peculiar property of total internal reflection is useful in advertising signs and some medical applications.

Typical uses include illuminated notices, control panels, dome-lights, lighting diffusers, baths, face guards, nameplates, lenses and display models.

Polystyrene (PS). Polystyrene is available in a range of grades which generally vary in impact strength from brittle to very tough. The non-pigmented grades have crystal clarity and overall their low cost coupled with ease of processing makes them used for such things as model aircraft kits, vending cups, yoghurt containers, light fittings, coils, relays, disposable syringes and casings for ballpoint pens. Polystyrene is also available in an expanded form which is used for such things as ceiling tiles and is excellent as a packaging material and thermal insulator.

Acrylonitrile-butadiene-styrene (ABS). ABS materials have superior strength, stiffness and toughness properties to many plastics and so they are often considered in the category of engineering plastics. They compare favourably with nylon and acetal in many applications and are generally less expensive. However, they are susceptible to chemical attack by chlorinated solvents, esters, ketones, acids and alkalis.

Typical applications are housings for TV sets, telephones, fascia panels, hair brush handles, luggage, helmets and linings for refrigerators.

Polycarbonates. These materials also come within the category of engineering plastics and their outstanding feature is extreme toughness. They are transparent and have good temperature resistance but are attacked by alkaline solutions and hydrocarbon solvents. Typical applications include vandal-proof street lamp covers, baby feeding bottles, machine housings and guards, camera parts, electrical components, safety equipment and compact discs.

Polyethersulphone. This material is one of the new high temperature plastics. It is recommended for load bearing applications up to 180°C. Even without flame retardants it offers low flammability and there is little change in dimensions of electrical properties in the temperature range 0–200°C. It is easily processed on conventional moulding equipment. Applications include aircraft heating ducts, terminal blocks, engine manifolds, bearings, grilles, tool handles, non-stick coatings.

Modified Polyphenylene Oxide (PPO). The word *modified* in this material refers to the inclusion of high impact polystyrene to improve processability and reduce the cost of the basic PPO. This material offers a range of properties which make it attractive for a whole range of applications. For example, it may be used at 100–150°C where it is rigid, tough and strong with good creep resistance and hydrolytic stability. Water absorption is very small and there is excellent dimensional stability. Applications include business machine parts, flow values, headlight parts, engine manifolds, fascia panels, grilles, pump casings, hair dryer housings, etc.

(c) Thermoplastic Rubbers

There are five types of thermoplastic rubbers currently available. These are based on (i) Olefinics (e.g. *Alcryn, Santoprene*) (ii) Polyurethanes (e.g. *Elastollan, Caprolan, Pellethane*) (iii) Polyesters (e.g. *Hytrel, Arnitel*) (iv) Styrenics (e.g. *Solprene, Cariflex*) and (v) Polyamides (e.g. *Pebax, Dinyl*) Some typical properties are given in Table 1.4.

Table 1.4

Physical characteristics of thermoplastic rubbers

Type	Olefinic	Polyurethane	Polyester	Styrenic	Polyamide
Hardness (Shore A–D)	60A to 60D	60A to 60D	40D to 72D	30A to 45D	40D to 63D
Resilience (%)	30 to 40	40 to 50	43 to 62	60 to 70	–
Tensile strength					
$\quad$ (MN/m^2)	8 to 20	30 to 55	21 to 45	25 to 45	–
Resistances					
$\quad$ Chemicals	F	P/G	E	E	P/E
$\quad$ Oils	F	E	E	F	–
$\quad$ Solvents	P/F	F	G	P	P/E
$\quad$ Weathering	E	G	E	P/E	E
Specific gravity	0.97–1.34	1.11–1.21	1.17–1.25	0.93–1.0	1.0–1.12
Service temperature (°C)	−50–130	−40–130	−65–130	−30–120	−65–130

KEY: P = poor, F = fair, G = good, E = excellent

(d) Thermosetting Plastics

Aminos. There are two basic types of amino plastics – urea formaldehyde and melamine formaldehyde. They are hard, rigid materials with good abrasion resistance and their mechanical characteristics are sufficiently good for continuous use at moderate temperatures (up to 100°C). Urea formaldehyde is relatively inexpensive but moisture absorption can result in poor dimensional stability. It is generally used for bottle caps, electrical switches, plugs, utensil handles and trays. Melamine formaldehyde has lower water absorption and improved temperature and chemical resistance. It is typically used for tableware, laminated worktops and electrical fittings.

Phenolics. Phenol-formaldehyde *(Bakelite)* is one of the oldest synthetic materials available. It is a strong, hard, brittle material with good creep resistance and excellent electrical properties. Unfortunately the material is only available in dark colours and it is susceptible to attack by alkalis and oxidising agents. Typical applications are domestic electrical fittings, saucepan handles, fan blades, smoothing iron handles and pump parts.

Polyurethanes. This material is available in three forms – rigid foam, flexible foam and elastomer. They are characterised by high strength and good chemical and abrasion resistance. The rigid foam is widely used as an insulation material, the flexible foam is an excellent cushion material for furniture and the elastomeric material is used in solid tyres and shock absorbers.

Polyesters. The main application of this material is as a matrix for glass fibre reinforcement. This can take many forms and is probably most commonly known as a DIY type material used for the manufacture of small boats, chemical containers, tanks and repair kits for cars, etc.

Epoxides. Epoxy resins are more expensive than other equivalent thermosets (e.g. polyesters) but they an generally out-perform these materials due to better toughness, less shrinkage during curing, better weatherability and lower moisture absorption. A major area of application is in the aircraft industry because of the combination of properties offered when they are reinforced with fibres. They have an operating temperature range of −25 to 150°C.

1.4 Selection of Plastics

The previous section has given an indication of the range of plastics available to the design engineer. The important question then arises *How do we decide which plastic, if any, is best for a particular application?* Material selection is not as difficult as it might appear but it does require an awareness of the general behaviour of plastics as a group, as well as a familiarity with the special characteristics of individual plastics.

The first and most important steps in the design process are to define clearly the purpose and function of the proposed product and to identify the service environment. Then one has to assess the suitability of a range of candidate materials. The following are generally regarded as the most important characteristics requiring consideration for most engineering components.

(1) mechanical properties – strength, stiffness, specific strength and stiffness, fatigue and toughness, and the influence of high or low temperatures on these properties;

(2) corrosion susceptibility and degradation

(3) wear resistance and frictional properties;

(4) special properties, for example, thermal, electrical, optical and magnetic properties, damping capacity, etc;

(5) moulding and/or other methods of fabrication.

(6) total costs attributable to the selected material and manufacturing route.

In the following sections these factors will be considered briefly in relation to plastics.

1.4.1 Mechanical Properties

Strength and Stiffness. Thermoplastic materials are viscoelastic which means that their mechanical properties reflect the characteristics of both viscous liquids and elastic solids. Thus when a thermoplastic is stressed it responds by exhibiting viscous flow (which dissipates energy) and by elastic displacement (which stores energy). The properties of viscoelastic materials are time, temperature and strain rate dependent. Nevertheless the conventional stress-strain test is frequently used to describe the (short-term) mechanical properties of plastics. It must be remembered, however, that as described in detail in Chapter 2 the information obtained from such tests may only be used for an initial sorting of materials. It is not suitable, or intended, to provide design data which must usually be obtained from long term tests.

In many respects the stress–strain graph for a plastic is similar to that for a metal (see Fig. 1.2).

At low strains there is an elastic region whereas at high strains there is a non-linear relationship between stress and strain and there is a permanent element to the strain. In the absence of any specific information for a particular plastic, design strains should normally be limited to 1%. Lower values ($\simeq$0.5%) are recommended for the more brittle thermoplastics such as acrylic, polystyrene and values of 0.2–0.3% should be used for thermosets.

The effect of material temperature is illustrated in Fig. 1.3. As temperature is increased the material becomes more flexible and so for a given stress the

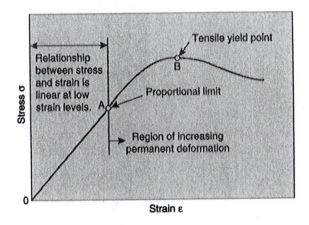

Fig. 1.2   Typical stress–strain graph for plastics

Fig. 1.3   Effect of material temperature on stress–strain behaviour of plastics

material deforms more. Another important aspect to the behaviour of plastics is the effect of strain rate. If a thermoplastic is subjected to a rapid change in strain it appears stiffer than if the same maximum strain were applied but at a slower rate. This is illustrated in Fig. 1.4.

Fig. 1.4   Effect of strain rate on stress–strain behaviour of plastics

It is important to realise also that within the range of grades that exist for a particular plastic, there can be significant differences in mechanical properties. For example, with polypropylene for each 1 kg/m^3 change in density there is a corresponding 4% change in modulus. Fig. 1.5 illustrates the typical variation which occurs for the different grades of ABS. It may be seen that very often a grade of material selected for some specific desirable feature (e.g. high impact strength) results in a decrease in some other property of the material (e.g. tensile strength).

The stiffness of a plastic is expressed in terms of a modulus of elasticity. Most values of elastic modulus quoted in technical literature represent the slope of a tangent to the stress–strain curve at the origin (see Fig. 1.6). This is often referred to as Youngs modulus, E, but it should be remembered that for a plastic this will not be a constant and, as mentioned earlier, is only useful for quality

Fig. 1.5   Effect of grade on mechanical properties of ABS

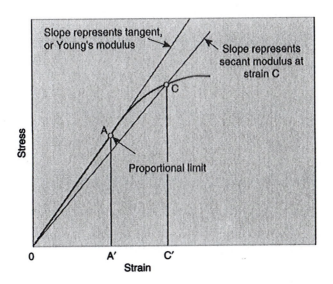

Fig. 1.6   Tangent and secant modulus

control purposes, not for design. Since the tangent modulus at the origin is sometimes difficult to determine precisely, a secant modulus is often quoted to remove any ambiguity. A selected strain value of, say 2% (point C', Fig. 1.6) enables a precise point, C, on the stress–strain curve to be identified. The slope of a line through C and O is the secant modulus. Typical short-term mechanical properties of plastics are given in Table 1.5. These are given for illustration purposes. For each type of plastic there are many different grades and a wide variety of properties are possible. The literature supplied by the manufacturers should be consulted in specific instances.

Table 1.5

Short-term properties of some important plastics

Material	Density (kg/m^3)	Tensile strength (MN/m^2)	Flexural modulus (GN/m^2)	% elongation at break	Price*
ABS (high impact)	1040	38	2.2	8	2.1
Acetal (homopolymer)	1420	68	2.8	40	3.5
Acetal (copolymer)	1410	70	2.6	65	3.3
Acrylic	1180	70	2.9	2	2.5
Cellulose acetate	1280	30	1.7	30	3.2
CAB	1190	25	1.3	60	–
Epoxy	1200	70	3.0	3	8.3
Modified PPO	1060	45	2.3	70	–
Nylon 66	1140	70	2.8	60	3.9
Nylon 66 (33% glass)	1380	115	5.1	4	4.0
PEEK	1300	62	3.8	4	42
PEEK (30% carbon)	1400	240	14	1.6	44
PET	1360	75	3	70	3.0
PET (36% glass)	1630	180	12	3	3.5
Phenolic (mineral filled)	1690	55	8.0	0.8	1.25
Polyamide-imide	1400	185	4.5	12	67
Polycarbonate	1150	65	2.8	100	4.2
Polyetherimide	1270	105	3.3	60	–
Polyethersulphone	1370	84	2.6	60	13.3
Polyimide	1420	72	2.5	8	150
Polypropylene	905	33	1.5	150	1
Polysulphone	1240	70	2.6	80	11
Polystyrene	1050	40	3.0	1.5	1.1
Polythene (LD)	920	10	0.2	400	0.83
Polythene (HD)	950	32	1.2	150	1.1
PTFE	2100	25	0.5	200	13.3
PVC (rigid)	1400	50	3.0	80	0.88
PVC (flexible)	1300	14	0.007	300	0.92
SAN	1080	72	3.6	2	1.8
DMC (polyester)	1800	40	9.0	2	1.5
SMC (polyester)	1800	70	11.0	3	1.3

*On a weight basis, relative to polypropylene.

Material Selection for Strength

If, in service, a material is required to have a certain strength in order to perform its function satisfactorily then a useful way to compare the structural efficiency of a range of materials is to calculate their strength desirability factor.

Consider a structural member which is essentially a beam subjected to bending (Fig. 1.7). Irrespective of the precise nature of the beam loading the

Fig. 1.7 Beam subjected to bending

maximum stress, σ, in the beam will be given by

$$\sigma = \frac{M_{max}(d/2)}{I} = \frac{M_{max}(d/2)}{bd^2/12} \tag{1.1}$$

Assuming that we are comparing different materials on the basis that the mean length, width and loading is fixed but the beam depth is variable then equation (1.1) may be written as

$$\sigma = \beta_1/d^2 \tag{1.2}$$

where β_1 is a constant.

But the weight, w, of the beam is given by

$$w = \rho b d L \tag{1.3}$$

So substituting for d from (1.2) into (1.3)

$$w = \beta_2 \rho/\sigma^{1/2} \tag{1.4}$$

where β_2 is the same constant for all materials.

Hence, if we adopt loading/weight as a desirability factor, D_f, then this will be given by

$$D_f = \frac{\sigma_y^{1/2}}{\rho} \tag{1.5}$$

where σ_y and ρ are the strength and density values for the materials being compared.

Similar desirability factors may be derived for other geometries such as struts, columns etc. This concept is taken further later where material costs are taken into account and Tables 1.11 and 1.12 give desirability factors for a range of loading configurations and materials.

Material Selection for Stiffness

If in the service of a component it is the deflection, or stiffness, which is the limiting factor rather than strength, then it is necessary to look for a different desirability factor in the candidate materials. Consider the beam situation described above. This time, irrespective of the loading, the deflection, δ, will be given by

$$\delta = \alpha_1 \left(\frac{WL^3}{EI} \right) \tag{1.6}$$

where α_1 is a constant and W represents the loading.

The stiffness may then be expressed as

$$\frac{W}{\delta} = \left(\frac{1}{\alpha_1} \right) \frac{EI}{L^3}$$

$$\frac{W}{\delta} = \alpha_2 (Ed^3) \tag{1.7}$$

where α_2 is a constant and again it is assumed that the beam width and length are the same in all cases.

Once again the beam weight will be given by equation (1.3) so substituting for d from equation (1.7)

$$w = \alpha_3 \rho / E^{1/3} \tag{1.8}$$

Hence, the desirability factor, D_f, expressed as maximum stiffness for minimum weight will be given by

$$D_f = \frac{E^{1/3}}{\rho} \tag{1.9}$$

where E is the elastic modulus of the material in question and ρ is the density. As before a range of similar factors can be derived for other structural elements and these are illustrated in Section 1.4.6. (Tables 1.11 and 1.12) where the effect of material cost is also taken into account. Note also that since for plastics the modulus, E, is not a constant it is often necessary to use a long-term (creep) modulus value in equation (1.9) rather than the short-term quality control value usually quoted in trade literature.

Ductility. A load-bearing device or component must not distort so much under the action of the service stresses that its function is impaired, nor must it fail by rupture, though local yielding may be tolerable. Therefore, high modulus and high strength, with ductility, is the desired combination of attributes. However, the inherent nature of plastics is such that high modulus tends to be associated with low ductility and steps that are taken to improve the one cause the other to deteriorate. The major effects are summarised in Table 1.6. Thus it may be seen that there is an almost inescapable rule by which increased modulus is accompanied by decreased ductility and vice versa.

Creep and Recovery Behaviour. Plastics exhibit a time-dependent strain response to a constant applied stress. This behaviour is called creep. In a similar fashion if the stress on a plastic is removed it exhibits a time dependent recovery of strain back towards its original dimensions. This is illustrated in

Table 1.6

Balance between stiffness and ductility in thermoplastics

	Effect on	
	Modulus	Ductility
Reduced temperature	increase	decrease
Increased straining rate	increase	decrease
Multiaxial stress field	increase	decrease
Incorporation of plasticizer	decrease	increase
Incorporation of rubbery phase	decrease	increase
Incorporation of glass fibres	increase	decrease
Incorporation of particulate filler	increase	decrease

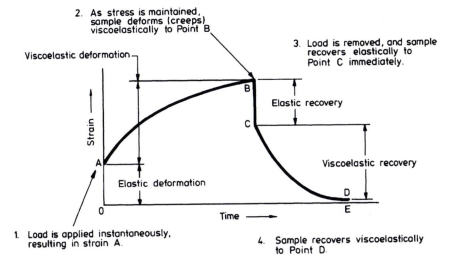

Fig. 1.8   Typical Creep and recovery behaviour of a plastic

Fig. 1.8 and because of the importance of these phenomena in design they are dealt with in detail in Chapter 2.

Stress Relaxation. Another important consequence of the viscoelastic nature of plastics is that if they are subjected to a particular strain and this strain is held constant it is found that as time progresses, the stress necessary to maintain this strain decreases. This is termed stress relaxation and is of vital importance in the design of gaskets, seals, springs and snap-fit assemblies. This subject will also be considered in greater detail in the next chapter.

Creep Rupture. When a plastic is subjected to a constant tensile stress its strain increases until a point is reached where the material fractures. This is called creep rupture or, occasionally, static fatigue. It is important for designers to be aware of this failure mode because it is a common error, amongst those accustomed to dealing with metals, to assume that if the material is capable of withstanding the applied (static) load in the short term then there need be no further worries about it. This is not the case with plastics where it is necessary to use long-term design data, particularly because some plastics which are tough at short times tend to become embrittled at long times.

Fatigue. Plastics are susceptible to brittle crack growth fractures as a result of cyclic stresses, in much the same way as metals are. In addition, because of their high damping and low thermal conductivity, plastics are also prone to thermal softening if the cyclic stress or cyclic rate is high. The plastics with the best fatigue resistance are polypropylene, ethylene-propylene copolymer and PVDF. The fatigue failure of plastics is described in detail in Chapter 2.

Toughness. By toughness we mean the resistance to fracture. Some plastics are inherently very tough whereas others are inherently brittle. However, the picture is not that simple because those which are nominally tough may become embrittled due to processing conditions, chemical attack, prolonged exposure to constant stress, etc. Where toughness is required in a particular application it is very important therefore to check carefully the service conditions in relation to the above type of factors. At room temperature the toughest unreinforced plastics include nylon 66, LDPE, LLDPE, EVA and polyurethane structural foam. At sub-zero temperatures it is necessary to consider plastics such as ABS, polycarbonate and EVA. The whole subject of toughness will be considered more fully in Chapter 2.

1.4.2 Degradation

Physical or Chemical Attack. Although one of the major features which might prompt a designer to consider using plastics is corrosion resistance, nevertheless plastics are susceptible to chemical attack and degradation. As with metals, it is often difficult to predict the performance of a plastic in an unusual environment so it is essential to check material specifications and where possible carry out proving trials. Clearly, in the space available here it is not possible to give precise details on the suitability of every plastic in every possible environment. Therefore the following sections give an indication of the general causes of polymer degradation to alert the designer to a possible problem.

The degradation of a plastic occurs due to a breakdown of its chemical structure. It should be recognised that this breakdown is not necessarily caused by concentrated acids or solvents. It can occur due to apparently innocuous mediums such as water (**hydrolysis**), or oxygen (**oxidation**). Degradation of plastics is also caused by heat, stress and radiation. During moulding the material is subjected to the first two of these and so it is necessary to incorporate stabilisers and antioxidants into the plastic to maintain the properties of the material. These additives also help to delay subsequent degradation for an acceptably long time.

As regards the general behaviour of polymers, it is widely recognised that crystalline plastics offer better environmental resistance than amorphous plastics. This is as a direct result of the different structural morphology of these two classes of material (see Appendix A). Therefore engineering plastics which are also crystalline e.g. Nylon 66 are at an immediate advantage because they can offer an attractive combination of load-bearing capability and an inherent chemical resistance. In this respect the arrival of crystalline plastics such as PEEK and polyphenylene sulfide (PPS) has set new standards in environmental resistance, albeit at a price. At room temperature there is no known solvent for PPS, and PEEK is only attacked by 98% sulphuric acid.

Weathering. This generally occurs as a result of the combined effect of water absorption and exposure to ultra-violet radiation (u-v). Absorption of water can have a plasticizing action on plastics which increases flexibility but ultimately (on elimination of the water) results in embrittlement, while u-v causes breakdown of the bonds in the polymer chain. The result is general deterioration of physical properties. A loss of colour or clarity (or both) may also occur. Absorption of water reduces dimensional stability of moulded articles. Most thermoplastics, in particular cellulose derivatives, are affected, and also polyethylene, PVC, and nylons.

Oxidation. This is caused by contact with oxidising acids, exposure to u-v, prolonged application of excessive heat, or exposure to weathering. It results in a deterioration of mechanical properties (embrittlement and possibly stress cracking), increase in power factor, and loss of clarity. It affects most thermoplastics to varying degrees, in particular polyolefins, PVC, nylons, and cellulose derivatives.

Environmental Stress Cracking (ESC). In some plastics, brittle cracking occurs when the material is in contact with certain substances whilst under stress. The stress may be externally applied in which case one would be prompted to take precautions. However, internal or residual stresses introduced during processing are probably the more common cause of ESC. Most organic liquids promote ESC in plastics but in some cases the problem can be caused by a liquid which one would not regard as an aggressive chemical. The classic example of ESC is the brittle cracking of polyethylene washing-up bowls due to the residual stresses at the moulding gate (see injection moulding, Chapter 4) coupled with contact with the aqueous solution of washing-up liquid. Although direct attack on the chemical structure of the plastic is not involved in ESC the problem can be alleviated by controlling structural factors. For example, the resistance of polyethylene is very dependent on density, crystallinity, melt flow index (MFI) and molecular weight. As well as polyethylene, other plastics which are prone to ESC are ABS and polystyrene.

The mechanism of ESC is considered to be related to penetration of the promoting substance at surface defects which modifies the surface energy and promotes fracture.

1.4.3 Wear Resistance and Frictional Properties

There is a steady rate of increase in the use of plastics in bearing applications
and in situations where there is sliding contact e.g. gears, piston rings, seals,
cams, etc. The advantages of plastics are low rates of wear in the absence of
conventional lubricants, low coefficients of friction, the ability to absorb shock
and vibration and the ability to operate with low noise and power consumption.
Also when plastics have reinforcing fibres they offer high strength and load
carrying ability. Typical reinforcements include glass and carbon fibres and
fillers include PTFE and molybdenum disulphide in plastics such as nylon,
polyethersulphone (PES), polyphenylene sulfide (PPS), polyvinylidene fluoride
(PVDF) and polyetheretherketone (PEEK).

The friction and wear of plastics are extremely complex subjects which
depend markedly on the nature of the application and the properties of the
material. The frictional properties of plastics differ considerably from those of
metals. Even reinforced plastics have modulus values which are much lower
than metals. Hence metal/thermoplastic friction is characterised by adhesion
and deformation which results in frictional forces that are not proportional to
load but rather to speed. Table 1.7 gives some typical coefficients of friction
for plastics.

Table 1.7

Coefficients of friction and relative wear rates for plastics

Material	Coefficient of friction		Relative wear rate
	Static	Dynamic	
Nylon	0.2	0.28	33
Nylon/glass	0.24	0.31	13
Nylon/carbon	0.1	0.11	1
Polycarbonate	0.31	0.38	420
Polycarbonate/glass	0.18	0.20	5
Polybutylene terephthalate (PBT)	0.19	0.25	35
PBT/glass	0.11	0.12	2
Polyphenylene sulfide (PPS)	0.3	0.24	90
PPS/glass	0.15	0.17	19
PPS/carbon	0.16	0.15	13
Acetal	0.2	0.21	–
PTFE	0.04	0.05	–

The wear rate of plastics is governed by several mechanisms. The primary
one is adhesive wear which is characterised by fine particles of polymer being
removed from the surface. This is a small-scale effect and is a common
occurrence in bearings which are performing satisfactorily. However, the other
mechanism is more serious and occurs when the plastic becomes overheated
to the extent where large troughs of melted plastic are removed. Table 1.7

shows typical primary wear rates for different plastics, the mechanism of wear is complex the relative wear rates may change depending on specific circumstances.

In linear bearing applications the suitability of a plastic is usually determined from its PV rating. This is the product of P (the bearing load divided by the projected bearing area) and V (the linear shaft velocity). Fig. 1.9 shows the limiting PV lines for a range of plastics – combinations of P and V above the lines are not permitted. The PV ratings may be increased if the bearing is lubricated or the mode of operation is intermittent. The PV rating will be decreased if the operating temperature is increased. Correction factors for these variations may be obtained from material/bearing manufacturers. The plastics with the best resistance to wear are ultra high molecular weight polyethylene (used in hip joint replacements) and PTFE lubricated versions of nylon, acetal and PBT. It is not recommended to use the same plastic for both mating surfaces in applications such as gear wheels.

Fig. 1.9 Typical P–V ratings for plastics rubbing on steel

1.4.4 Special Properties

Thermal Properties. Before considering conventional thermal properties such as conductivity it is appropriate to consider briefly the effect of temperature on the mechanical properties of plastics. It was stated earlier that the properties of plastics are markedly temperature dependent. This is as a result of their molecular structure. Consider first an amorphous plastic in which the molecular chains have a random configuration. Inside the material, even though it is not possible to view them, we know that the molecules are in a state of continual motion. As the material is heated up the molecules receive more energy and there is an increase in their relative movement. This makes the material more flexible. Conversely if the material is cooled down then molecular mobility decreases and the material becomes stiffer.

With plastics there is a certain temperature, called the **glass transition temperature**, T_g, below which the material behaves like glass i.e. it is hard and rigid. As can be seen from Table 1.8 the value for T_g for a particular plastic is not necessarily a low temperature. This immediately helps to explain some of the differences which we observe in plastics. For example, at room temperature polystyrene and acrylic are below their respective T_g values and hence we observe these materials in their glassy state. Note, however, that in contrast, at room temperature, polyethylene is above its glass transition temperature and so we observe a very flexible material. When cooled below its T_g it then becomes a hard, brittle solid. Plastics can have several transitions.

The main T_g is called the glass–rubber transition and signifies a change from a flexible, tough material to a glassy state in which the material exhibits stiffness, low creep and toughness although with a sensitivity to notches. At lower temperatures there is then a secondary transition characterised by a change to a hard, rigid, brittle state.

It should be noted that although Table 1.8 gives specific values of T_g for different polymers, in reality the glass–transition temperature is not a material constant. As with many other properties of polymers it will depend on the testing conditions used to obtain it.

In the so-called crystalline plastics the structure consists of both crystalline (ordered) regions and amorphous (random) regions. When these materials are heated there is again increased molecular mobility but the materials remain relatively stiff due to the higher forces between the closely packed molecules. When the crystalline plastics have their temperature reduced they exhibit a glass transition temperature associated with the amorphous regions. At room temperature polypropylene, for example, is quite rigid and tough, not because it is below its T_g but because of the strong forces between the molecules in the crystalline regions. When it is cooled below $-10°C$ it becomes brittle because the amorphous regions go below their T_g.

In the past a major limitation to the use of plastics materials in the engineering sector has been temperature. This limitation arises not only due to the

Table 1.8

Typical Thermal properties of materials

Material	Density (kg/m^3)	Specific heat (kJ/kg K)	Thermal conductivity (W/m/K)	Coeff. of therm exp (μm/m/°C)	Thermal diffusivity (m^2/s) $\times 10^{-7}$	Glass transition Temp, T_g(°C)	Max. operating, Temp (°C)
ABS	1040	1.3	0.25	90	1.7	115	70
Acetal (homopolymer)	1420	1.5	0.2	80	0.7	−85	85
Acetal (copolymer)	1410	1.5	0.2	95	0.72	−85	90
Acrylic	1180	1.5	0.2	70	1.09	105	50
Cellulose acetate	1280	1.6	0.15	100	1.04	−	60
CAB	1190	1.6	0.14	100	1.27	−	60
Epoxy	1200	0.8	0.23	70	−	−	130
Modified PPO	1060	−	0.22	60	−	−	120
Nylon 66	1140	1.7	0.24	90	1.01	56	90
Nylon 66 (33% glass)	1380	1.6	0.52	30	1.33	−	100
PEEK	1300	−	−	48	−	143	204
PEEK (30% carbon)	1400	−	−	14	−	−	255
PET	1360	1.0	0.2	90	−	75	110
PET (36% glass)	1630	−	−	40	−	−	150
Phenolic (glass filled)	1700	−	0.5	18	−	−	185
Polyamide-imide	1400	−	0.25	36	−	260	210
Polycarbonate	1150	1.2	0.2	65	1.47	149	125
Polyester	1200	1.2	0.2	100	−	−	−
Polyetherimide	1270	−	0.22	56	−	200	170
Polyethersulphone	1370	−	1.18	55	−	230	180
Polyimide	1420	−	−	45	−	400	260
Polyphenylene sulfide	1340	−	−	49	−	85	150
Polypropylene	905	2.0	0.20	100	0.65	−10	100
Polysulphone	1240	1.3	−	56	−	180	170
Polystyrene	1050	1.3	0.15	80	0.6	100	50
Polythene (LD)	920	2.2	0.24	200	1.17	−120	50
Polythene (HD)	950	2.2	0.25	120	1.57	−120	55
PTFE	2100	1.0	0.25	140	0.7	−113	250
PVC (rigid)	1400	0.9	0.16	70	1.16	80	50
PVC (flexible)	1300	1.5	0.14	140	0.7	80	50
SAN	1080	1.3	0.17	70	0.81	115	60
DMC (polyester)	1800	−	0.2	20	−	−	130
SMC (polyester)	1800	−	0.2	20	−	−	130
Polystyrene foam	32	−	0.032	−	−	−	−
PU foam	32	−	0.032	−	−	−	−
Stainless steel	7855	0.49	90	10	−	−	800
Nickel chrome alloy	7950	−	12	14	−	−	900
Zinc	7135	0.39	111	39	−	−	−
Copper	8940	0.39	400	16	−	−	−

reduction in mechanical properties at high temperatures, including increased propensity to creep, but also due to limitations on the continuous working temperature causing permanent damage to the material as a result of thermal and oxidative degradation. Significant gains in property retention at high temperatures with crystalline polymers have been derived from the incorporation of fibrous reinforcement, but the development of new polymer matrices is the key to further escalation of the useful temperature range.

Table 1.8 indicates the service temperatures which can be used with a range of plastics. It may be seen that there are now commercial grades of unreinforced plastics rated for continuous use at temperatures in excess of 200°C. When glass or carbon fibres are used the service temperatures can approach 300°C.

The other principal thermal properties of plastics which are relevant to design are thermal conductivity and coefficient of thermal expansion. Compared with most materials, plastics offer very low values of thermal conductivity, particularly if they are foamed. Fig. 1.10 shows comparisons between the thermal conductivity of a selection of metals, plastics and building materials. In contrast to their low conductivity, plastics have high coefficients of expansion when compared with metals. This is illustrated in Fig. 1.11 and Table 1.8 gives fuller information on the thermal properties of plastics and metals.

25mm Polyurethane
40mm Polystyrene
45mm Mineral wool
50mm Cork
65mm Fibreboard
140mm Softwood
380mm Concrete blocks
860mm Bricks

Equivalent thickness of common
building and insulation materials required
to achieve the same degree of insulation

Fig. 1.10   Comparative Thermal conductivities for a range of materials

Electrical Properties Traditionally plastics have established themselves in applications which require electrical insulation. PTFE and polyethylene are among the best insulating materials available. The material properties which are particularly relevant to electrical insulation are *dielectric strength, resistance and tracking*.

The insulating property of any insulator will break down in a sufficiently strong electric field. The dielectric strength is defined as the electric strength (V/m) which an insulating material can withstand. For plastics the dielectric strength can vary from 1 to 1000 MV/m. Materials may be compared on the basis of their relative permittivity (or dielectric constant). This is the ratio of the permittivity of the material to the permittivity of a vacuum. The ability of a

Fig. 1.11   Typical thermal properties of plastics

material to resist the flow of electricity is determined by its volume resistivity, measured in ohm m. Insulators are defined as having volume resistivities greater than about 10^4 ohm m. Plastics are well above this, with values ranging from about 10^8 to 10^{16} ohm m. These compare with a value of about 10^{-8} ohm m for copper. Although plastics are good insulators, local breakdown may occur due to tracking. This is the name given to the formation of a conducting path (arc) across the surface of the polymer. It can be caused by surface contamination (for example dust and moisture) and is characterised by the development of carbonised destruction of the surface carrying the arc. Plastics differ greatly in their propensity to tracking – PTFE, acetal, acrylic and PP/PE copolymers offer very good resistance.

It is interesting to note that although the electrical insulation properties of plastics have generally been regarded as one of their major advantages, in recent years there has been a lot of research into the possibility of conducting plastics. This has been recognised as an exciting development area for plastics because electrical conduction if it could be achieved would offer advantages in designing against the build up of static electricity and in shielding of computers, etc from electro-magnetic interference (EMI). There have been two approaches – coating or compounding. In the former the surface of the plastic is treated with a conductive coating (e.g. carbon or metal) whereas in the second, fillers such as brass, aluminium or steel are incorporated into the plastic. It is important that the filler has a high aspect ratio (length:diameter) and so fibres or flakes of metal are used. There has also been some work done using glass fibres which are coated with a metal before being incorporated into the plastic. Since the fibre aspect ratio is critical in the performance of conductive plastics there can be problems due to breaking up of fibres during processing. In this regard thermosetting plastics have an advantage because their simpler processing methods cause less damage to the fibres. Conductive grades of DMC are now available with resistivities as low as 7×10^{-3} ohm m.

Optical Properties. The optical properties of a plastic which are important are refraction, transparency, gloss and light transfer. The reader is referred to BS 4618:1972 for precise details on these terms. Table 1.9 gives data on the optical properties of a selection of plastics. Some plastics may be optically clear (e.g. acrylic, cellulosics and ionomers) whereas others may be made transparent. These include epoxy, polycarbonate, polyethylene, polypropylene, polystyrene, polysulphone and PVC.

Table 1.9

Typical properties of plastics

Material	Refractive index	Light transmission	Dispersive power
Acrylic	1.49	92	58
Polycarbonate	1.59	89	30–35
Polystyrene	1.59	88	31
CAB	1.49	85	–
SAN	1.57	–	36
Nylon 66	1.54	0	–

Flammability. The fire hazard associated with plastics has always been difficult to assess and numerous tests have been devised which attempt to grade materials as regards flammability by standard small scale methods under controlled but necessarily artificial conditions. Descriptions of plastics as *self-extinguishing, slow burning, fire retardant* etc. have been employed to describe their behaviour under such standard test conditions, but could never be regarded as predictions of the performance of the material in real fire situations, the nature and scale of which can vary so much.

Currently there is a move away from descriptions such as *fire-retardant* or *self-extinguishing* because these could imply to uninformed users that the material would not burn. The most common terminology for describing the flammability characteristics of plastics is currently the **Critical Oxygen Index (COI)**. This is defined as the minimum concentration of oxygen, expressed as volume per cent, in a mixture of oxygen and nitrogen that will just support combustion under the conditions of test. Since air contains 21% oxygen, plastics having a COI of greater than 0.21 are regarded as self-extinguishing. In practice a higher threshold (say 0.27) is advisable to allow for unforeseen factors in a particular fire hazard situation. Fig. 1.12 shows the typical COI values for a range of plastics.

Fig. 1.12 Oxygen Index Values for Plastics

Permeability. The low density of plastics is an advantage in many situations but the relatively loose packing of the molecules means that gases and liquids can permeate through the plastic. This can be important in many applications such as packaging or fuel tanks. It is not possible to generalise about the performance of plastics relative to each other or in respect to the performance of a specific plastic in contact with different liquids and gases.

Some plastics are poor at offering resistance to the passage of fluids through them whereas others are excellent. Their relative performance may be quantified in terms of a permeation constant, k, given by

$$k = \frac{Qd}{Atp} \tag{1.10}$$

where Q = volume of fluid passing through the plastic
d = thickness of plastic
A = exposed area
t = time
p = pressure difference across surfaces of plastic.

The main fluids of interest with plastics are oxygen and water vapour (for packaging applications) and CO_2 (for carbonated drinks applications). Fig. 1.13 and Fig. 1.14 illustrate the type of behaviour exhibited by a range of plastics. In some cases it is necessary to use multiple layers of plastics because no single plastic offers the combination of price, permeation resistance, printability, etc. required for the application. When multi-layers are used, an overall permeation constant for the composite wall may be obtained from

$$\frac{1}{k} = \frac{1}{d} \sum_{i=1}^{i=N} \frac{d_i}{k_i} \tag{1.11}$$

1.4.5 Processing

A key decision in designing with plastics is the processing method employed. The designer must have a thorough knowledge of processing methods because

Fig. 1.13   Permeability data for a range of plastics

Fig. 1.14   Permeability data for a range of plastics

plastics are unique in design terms in that they offer a wider choice of conversion techniques than are available for any other material. A simple container for example could be made by injection moulding, blow moulding or rotational moulding; light fittings could be thermoformed or injection moulded. In this brief introduction to designing with plastics it is not possible to do justice to the range of processing methods available for plastics. Therefore Chapter 4 will be devoted to processing. This describes the suitability of particular plastics for each moulding method and considers the limitations which these place on the designer.

1.4.6 Costs

It is a popular misconception that plastics are cheap materials. They are not. On a weight basis most plastics are more expensive than steel and only slightly less expensive than aluminium. Prices for plastics can range from about £600 per tonne for polypropylene to about £25,000 per tonne for carbon fibre reinforced PEEK. Table 1.5 compares the costs of a range of plastics.

However, it should always be remembered that it is bad design practice to select materials on the basis of cost per unit weight. In the mass production industries, in particular, the raw material cost is of relatively little importance. It is the *in-position* cost which is all important. The in-position cost of a component is the sum of several independent factors i.e. raw material costs, fabrication costs and performance costs.

It is in the second two of these cost components that, in relation to other materials, plastics can offer particular advantages. Fabrication costs include power, labour, consumables, etc and Table 1.10 shows that, in terms of the overall energy consumption, plastics come out much better than metals. Performance costs relate to servicing, warranty claims, etc. On this basis plastics can be very attractive to industries manufacturing consumer products because they can offer advantages such as colour fastness, resilience, toughness, corrosion resistance and uniform quality – all features which help to ensure a reliable product.

However, in general these fabrication and performance advantages are common to all plastics and so a decision has to be made in regard to which plastic would be best for a particular application. Rather than compare the basic raw material costs it is better to use a cost index on the basis of the cost to achieve a certain performance. Consider again the material selection procedures illustrated in Section 1.4.1 in relation to strength and stiffness.

Selection for Strength at Minimum Cost

If the cost of a material is C per unit weight then from equation (1.3) the cost of the beam considered in the analysis would be

$$C_b = (\rho bdL)C \qquad (1.12)$$

Substituting for d from (1.2) then the cost of the beam on a strength basis would be

$$C_b = \beta_3 \left(\frac{\rho C}{\sigma_y^{1/2}} \right) \qquad (1.13)$$

Table 1.10

The energy required to manufacture and process a range of materials at typical design thickness

Material	Design thickness -mm	Energy J/m² × 10⁶
Magnesium	1.9	
Aluminium sheet	1.3	
Zinc die casting	1.3	
Aluminium casting	1.3	
Steel	0.8	
Acetal	2.0	
Modified PPO	2.0	
SMC	3.3	
Rigid polyurethane foam	5.3	
Polycarbonate	2.0	
Acrylic	2.0	
Nylon 6	1.8	
Nylon 66	1.8	
LDPE	2.7	
HDPE	2.4	
Polystyrene	2.0	
RRIM polyurethane	3.2	
Polypropylene	2.5	
ABS	1.9	
PVC	2.0	

(scale: 160, 320, 480, 640, 800, 960, 1120, 1280)

▨ Feedstock   ▨ Fuel   ▢ Process

where β_3 is a constant, which will be the same for all materials. Therefore we can define a cost factor, C_f, where

$$C_f = \left(\frac{\rho C}{\sigma_y^{1/2}} \right) \tag{1.14}$$

which should be minimised in order to achieve the best combination of price and performance. Alternatively we may take the reciprocal of C_f to get a desirability factor, D_f,

$$D_f = \left(\frac{\sigma_y^{1/2}}{\rho C} \right) \tag{1.15}$$

and this may be compared to D_f given by equation (1.5).

Selection for Stiffness at Minimum Cost

Using equation (1.7) and an analysis similar to above it may be shown that on the basis of stiffness and cost, the desirability factor, D_f, is given by

$$D_f = \left(\frac{E^{1/3}}{\rho C} \right) \tag{1.16}$$

Table 1.11

Desirability factors for some common loading configurations

Component	Desirability factor, D_f	
	Strength basis	Stiffness basis
Rectangular beam with fixed width	$\sigma_y^{1/2}/\rho C$	$E^{1/3}/\rho C$
Struts or ties	$\sigma_y/\rho C$	$E/\rho C$
Thin wall cylinders under pressure	$\sigma_y/\rho C$	–
Thin wall shafts in tension	$\tau_m/\rho C$	$G/\rho C$
Long rods in compression (buckling)	–	$E^{1/2}/\rho C$

Table 1.12

Desirability factors for a range of materials

Material	Density, ρ kg/m^3	Proof or fracture stress σ_y (MN/m^2)	Modulus E (GN/m^2)	$\dfrac{\sigma_y}{\rho}$	$\dfrac{E}{\rho}$	$\dfrac{\sigma_y^{1/2}}{\rho}$ ($\times 10^{-3}$)	$\dfrac{E^{1/2}}{\rho}$ ($\times 10^{-3}$)	$\dfrac{E^{1/3}}{\rho}$ ($\times 10^{-3}$)
Aluminium (pure)	2700	90	70	0.033	0.026	3.51	3.12	1.53
Aluminium alloy	2810	500	71	0.178	0.025	7.95	3.0	1.47
Stainless steel	7855	980	185	0.125	0.024	4.0	1.73	0.73
Titanium alloy	4420	900	107	0.204	0.024	6.78	2.34	1.07
Spruce	450	35	9	0.078	0.020	13.15	6.67	4.62
GRP (80% unidirectional glass in polyester)	2000	1240	48	0.62	0.024	17.6	3.46	1.82
CFRP (60% unidirectional fibres in epoxy)	1500	1050	189	0.7	0.126	21.6	9.16	3.82
Nylon 66	1140	70	0.78*	0.061	6.8×10^{-4}	8.34	0.77	0.81
ABS	1040	35	1.2*	0.034	11.5×10^{-4}	5.68	1.05	1.02
Polycarbonate	1150	60	2.0*	0.052	17.4×10^{-4}	6.73	1.23	1.09
PEEK (+ 30% C)	1450	215	15.5	0.19	0.011	10.1	2.7	1.72

*1500 h creep modules

Tables 1.11 and 1.12 give desirability factors for configurations other than the beam analysed above and typical numerical values of these factors for a range of materials.

Bibliography

Waterman, N.A. *The Selection and Use of Engineering Materials*, Design Council, London (1979)

Crane, F.A.C. and Charles, J.A. *Selection and Use of Engineering Materials* Butterworths, London (1984)

Crawford, R.J. *Plastics and Rubber–Engineering Design and Application*, MEP, London (1985)

Powell, P.C. *Engineering with Polymers*, Chapman and Hall, London (1984)

Hall, C. *Polymer Materials*, Macmillan, London (1981)

Birley, A.W. and Scott, M.J. *Plastic Materials: Properties and Applications*, Leonard Hall, Glasgow (1982)

Benham, P.P. Crawford, R.J. and Armstrong, C.G *Mechanics of Engineering Materials*, Longmans (1996)

Lancaster, J.K. *Friction and Wear of Plastics*, Chapter 14 in Polymer Science edited by A.D. Jenkins (vol 2), North-Holland Publ. Co., London (192)

Bartenev, G.M. and Lavrentev, V.V. *Friction and Wear in Polymers*, Elsevier Science Publ. Co., Amsterdam (1981)

Schwartz, S.S. and Goodman, S.H. *Plastics Materials and Processes*, Van Nostrand Reinhold, New York (1982)

Blythe, A.R. *Electrical Properties of Polymers*, Cambridge Univ. Press (1980)

Van Krevelen, D.W. *Properties of Polymers* 2nd Edition, Elsevier, Amsterdam (1976)

Mills, N. *Plastics*, Edward Arnold, London (1986)

Kemmish, D.J. High performance engineering plastics, *RAPRA Review Reports*, **8**, 2 (1995)

Oswald, T.A. and Menges, G. *Materials Science of Polymers for Engineers*, Hanser, Munich (1995)

Birley, A.W. Haworth B. and Batchelor, J. *Physics of Plastics*, Hanser, Munich (1992)

Belofsky, H. *Plastics: Product Design and Process Engineering*, Hanser, Munich (1995)

Gruenwald, G. *Plastics: How Structure Determines Properties*, Hanser, Munich (1992)

Dominghaus, H. *Plastics for Engineers*, Hanser, New York (1993)

Charrier, J.M. *Polymeric Materials and Processing*, Hanser, New York (1990)

Progelhof, R.C. and Throne, J.L. *Polymer Engineering Principles*, Hanser, New York (1993)

CHAPTER 2 – Mechanical Behaviour of Plastics

2.1 Introduction

In Chapter 1 the general mechanical properties of plastics were introduced. In order to facilitate comparisons with the behaviour of other classes of materials the approach taken was to refer to standard methods of data presentation, such as stress–strain graphs, etc. However, it is important to note that when one becomes involved in engineering design with plastics, such graphs are of limited value. The reason is that they are the results of relatively short-term tests and so their use is restricted to quality control and, perhaps, the initial sorting of materials in terms of stiffness, strength etc. Designs based on, say, the modulus obtained from a short-term test would not predict accurately the long-term behaviour of plastics because they are viscoelastic materials. This viscoelasticity means that quantities such as modulus, strength, ductility and coefficient of friction are sensitive to straining rate, elapsed time, loading history, temperature, etc. It will also be shown later that the manufacturing method used for the plastic product can create changes in the structure of the material which have a pronounced effect on properties. The behaviour of the moulded product may therefore be different from the behaviour of a moulded test-piece of the same material.

The time-dependent change in the dimensions of a plastic article when subjected to a constant stress is called **creep**. As a result of this phenomenon the modulus of a plastic is not a constant, but provided its variation is known then the creep behaviour of plastics can be allowed for using accurate and well established design procedures. Metals also display time dependent properties at high temperatures so that designers of turbine blades, for example, have to allow for creep and guard against creep rupture. At room temperature the creep behaviour of metals is negligible and so design procedures are simpler in that

41

the modulus may be regarded as a constant. In contrast, thermoplastics at room temperature behave in a similar fashion to metals at high temperatures so that design procedures for relatively ordinary load-bearing applications must always take into account the viscoelastic behaviour of plastics.

For most traditional materials, the objective of the design method is to determine stress values which will not cause fracture. However, for plastics it is more likely that excessive deformation will be the limiting factor in the selection of working stresses. Therefore this chapter looks specifically at the deformation behaviour of plastics and fracture will be treated separately in the next chapter.

2.2 Viscoelastic Behaviour of Plastics

For a component subjected to a uniaxial force, the engineering stress, σ, in the material is the applied force (tensile or compressive) divided by the original cross-sectional area. The engineering strain, ε, in the material is the extension (or reduction in length) divided by the original length. In a perfectly elastic (Hookean) material the stress, σ, is directly proportional to be strain, ε, and the relationship may be written, for uniaxial stress and strain, as

$$\sigma = \text{constant} \times \varepsilon \tag{2.1}$$

where the constant is referred to as the modulus of the material.

In a perfectly viscous (Newtonian) fluid the shear stress, τ is directly proportional to the rate of strain $(d\gamma/dt$ or $\dot{\gamma})$ and the relationship may be written as

$$\tau = \text{constant} \times \dot{\gamma} \tag{2.2}$$

where the constant in this case is referred to as the **viscosity** of the fluid.

Polymeric materials exhibit mechanical properties which come somewhere between these two ideal cases and hence they are termed **viscoelastic**. In a viscoelastic material the stress is a function of strain and time and so may be described by an equation of the form

$$\sigma = f(\varepsilon, t) \tag{2.3}$$

This type of response is referred to as non-linear viscoelastic but as it is not amenable to simple analysis it is often reduced to the form

$$\sigma = \varepsilon \cdot f(t) \tag{2.4}$$

This equation is the basis of linear viscoelasticity and simply indicates that, in a tensile test for example, for a fixed value of elapsed time, the stress will be directly proportional to the strain. The different types of response described are shown schematically in Fig. 2.1.

The most characteristic features of viscoelastic materials are that they exhibit a time dependent strain response to a constant stress (**creep**) and a time dependent stress response to a constant strain (**relaxation**). In addition when the

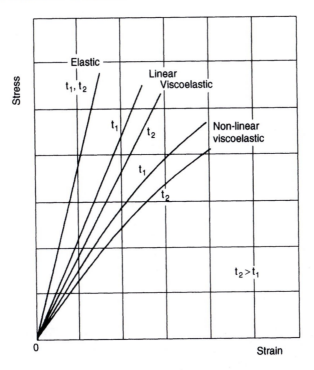

Fig. 2.1   Stress−strain behaviour of elastic and viscoelastic materials at two values of elapsed
time, *t*

applied stress is removed the materials have the ability to **recover** slowly over
a period of time. These effects can also be observed in metals but the difference
is that in plastics they occur at room temperature whereas in metals they only
occur at very high temperatures.

2.3 Short-Term Testing of Plastics

The simple tensile test is probably the most popular method for characterising metals and so it is not surprising that it is also widely used for plastics.
However, for plastics the tensile test needs to be performed very carefully
and the results of the single test should only be used as a means of quality
control − not as design data. This is because, with plastics it is possible to
obtain quite different results from the same material simply by changing the
test conditions. Fig. 2.2 shows that at high extension rates (>1 mm/s) unplasticised PVC is almost brittle with a relatively high modulus and strength. At
low extension rates (<0.05 mm/s) the same material exhibits a lower modulus
and strength but its ductility is now very high. Therefore a single tensile test
could be quite misleading if the results were used in design formulae but the
test conditions were not similar to the service conditions.

Fig. 2.2   Typical tensile behaviour of unplasticised PVC

Fig. 2.2 also illustrates an interesting phenomenon observed in some plastics. This is *cold drawing* and it occurs because at low extension rates the molecular chains in the plastic have time to align themselves under the influence of the applied stress. Thus the the material is able to flow at the same rate as it is being strained. Nowadays this phenomena is utilised in some forming methods for plastics. It creates highly aligned molecular structures which exhibit excellent strength and stiffness properties in the alignment direction.

Occasionally, materials are tested in tension by applying the loads in increments. If this method is used for plastics then special caution is needed because during the delay between applying the load and recording the strain, the material creeps. Therefore if the delay is not uniform there may appear to be excessive scatter or non-linearity in the material. In addition, the way in which the loads are applied constitutes a loading history which can affect the performance of the material. A test in which the increments are large would quite probably give results which are different from those obtained from a test in which the increments were small or variable.

As a result of these special effects in plastics it is not reasonable to quote properties such as modulus, yield strength, etc as a single value without qualifying these with details of the test method. Standard short-term test methods for plastics are described in BS 2782, ASTM D638 and ASTM D790. These relate to both tensile and flexural short-term tests. It should be noted that, as pointed out above, in uniaxial loading the stress, σ, is simply the force divided by the cross-sectional area and the strain, ε, is the deformation divided by the original length. The modulus of the material is then the ratio of stress/strain.

Hence,

$$\text{stress,} \quad \sigma = \frac{W}{A}$$

$$\text{strain,} \quad \varepsilon = \frac{\delta}{L_0}$$

$$\text{modulus,} \quad E = \frac{\sigma}{\varepsilon} = \frac{WL_0}{A\delta}$$

In flexure (bending) situations, these equations do not apply. For the three-point loading shown in Fig. 2.3, the relevant equations are

$$\text{stress, } \sigma = \frac{My}{I}$$

where M = bending moment at loading point ($= WL/4$)
  y = half depth of beam ($= d/2$)
  I = second moment of area ($bd^3/12$)

hence stress, $\qquad\qquad \sigma = \dfrac{3WL}{2bd^2}$ $\qquad\qquad$ (2.5)

Also,

$$\text{deflection } \delta = \frac{WL^3}{48EI} \qquad \text{(see Benham } et\ al.\text{)}$$

$$= \sigma\left(\frac{L^2}{6Ed}\right) = \varepsilon\left(\frac{L^2}{6d}\right)$$

hence strain, $\varepsilon = \dfrac{6\delta d}{L^2}$ $\qquad\qquad$ (2.6)

modulus, $E = \dfrac{\sigma}{\varepsilon} = \dfrac{WL^3}{4bd^3\delta}$ $\qquad\qquad$ (2.7)

Note that these stress, strain and modulus equations are given for illustration purposes. They apply to three-point bending as shown in Fig. 2.3. Other types of bending can occur (e.g. four-point bending, cantilever, etc.) and different equations will apply. Some of these are illustrated in the Worked Examples later in this chapter and the reader is referred to Benham *et al.* for a greater variety of bending equations.

2.4 Long-Term Testing of Plastics

Since the tensile test has disadvantages when used for plastics, creep tests have evolved as the best method of measuring the deformation behaviour of

Fig. 2.3   Three-point bending

polymeric materials. In these tests a constant load is applied to the material and
the variation of strain with time is recorded as shown in Fig. 2.4(a). Normally
a logarithmic time scale is used as shown in Fig. 2.4(b) so that the time dependence after long periods can be included and as an aid to extrapolation. This
figure shows that there is typically an almost instantaneous strain followed by
a gradual increase. If a material is linearly viscoelastic then at any selected
time each line in a family of creep curves should be offset along the strain
axis by the same amount. Although this type of behaviour may be observed
for plastics at low strains and short times, in the majority of cases the response
is non-linear as indicated in Fig. 2.5.

Occasionally creep curves are plotted as log (strain) against log (time). This
is convenient because quite often this results in straight line plots suggesting
that the creep behaviour can be described by an equation of the type

$$\varepsilon(t) = At^n \tag{2.8}$$

where 'n' is a material constant and 'A' is a constant which depends on the
stress level.

To be strictly accurate, the right hand side should have the form $(A_0 + At^n)$
to allow for the instantaneous strain at zero time. However, for long creep
times, sufficient accuracy can be obtained by ignoring A_0.

Fig. 2.4   Typical creep curves

Fig. 2.5   Creep curves for polypropylene at 20°C (density 909 kg/m³)

A variety of other creep-strain equations have been proposed by Sterrett *et al.* These take the form

1. $\varepsilon = a + bt^c$ Parabolic

2. $\varepsilon = a + bt^{1/3} + ct$ Power

3. $\varepsilon = \dfrac{at}{1 + bt}$ Hyperbolic

4. $\varepsilon = \dfrac{1}{1 + at^b}$ Hyperbolic

5. $\varepsilon = a + \ln(b + t)$ Logarithmic

6. $\varepsilon = \dfrac{a(b + \ln t)}{c + d \ln t}$ Logarithmic

7. $\varepsilon = a + b(1 - e^{-ct})$ Exponential

8. $\varepsilon = \sinh at$ Hyperbolic sine

9. $\varepsilon = a + b \sinh ct$ Hyperbolic sine

10. $\varepsilon = a + b \sinh ct^{1/3}$ Hyperbolic sine

11. $\varepsilon = a(1 + bt^{1/3})e^{-ct}$ Power/exponential

The mechanism of creep is not completely understood but some aspects have been explained based on the structures described in Appendix A. For example, in a glassy plastic a particular atom is restricted from changing its position as a result of attractions and repulsions between it and (a) atoms in the same chain, (b) atoms in adjacent chains. It is generally considered that for an atom to change its position it must overcome an energy barrier and the probability of it achieving the necessary energy is improved when a stress is applied. In a semi-crystalline plastic there is an important structural difference in that the crystalline regions are set in an amorphous matrix. Movement of atoms can occur in both regions but in the majority of cases atom mobility is favoured in the non-crystalline material between the spherulites.

Plastics also have the ability to recover when the applied stress is removed and to a first approximation this can often be considered as a reversal of creep. This was illustrated in Fig. 1.8 and will be studied again in Section 2.7. At present it is proposed to consider the design methods for plastics subjected to steady forces.

2.5 Design Methods for Plastics using Deformation Data

Isochronous and Isometric Graphs

The most common method of displaying the interdependence of stress, strain and time is by means of creep curves. However, it should be realised that these represent a two-dimensional view of (or slices through) the three-dimensional stress–strain–time behaviour of plastics shown in Fig. 2.6. It would be equally sensible to consider the other two-dimensional views which could be taken of Fig. 2.6. These are the stress–strain plot (sometimes called the **isochronous curve** because it represents a constant time section) and the stress–time plot (sometimes called the **isometric curve** because it represents constant strain section). These two 'views' are shown in Fig. 2.7 and 2.8.

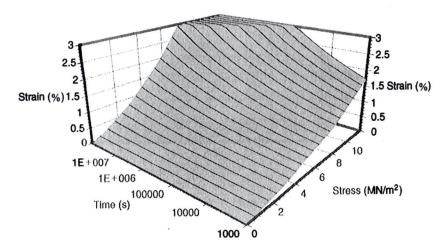

Fig. 2.6 Stress–strain–time curves for polypropylene

In practice, of course, it is most common to have the creep curves (strain–time curve) since these can be obtained relatively simply by experiment. The other two curves can then be derived from it. For example, the isometric graph is obtained by taking a constant strain section through the creep curves and re-plotting as stress versus time (see Fig. 2.10(a)). This is termed an **isometric graph** and is an indication of the relaxation of stress in the material when the strain is kept constant. This data is often used as a good approximation of stress relaxation in a plastic since the latter is a less common experimental procedure than creep testing. In addition, if the vertical axis (stress) is divided by the strain, ε, then we obtain a graph of modulus against time (Fig. 2.10(b)). This is a good illustration of the time dependent variation of modulus which was referred to earlier. It should be noted that this is a *Relaxation Modulus* since it relates to a constant strain situation. It will be slightly different to the *Creep Modulus* which could be obtained by dividing the constant creep stress by the strain at various times. Thus

$$\text{creep modulus, } E(t) = \frac{\sigma}{\varepsilon(t)} \qquad (2.9)$$

Fig. 2.7   Stress–strain curves for polypropylene

Fig. 2.8   Stress–time curves for polypropylene

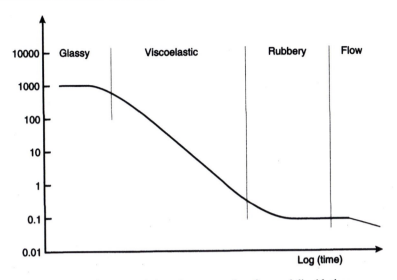

Fig. 2.9   Typical variation of creep or relaxation moduli with time

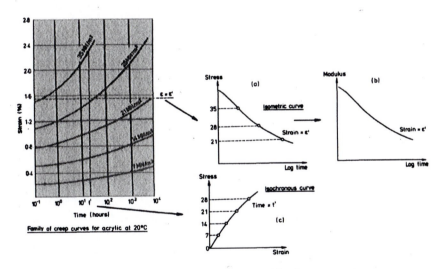

Fig. 2.10   Construction of isometric and isochronous graphs

The creep modulus will vary with time, i.e. decrease as time increases, in a manner similar to that shown for the relaxation modulus. The classical variation of these moduli is illustrated in Fig. 2.9. On log–log scales it is observed that there is a high value of creep or relaxation modulus at short times. This is referred to as the *Unrelaxed Modulus* and is independent of time. Similarly at long times there is a low value *Relaxed Modulus* which is also independent of time.

Occasionally in creep analysis it is convenient to use a *Creep Compliance* instead of the creep modulus. This is simply given by

$$\text{Creep compliance, } C(t) = \frac{1}{E(t)} = \frac{\varepsilon(t)}{\sigma} \tag{2.10}$$

An alternative to the isometric graph may be obtained by taking a constant time section through the creep curves and plotting stress versus strain as shown in Fig. 2.10(c). This **Isochronous Graph** can also be obtained experimentally by performing a series of mini-creep and recovery tests on a plastic. In this experiment a stress is applied to a plastic test-piece and the strain is recorded after a time, *t* (typically 100 seconds). The stress is then removed and the plastic allowed to recover, normally for a period of 4*t*. A larger stress is then applied to the same specimen and after recording the strain at time *t*, this stress is removed and the material allowed to recover. This procedure is repeated until sufficient points have been obtained for the isochronous graph to be plotted.

These latter curves are particularly important when they are obtained experimentally because they are less time consuming and require less specimen preparation than creep curves. Isochronous graphs at several time intervals can also be used to build up creep curves and indicate areas where the main experimental creep programme could be most profitably concentrated. They are also popular as evaluations of deformational behaviour because the data presentation is similar to the conventional tensile test data referred to in Section 2.3. It is interesting to note that the isochronous test method only differs from that of a conventional incremental loading tensile test in that (a) the presence of creep is recognised, and (b) the memory which the material has for its stress history is accounted for by the recovery periods.

Quite often isochronous data is presented on log–log scales. One of the reasons for this is that on linear scales any slight, but possibly important, non-linearity between stress and strain may go unnoticed whereas the use of log–log scales will usually give a straight-line graph, the slope of which is an indication of the linearity of the material. If it is perfectly linear the slope will be 45°. If the material is non-linear the slope will be less than this.

As indicated above, the stress–strain presentation of the data in isochronous curves is a format which is very familiar to engineers. Hence in design situations it is quite common to use these curves and obtain a secant modulus (see Section 1.4.1, Fig. 1.6) at an appropriate strain. Strictly speaking this will be different to the creep modulus or the relaxation modulus referred to above since the secant modulus relates to a situation where both stress and strain are changing. In practice the values are quite similar and as will be shown in the following sections, the values will coincide at equivalent values of strain and time. That is, a 2% secant modulus taken from a 1 year isochronous curve will be the same as a 1 year relaxation modulus taken from a 2% isometric curve.

Pseudo-Elastic Design Method for Plastics

Throughout this chapter the viscoelastic behaviour of plastics has been described and it has been shown that deformations are dependent on such factors as the time under load and the temperature. Therefore, when structural components are to be designed using plastics, it must be remembered that the classical equations which are available for the design of springs, beams, plates, cylinders, etc., have all been derived under the assumptions that

 (i) the strains are small
 (ii) the modulus is constant
 (iii) the strains are independent of loading rate or history and are immediately reversible
 (iv) the material is isotropic
 (v) the material behaves in the same way in tension and compression

Since these assumptions are not always justified for plastics, the classical equations cannot be used indiscriminately. Each case must be considered on its merits and account taken of such factors as mode of deformation, service temperature, fabrication method, environment and so on. In particular it should be noted that the classical equations are derived using the relation.

$$\text{stress} = \text{modulus} \times \text{strain}$$

where the modulus is a constant. From the foregoing sections it should be clear that the modulus of a plastic is not a constant. Several approaches have been used to allow for this and some give very accurate results. The drawback is that the methods can be quite complex, involving Laplace transforms or numerical methods and they are certainly not attractive to designers. However, one method that has been widely accepted is the so called **Pseudo Elastic Design Method**. In this method, appropriate values of time dependent properties, such as modulus, are selected and substituted into the classical equations. It has been found that this approach gives sufficient accuracy in most cases provided that the value chosen for the modulus takes into account the service life of the component and the limiting strain of the plastic. This of course assumes that the limiting strain for the material is known. Unfortunately this is not just a straightforward value which applies for all plastics or even for one plastic in all applications. It is often arbitrarily chosen although several methods have been suggested for arriving at a suitable value. One method is to plot a secant modulus which is 0.85 of the initial tangent modulus (see Fig. 1.6) and note the strain at which this intersects the stress–strain characteristic. However, for many plastics (particularly crystalline ones) this is too restrictive and so in most practical situations the limiting strain is decided in consultations between the designer and the material manufacturers.

Once the limiting strain is known, design methods based on the creep curves are quite straightforward and the approach is illustrated in the following examples. When using this pseudo-elastic design approach it should be remembered that the creep curves used to derive modulus values have normally been obtained on test pieces which are essentially isotropic. In practice the manufacture of the end-product by injection moulding or extrusion, etc. will have resulted in some degree of anisotropy. This may make the predictions inaccurate because the creep data for the material is no longer appropriate for the structural morphology introduced by the moulding method. Similar comments could, of course, also be made about metals in that the test data may have been obtained on specimens of the material which do not accurately reflect the nature of the material in the end-product. Therefore, pseudo-elastic design is a valid analytical procedure but one should always be cautious about the way in which the manufacturing method has affected the behaviour of the material.

Example 2.1 A ball-point pen made from polypropylene has the clip design shown in Fig. 2.11. When the pen is inserted into a pocket, the clip is subjected to a deflection of 2 mm at point A. If the limiting strain in the material is to be 0.5% calculate (i) a suitable thickness, d, for the clip (ii) the initial stress in the clip when it is first inserted into the pocket and (iii) the stress in the clip when it has been in the pocket for 1 week. The creep curves in Fig. 2.5 may be used and the short-term modulus of polypropylene is 1.6 GN/m^2.

Fig. 2.11   Ball-point pen clip design

Solution Strain, ε, is given by the ratio of stress, σ, to modulus, E. In the case of the pen clip, it is effectively a cantilever of length 40 mm.

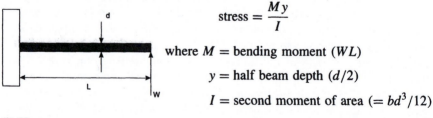

$$\text{stress} = \frac{M y}{I}$$

where M = bending moment (WL)

y = half beam depth $(d/2)$

I = second moment of area $(= bd^3/12)$

(i) Hence

$$\text{strain, } \varepsilon = \frac{WLd}{2EI} \tag{2.11}$$

Also, the classical elastic equation for the end deflection of a cantilever is:

$$\text{deflection, } \delta = \frac{WL^3}{3EI} \tag{2.12}$$

Combining (2.11) and (2.12) gives

$$\text{strain, } \varepsilon = \frac{3\delta d}{2L^2} \tag{2.13}$$

So

$$d = \frac{2(40)^2 \times 0.005}{3 \times 2} = 2.7 \text{ mm}$$

(ii) The short-term stress in the material is obtained from the short-term modulus which is given in this question (or could be obtained from the creep/isometric curves, i.e. at 10 seconds, $E = 8 \times 10^6/0.5\% = 1.6$ GN/m^2 or from the appropriate isometric curve).

$$\text{stress} = E\varepsilon = 1.6 \times 10^9 \times 0.005 = 8 \text{ MN/m}^2$$

(iii) After 1 week (6.1×10^5 seconds), the isometric curves (Fig. 2.8) derived from the creep curves show that at a strain of 0.5% the stress would have decayed to about 3.3 MN/m^2.

Example 2.2 A polypropylene beam is 100 mm long, simply supported at each end and is subjected to a load W at its mid-span. If the maximum permissible strain in the material is to be 1.5%, calculate the largest load which may be applied so that the deflection of the beam does not exceed 5 mm in a service life of 1 year. For the beam $I = 28$ mm^4 and the creep curves in Fig. 2.5 should be used.

Solution The central deflection in a beam loaded as shown in Fig. 2.12 is given by

$$\delta = \frac{WL^3}{48EI}$$

$$W = \frac{48EI\delta}{L^3}$$

Fig. 2.12 Simply supported beam with central load

Fig. 2.13   Isochronous curve for polypropylene (1 year)

The only unknown on the right hand side is a value for modulus E. For the plastic this is time-dependent but a suitable value may be obtained by reference to the creep curves in Fig. 2.5. A section across these curves at the service life of 1 year gives the isochronous graph shown in Fig. 2.13. The maximum strain is recommended as 1.5% so a secant modulus may be taken at this value and is found to be 347 MN/m². This is then used in the above equation.

So
$$W = \frac{48 \times 347 \times 28 \times 5}{(100)^3} = 2.33 \text{ N}$$

As before, a similar result could have been achieved by taking a section across the creep curves at 1.5% strain, plotting an isometric graph (or a 1.5% modulus/time graph) and obtaining a value for modulus at 1 year (see Fig. 2.8)

In this example it has been assumed that the service temperature is 20 °C. If this is not the case, then curves for the appropriate temperature should be used. If these are not available then a linear extrapolation between temperatures which are available is usually sufficiently accurate for most purposes. If the beam in the above example had been built-in at both ends at 20 °C, and subjected to service conditions at some other temperature, then allowance would need to be made for the *thermal strains* set up in the beam. These could be obtained from a knowledge of the coefficient of thermal expansion of the beam material. This type of situation is illustrated later.

For some plastics, particularly nylon, the moisture content can have a significant effect on the creep behaviour. For such plastics, creep curves are normally available in the wet and dry states and a knowledge of the service conditions enables the appropriate data to be used.

For convenience so far we have referred generally to *creep curves* in the above examples. It has been assumed that one will be using the correct curves for the particular loading configuration. In practice, creep curves obtained under tensile and flexural loading conditions are quite widely available. Obviously it is important to use the creep curves which are appropriate to the particular loading situation. Occasionally it is possible to obtain creep curves for compressive or shear loading but these are less common.

If only one type of data is available (e.g. tensile creep curves) then it is possible to make conversions to the other test modes. It should always be remembered, however, that these may not always be absolutely accurate for plastics under all situations.

Generally there is a stiffening effect in compression compared to tension. As a first approximation one could assume that tension and compression behaviour are the same. Thomas has shown that typically for PVC, the compression modulus is about 10% greater than the tensile modulus. However, one needs to be careful when comparing the experimental data because normally no account is taken of the changes in cross-sectional area during testing. In tension, the area will decrease so that the true stress will increase whereas in compression the opposite effect will occur.

The classical relationship between moduli in tension, compression and flexure is

$$E_{\text{flex}} = E_{\text{T}} \left[\frac{2\sqrt{M_{\text{R}}}}{1 + \sqrt{M_{\text{R}}}} \right]^2 \qquad (2.14)$$

where

$$M_{\text{R}} = E_{\text{c}}/E_{\text{T}}$$

It may be seen that if $E_{\text{c}} = E_{\text{T}}$ then $E_{\text{flex}} = E_{\text{T}}$. However, if $E_{\text{c}} = 1.1E_{\text{T}}$ (for example), then $E_{\text{flex}} = 1.05E_{\text{T}}$.

The classical relationship between the *shear modulus G*, and the tensile modulus, E, for an isotropic material is

$$G = \frac{E}{2(1 + v)} \qquad (2.15)$$

where v = Poissons ratio.

Finally, although it is less commonly used for plastics, the *bulk modulus, K*, is given by

$$K = \frac{E}{3(1 - 2v)} \qquad (2.16)$$

The bulk modulus is appropriate for situations where the material is subjected to hydrostatic stresses. The proof of equations (2.15) and (2.16) is given by Benham *et al.*

Example 2.3 A cylindrical polypropylene tank with a mean diameter of 1 m is to be subjected to an internal pressure of 0.2 MN/m^2. If the maximum strain in the tank is not to exceed 2% in a period of 1 year, estimate a suitable value for its wall thickness. What is the ratio of the hoop strain to the axial strain in the tank. The creep curves in Fig. 2.5 may be used.

Solution The maximum strain in a cylinder which is subjected to an internal pressure, p, is the hoop strain and the classical elastic equation for this is

$$\varepsilon_\theta = \frac{pR}{2hE}(2 - v)$$

where E is the modulus, R is the cylinder radius and h is the wall thickness (See Appendix C).

The modulus term in this equation can be obtained in the same way as in the previous example. However, the difference in this case is the term v. For elastic materials this is called **Poissons Ratio** and is the ratio of the transverse strain to the axial strain (See Appendix C). For any particular metal this is a constant, generally in the range 0.28 to 0.35. For plastics v is not a constant. It is dependent on time, temperature, stress, etc and so it is often given the alternative names of **Creep Contraction Ratio** or **Lateral Strain Ratio**. There is very little published information on the creep contraction ratio for plastics but generally it varies from about 0.33 for hard plastics (such as acrylic) to almost 0.5 for elastomers. Some typical values are given in Table 2.1 but do remember that these may change in specific loading situations.

Using the value of 0.4 for polypropylene,

$$h = \frac{pR}{2\varepsilon E}(2 - v)$$

from Fig. 2.7, $E = \dfrac{6.5}{0.02} = 325$ MN/m^2

$$\therefore h = \frac{0.2 \times 0.5 \times 10^3 \times 1.6}{2 \times 0.02 \times 325} = 12.3 \text{ mm}$$

For a cylindrical tank the axial strain is given by

$$\varepsilon_x = \frac{pR}{2hE}(1 - 2v)$$

So
$$\frac{\varepsilon_\theta}{\varepsilon_x} = \left(\frac{2 - v}{1 - 2v}\right) = \frac{1.6}{0.2} = 8$$

Table 2.1

Typical tensile and shear moduli for a range of polymers

Material	Density (kg/m^3)	Tensile modulus $(E)^\dagger$ (GN/m^2)	Shear modulus (G) (GN/m^2)	Poisson's ratio (v)
Polystyrene (PS)	1050	2.65	0.99	0.33
Polymethyl Methacrylate (PMMA)	1180	3.10	1.16	0.33
Polyvinyl Chloride (PVC) (Unplasticised)	1480	3.15	1.13	0.39
Nylon 66 (at 65% RH)	1140	0.99	0.34	0.44
Acetal Homopolymer (POM)	1410	3.24	1.15	0.41
Acetal Copolymer (POM)	1410	2.52	0.93	0.39
Polyethylene – High Density (HDPE)	955	1.05	0.39	0.34
Polyethylene – Low Density (LDPE)	920	0.32	0.11	0.45
Polypropylene Homopolymer (PP)	910	1.51	0.55	0.36
Polypropylene Copolymer (PP)	902	1.13	0.40	0.40
Polyethersulphone	1390	2.76	0.98	0.41

† 100 second modulus at 20°C for small strains (<0.2%)

Note that the ratio of the ratio of the hoop stress (pR/h) to the axial stress $(pR/2h)$ is only 2. From the data in this question the hoop stress will be 8.12 MN/m^2. A plastic cylinder or pipe is an interesting situation in that it is an example of creep under biaxial stresses. The material is being stretched in the hoop direction by a stress of 8.12 MN/m^2 but the strain in this direction is restricted by the perpendicular axial stress of 0.5(8.12) MN/m^2. Reference to any solid mechanics text will show that this situation is normally dealt with by calculating an equivalent stress, σ_e. For a cylinder under pressure σ_e is given by $0.5\sigma_\theta\sqrt{3}$ where σ_θ is the hoop stress. This would permit the above question to be solved using the method outlined earlier.

Example 2.4 A glass bottle of sparkling water has an acetal cap as shown in Fig. 2.14. If the carbonation pressure is 375 kN/m^2, estimate the deflection at the centre of the cap after 1 month. The value of Poissons ratio for acetal may be taken as 0.33.

Solution The top of the bottle cap is effectively a plate with clamped edges. The central deflection in such a situation is given by Benham et al. as

$$\delta = \frac{pR^4}{64D} \quad \text{where } D = \frac{Eh^3}{12(1 - v^2)}$$

To calculate δ after 1 month it is necessary to know the 1 month creep modulus. The stresses at the centre of the cap are biaxial (radial and circumferential) both

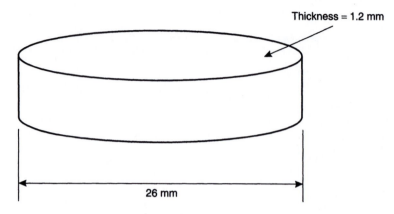

Fig. 2.14   Acetal bottle cap

given by

$$\sigma = \frac{3pR^2(1+v)}{8h^2} = \frac{3 \times 0.375 \times (13)^2(1.33)}{8(1.2)^2}$$

$$\sigma = 22 \text{ MN/m}^2$$

After 1 month (2.6×10^6 seconds) at this stress, the strain is obtained from Fig. 2.15 as 2.2%. Hence $E = 22 \times 10^6/0.022 = 1$ GN/m^2.

So,

$$\delta = \frac{pR^4}{64D} = \frac{0.375(13)^4 12(1 - 0.33^2)}{64 \times 1000 \times (1.2)^3}$$

$$\delta = 1 \text{ mm}$$

Example 2.5 In a small polypropylene pump the flange on the cover plate is 2 mm thick. When the rigid clamping screws are tightened, the flange is reduced in thickness by 0.03 mm. Estimate the initial stress in the plastic and the stress after 1 week

Solution The strain in the material is given by

$$\varepsilon = \frac{0.03}{2} \times 100 = 1.5\%$$

This is a stress relaxation problem and strictly speaking stress relaxation data should be used. However, for most purposes isometric curves obtained from the creep curves are sufficiently accurate. By considering the 1.5% isometric curve shown in Fig. 2.8 it may be seen that the initial stress is 16 MN/m^2 and the stress after 1 week is 7 MN/m^2.

Fig. 2.15   Creep curves for acetal (20°C)

Accurately performed relaxation tests in which the strain in the material was maintained constant and the decaying stress monitored, would give slightly lower values than those values obtained from the isometric data.

It should also be noted that in this case the material was loaded in compresion whereas the tensile creep curves were used. The vast majority of creep data which is available is for tensile loading mainly because this is the simplest and most convenient test method. However, it should not be forgotten that the material will behave differently under other modes of deformation. In compression the material deforms less than in tension although the effect is small for strains up to 0.5%. If no compression data is available then the use of tensile data is permissible because the lower modulus in the latter case will provide a conservative design.

2.6 Thermal Stresses and Strains

It is quite common in modern engineering designs, for plastics to be used in conjunction with other materials, particularly metals. In such cases it is wise to consider the possibility of thermal stresses being set up due to the differences in the thermal expansion (or contraction) in each material.

The change in shape of a material when it is subjected to a change in temperature is determined by the coefficient of thermal expansion, α_T. Normally for isotropic materials the value of α_T will be the same in all directions. For convenience this is often taken to be the case in plastics but one always needs to bear in mind that the manufacturing method may have introduced anisotropy which will result in different thermal responses in different directions in the material.

The coefficient of thermal expansion, α_T, is given by

$$\alpha_T = \frac{\delta L}{L(\Delta T)} \qquad (2.17)$$

where δL is the change in length in the material
$\quad\quad L$ is the original length
$\quad\quad \Delta T$ is the change in temperature.

There are standard procedures for determining α_T (e.g. ASTM 696) and typical values for plastics are given in Table 1.2. It may be observed that the coefficients of thermal expansion for plastics are higher than those for metals. Thus if 50 mm lengths of polypropylene and stainless steel are each heated up by 60°C the changes in length would be

(a) polypropylene, $\quad \delta l = 100 \times 10^{-6} \times 50 \times 60 = 0.3$ mm
(b) stainless steel, $\quad \delta l = 10 \times 10^{-6} \times 50 \times 60 = 0.03$ mm

If these changes in length take place freely then we will have a thermally induced strain in the material ($= 0.3 \times 100/50 = 0.6\%$ in the polypropylene) but no stress. However, if the polypropylene was constrained in some way so that the 0.3 mm expansion could not happen when it is heated by 60°C, then there would be a thermally induced stress in the material, i.e.

$$\text{stress} = \text{modulus} \times \text{strain}$$

If the modulus of the material is 1.2 GN/m^2 at the final temperature, then the stress in the material would be given by

$$\text{stress} = 1.2 \times 10^9 \left(\frac{-0.6}{100} \right) = -7.2 \text{ MN/m}^2$$

Note that the stress is compressive because the material is effectively compressed by 0.3 mm.

Example 2.6 The bobbin shown in Fig. 2.16 has been manufactured by sliding the acetal ring on to the steel inner and then placing the end-plate in position. At 20°C there are no stresses in the acetal and the distance between the metal end-plates is equal to the length of the acetal ring. If the whole assembly is heated to 100°C, calculate the axial stress in the acetal. It may be assumed that there is no friction between the acetal and the steel. The coefficients of thermal expansion for the acetal and the steel are 80×10^{-6}°C^{-1} and 11×10^{-6}°C^{-1} respectively. The modulus of the acetal at 100°C is 1.5 GN/m^2.

Fig. 2.16   Metal bobbin with plastic sleeve

Solution Under free conditions, the acetal would expand more than the steel but in the configuration shown they will both expand to the same extent. Hence, the acetal will effectively be put in compression by an amount given by eqn (2.17)

$$\delta = (\alpha_a - \alpha_s)L \cdot \Delta T$$
$$= (80 - 11)10^{-6}(40)(80)$$
$$= 0.22 \text{ mm}$$
$$\text{strain} = -\frac{0.22}{40} \times 100 = -0.55\%$$
$$\text{stress} = E\varepsilon = -1.5 \times 10^9 \times \frac{0.55}{100}$$
$$= -8.3 \text{ MN/m}^2$$

Thus there will be a compressive stress of 8.3 MN/m² in the acetal. It should be noted that the above analysis ignores the effect of the constraining effect which the acetal has on the thermal expansion of the steel. However, as the modulus of the steel is over 100 times greater than the acetal, this constraining effect will be very small. Thus the above analysis is perfectly acceptable for engineering design purposes.

Example 2.7 A nylon ring with a nominal inside diameter of 30 mm, an outer diameter of 50 mm and a width of 5 mm is to be made an interference fit on a metal shaft of 30 mm diameter as shown in Fig. 2.17. The design condition is that the initial separation force is to be 1 kN. Calculate (a) the interference on radius needed between the ring and the shaft and (b) the temperature to which the nylon must be heated to facilitate easy assembly. What will be the maximum stress in the nylon when it is in position on the shaft? The coefficient of friction between nylon and steel is 0.25. The short-term modulus of the nylon is 1 GN/m^2, its Poisson's ratio is 0.4 and its coefficient of thermal expansion is 100×10^{-6}°C^{-1}.

Fig. 2.17   Interference fit of nylon ring on metal shaft

Solution

(a)   Force, $F = \mu N$

where $N = pA$

$\qquad\qquad = p\pi DW$

So, $F = \pi \mu p D W$

Therefore

$$p = \frac{1 \times 10^3}{\pi(0.25)(30)(5)} = 8.5 \text{ MN/m}^2$$

The nylon ring may be considered as a thick wall cylinder subjected to this internal pressure (see Appendix D). At the inner surface of the ring there will be a hoop stress, σ_θ, and a radial stress, σ_r. Benham *et al.* shows these to be

$$\text{Hoop stress, } \sigma_\theta = p \left\{ \frac{k^2 + 1}{k^2 - 1} \right\}$$

$$\text{Radial stress, } \sigma_r = -p$$

where k = ratio of outer to inner radius.

Hence the hoop strain, ε_θ, at the bore of the ring is given by

$$\varepsilon_\theta = \frac{\delta r}{r} = \frac{\sigma_\theta}{E} - \frac{v\sigma_r}{E}$$

So

$$\delta r = \frac{pr}{E} \left\{ \left(\frac{k^2 + 1}{k^2 - 1} \right) + v \right\}$$

$$= \frac{8.5(15)}{1 \times 10^3} \left\{ \left(\frac{1.67^2 + 1}{1.67^2 - 1} \right) + 0.4 \right\}$$

$$\delta r = 0.32 \text{ mm}$$

This is the interference needed on radius to give the desired separation force.

(b) To facilitate easy assembly, the nylon should be heated so that the inner radius expands by this amount. As shown in the previous section

$$\delta r = \alpha \cdot r \cdot \Delta T$$

$$\Delta T = \frac{0.32}{100 \times 10^{-6} \times 15} = 214°C$$

(c) The maximum stress in the nylon ring when it is on the shaft will be the hoop stress at the bore. This is given by

$$\sigma_\theta = p \left\{ \frac{k^2 + 1}{k^2 - 1} \right\} = 8.5 \left\{ \frac{1.67^2 + 1}{1.67^2 - 1} \right\}$$

$$\sigma_\theta = 18 \text{ MN/m}^2$$

As a practical point, it should be noted that the separation force will change with time. This is because the modulus, E, will decrease with time. Suppose that in this Example, the assembly is to be maintained in position for 1 year and that during this time the modulus decreases to half its initial value (the 1 year modulus would be obtained from the creep curves in the normal way). The above analysis shows that the interface pressure would then be half its initial value (because δr is fixed) and this in turn means that the separation force would become 500 N instead of 1 kN.

2.7 Multi-layer Mouldings

It is becoming common practice to have the cross-section of a plastic moulding made up of several different materials. This may be done to provide a permeation barrier whilst retaining attractive economics by having a less expensive material making up the bulk of the cross-section. To perform stress analysis in such cases, it is often convenient to convert the cross-section into an equivalent section consisting of only one material. This new section will behave in exactly the same way as the multi-layer material when the loads are applied. A very common example of this type of situation is where a solid skin and a foamed core are moulded to provide a very efficient stiffness/weight ratio. This type of situation may be analysed as follows:

Example 2.8 A polypropylene sandwich moulding is 12 mm thick and consists of a foamed core sandwiched between solid skin layers 2 mm thick. A beam 12 mm wide is cut from the moulding and is subjected to a point load, W, at mid-span when it is simply supported over a length of 200 mm. Estimate the depth of a solid beam of the same width which would have the same stiffness when loaded in the same way. Calculate also the weight saving by using the foam moulding. The density of the solid polypropylene is 909 kg/m^3 and the density of the foamed core is 600 kg/m^3.

Solution The first step in analysing the foamed sandwich type structure is to calculate the second moment of area of the cross-section. This is done by converting the cross section to an equivalent section of solid plastic. This is shown in Fig. 2.18.

The equivalent width of the flange in the I section is given by

$$b' = \frac{E_c}{E_s}(b_c) \tag{2.18}$$

where E_c and E_s refer to the modulus values for the core (c) and solid (s) material. In most cases there is very little information available on the modulus of foamed plastics but fortunately an empirical relationship has been found to

Fig. 2.18 Equivalent section for sandwich moulding

exist between density (ρ) and modulus (Moore *et al.*).

$$\frac{E_c}{E_s} = \left(\frac{\rho_c}{\rho_s}\right)^2 \tag{2.19}$$

From the equivalent section the second moment of area can then be calculated as

$$I = \frac{12(12)^3}{12} - \frac{12(8)^3}{12} + 12\left(\frac{600}{909}\right)^2\frac{(8)^3}{12}$$

$$= 1439 \text{ mm}^4$$

The solid polypropylene beam which would have the same stiffness when loaded in the same way would need to have the same 2nd moment of area. So if its depth is d then

$$\frac{12(d)^3}{12} = 1439, d = 11.3 \text{ mm}$$

The weight per unit length of the solid beam would be

$$W_s = 12 \times 11.3 \times 10^{-6} \times 909 \times 10^3 = 123 \text{ g}$$

The weight per unit length of the foamed beam is

$$W_f = (909 \times 2 \times 12 \times 2 \times 10^{-3}) + (600 \times 12 \times 8 \times 10^{-3})$$

$$= 101.2 \text{ g}$$

Hence the weight saving is 17.7%

Once the foamed plastic moulding has been converted to an equivalent section of solid plastic then the long term design procedures illustrated in the previous questions can be used. For example, to determine the 1 year deflection of the beam in this question, the appropriate 1 year modulus from the polypropylene creep curves could be used.

Throne has reported that the relationship between foam modulus and density can be generalised to other properties such as tensile strength, fatigue strength, creep properties as well as shear and compression modulus. Thus if X is the general material property then

$$\frac{X_c}{X_s} = \left(\frac{\rho_c}{\rho_s}\right)^2 \tag{2.20}$$

Example 2.9 A solid polyethylene beam is 10 mm thick and 15 mm wide. If it is to be replaced with a sandwich section with solid polyethylene in the two outer skins and polyethylene foam (density $= 200$ kg/m^3) in the centre, calculate the dimensions of the sandwich beam if it is to have optimum stiffness at the same weight as the solid beam. If the foam material costs 20% more than the solid material, calculate the increase or decrease in cost of the sandwich beam.

Solution The weight per unit length, W, for the solid beam is

$$W = 935 \times 10^3 \times 15 \times 10 \times 10^{-6} = 140 \text{ g/m}$$

For the sandwich section

$$W = \rho_c bh + \rho_s 2bd$$

$$h = \frac{W - 2bd\rho}{\rho_c b} = \frac{W}{b\rho_s D} - \frac{2d}{D}$$

where $D = $ density ratio (ρ_c/ρ_s).

The flexural stiffness of the beam will be proportional to EI. So converting the section to an equivalent 'I' beam as in the previous example

$$E_s I = E_s \left[\frac{b(2d + h)^3}{12} - \frac{bh^3}{12} + \left(\frac{\rho_c}{\rho_s}\right)^2 \frac{bh^3}{12}\right]$$

Substituting for h from above, the stiffness may be optimised for constant W, by differentiating this expression and equating to zero. This gives the optimum beam depth as

$$d = \frac{1}{2\left(24 + \dfrac{72}{D^2} - \dfrac{96}{D}\right)}\left[\frac{48W}{b\rho_s D^2} - \frac{48W}{b\rho_s D} + \frac{24W\sqrt{(D^2 - 1)(D - 1)}}{b\rho_s D^2}\right]$$

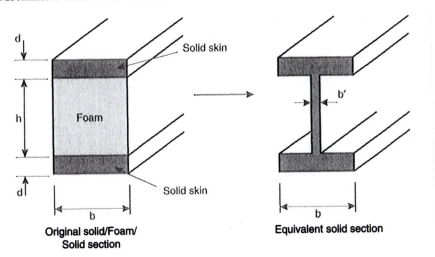

Fig. 2.19

Using the information given

$$D = 200/935 = 0.214, \qquad \rho_s = 935 \times 10^{-6} \text{ g/mm}^3, \qquad W = 0.14 \text{ g/mm}$$

So
$$d = 1.61 \text{ mm}$$

Using the expression for h from above, the thickness of the foam is given by

$$h = 31.6 \text{ mm}$$

The cost of the solid beam is given by

$$(\text{Cost})_s = 935 \times 10 \times 15 \times 10^{-6} \times C_s = 0.14 \, C_s$$

The cost of the sandwich beam is given by

$$(\text{Cost})_F = \rho_c bhC_c + \rho_s 2bdC_s = 0.159 \, C_s$$

Hence
$$\text{the additional cost} = \frac{0.159 - 0.14}{0.14} = 13.5\%$$

It is interesting to generalise this solution for any core: skin density ratio. Fig. 2.20 shows how the optimum skin thickness varies with D. This is independent of the solid material density or modulus but is based on a weight per unit length of 140 g/m (equivalent to a solid thickness of 10 mm). If a different solid thickness is of interest then the optimum skin thickness would need to be scaled accordingly. For example, for a 5 mm thick solid beam which is to be converted to a foam sandwich beam of the same weight per unit length, the optimum skin thickness would be half of the value shown in Fig. 2.20. The optimum skin/core thickness ratios are shown in Fig. 2.21.

Fig. 2.20   Variation of optimum skin thickness with core: skin density ratio

Fig. 2.21   Optimum skin to core ratio for constant weight

The stiffness ratios (i.e. stiffness of the foam sandwich beam relative to the original solid beam) are also given in Fig. 2.21. In both cases the values given are independent of the original solid material or its dimensions, so this provides a good design chart. The design of solid/foam sandwich structures is also considered in Chapter 3 in the laminate analysis.

2.8 Design of Snap Fits

A major attraction of plastics to designers is the ease with which fast assembly mechanisms can be incorporated into the end-product. A very good example of this is the snap fit. A typical design is shown in Fig. 2.22 although there are many variations. Snap fits exploit the fact that thin plastic sections can undergo relatively large flexural deflections for a short period of time and exhibit complete recovery. The design of snap fits is straightforward and does not involve creep curves since the time-scale of the deflection/stress is small. The point that will be illustrated here is that in a real design situations it is necessary to choose combinations of dimensions which provide the necessary function but which do not overstress the plastic. In the following example a

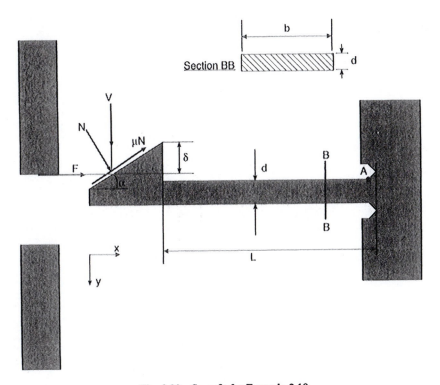

Fig. 2.22   Snap fit for Example 2.10

set of design curves are developed to show how the different combinations of dimensions might be selected.

Example 2.10 The polypropylene snap fit shown in Fig. 2.22 is to have a length of 10–30 mm. If the insertion force is not to exceed 4 N and the yield stress of the plastic is 30 MN/m^2, calculate suitable cross-sectional dimensions for the snap fit. The short-term modulus of the polypropylene is 900 MN/m^2 and the coefficient of friction is 0.3. The safety factor on stress is to be 2.

Solution

Resolving vertical forces gives

$$V = N \cos \alpha - \mu N \sin \alpha$$
$$= N(\cos \alpha - \mu \sin \alpha)$$

Resolving horizontal forces gives

$$F = N \sin \alpha + \mu N \cos \alpha$$
$$= N(\sin \alpha + \mu \cos \alpha)$$

Eliminating N from the above equations gives

$$V = F \left\{ \frac{1 - \mu \tan \alpha}{\mu + \tan \alpha} \right\}$$

The snap fit may be regarded as a cantilever and for this situation, the vertical deflection, δ, is given by

$$\delta = \frac{VL^3}{3EI} \qquad \text{where } I = \frac{bd^3}{12}$$

$$\delta = \frac{FL^3}{3EI} \left\{ \frac{1 - \mu \tan \alpha}{\mu + \tan \alpha} \right\}$$

Insertion force

$$F = \frac{Eb\delta}{4} \frac{1}{(L/d)^3} \left\{ \frac{1 - \mu \tan \alpha}{\mu + \tan \alpha} \right\} \tag{2.21}$$

For polypropylene in the situation given, Fig. 2.23 shows a set of design curves linking F, b and (L/d). For a limiting insertion force of 4 N and a typical maximum thickness, d, of 2 mm (to avoid excessive moulding times – Chapter 5), a length of 20 mm would give an L/d ratio of 10 which gives $b = 18$ mm.

Fig. 2.23 Variation of insertion force with beam dimensions

It is also necessary to check that the stress in the snap fit does not exceed the yield strength for the material (30 MN/m²). Allowing for a safety factor of 2, the maximum permissible stress will be 15 MN/m².

The stress in a cantilever beam of this type will be maximum at point A (Fig. 2.22) and is given by

$$\text{stress, } \sigma_A = \frac{My}{I}$$

where M = bending moment (= VL)
y = half depth of beam (= $d/2$)
I = second moment of area (= $bd^3/12$)

$$\sigma_A = \frac{VLd}{2I} = \frac{FLd}{2I} \left\{ \frac{1 - \mu \tan \alpha}{\mu + \tan \alpha} \right\}$$

$$\sigma_A = \frac{3E\delta d}{2L^2} \tag{2.22}$$

For the conditions given, $\sigma_A = 13.5$ MN/m² which is less than the permitted maximum of 15 MN/m² and so the chosen dimensions are acceptable. Note that the stress is independent of the width, b. Fig. 2.24 gives a set of design curves for the situation considered so that it may be seen that other combinations of L, d and b could be chosen to meet the design specification. For example, $L = 22$ mm, $d = 2.5$ mm and $b = 12$ mm gives an insertion force of 4 N and a maximum stress of 14 MN/m² which is again acceptable.

Fig. 2.24   Variation of stress with beam dimensions

2.9 Design of Ribbed Sections

It will be shown later (Chapters 4 and 5) that it is normally good practice to design plastic products with cross-sections which are as thin as possible. This is because thick sections cool very slowly and hence the moulding cycle times are long resulting in uneconomic production. Also, thick sections tend to shrink more and can lead to warpage and distortion. Of course thin sections tend to have low stiffness and so engineers usually adopt geometrical configurations which will enhance stiffness whilst retaining the required thinness in the wall section.

In any particular material, the flexural stiffness will be defined by the second moment of area, I, for the cross-section. As with a property such as area, the second moment of area is independent of the material – it is purely a function of geometry. If we consider a variety of cross-sections as follows, we can easily see the benefits of choosing carefully the cross-sectional geometry of a moulded plastic component.

All the sections have the same cross-sectional area (and hence the same weight).

(a)	–	Solid Rectangle	$(B = 10D)$
(b)	–	Solid Rectangle	$(B = 4D)$
(c)	–	Solid Rectangle	$(B = 2D)$
(d)	–	Square	$(B = D)$
(e)	–	Hollow rectangular box	$(B = 2D, B = 10h)$

(f)   –   Solid circular section
(g)   –   Thin wall tube $(D = 10h)$
(h)   –   Thick wall tube $(D = 4h)$
(i)   –   T-section $(B = 8h)$

Fig. 2.25 illustrates the various cross-sections and Fig. 2.26 compares their flexural stiffnesses. An important conclusion from this is that 'T' sections (or

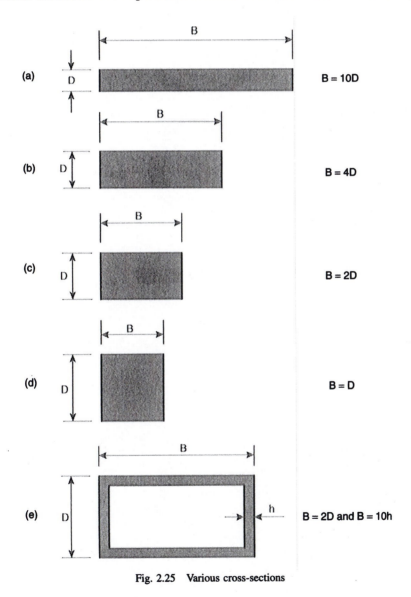

(a)   D   B = 10D

(b)   D   B = 4D

(c)   D   B = 2D

(d)   D   B = D

(e)   D   h   B = 2D and B = 10h

Fig. 2.25   Various cross-sections

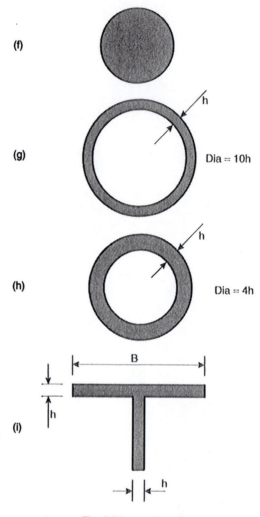

Fig. 2.25   (*continued*)

indeed 'L' or 'U' sections) are very efficient at providing excellent stiffness per unit weight. The following Example illustrates how such sections can be designed into plastic products.

Example 2.11 An aluminium cantilever beam is 50 mm wide, 80 mm long and 2 mm deep. The loading is 200 N spread uniformly over the cantilever. If the beam is to be replaced by one made from acetal and the design criteria is that the end deflections should be the same in each beam after one month, calculate the dimensions (a) of a solid acetal beam and (b) an acetal beam with unidirectional ribs. The modulus of the aluminium is 70 GN/m^2.

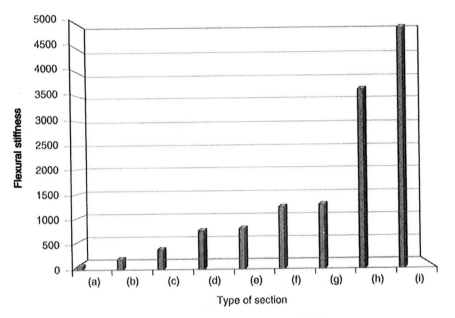

Fig. 2.26   Effect of geometry of flexural stiffness

Solution

(a) In order that the two beams (aluminium and acetal) will have the same stiffness it is necessary that

$$(EI)_{al} = (EI)_{acetal} \tag{2.23}$$

In order to get the 1 month modulus for acetal, the strain in the beam must be known.

$$\text{Strain,} \quad \varepsilon = \frac{\sigma}{E} = \frac{My}{EI} = \frac{W(L/2)(d_{al}/2)}{EI} = \frac{WLd_{al}}{4EI}$$

where W = total load (200 N) and d_{al} = depth of beam.

Also, deflection

$$\delta = \frac{WL^3}{8EI}$$

So, strain,

$$\varepsilon = \frac{2\delta d_{al}}{L^2}$$

Using the information for the aluminium

$$\delta = \frac{200 \times 80^3 \times 12}{8 \times 70 \times 10^3 \times 50 \times 2^3} = 5.5 \text{ mm}$$

Therefore,

$$\varepsilon = \frac{2 \times 5.5 \times 2}{80^2} = 0.34\%$$

From the creep curves for acetal, at 1 month (2.6×10^6 s) and 0.34% strain, the modulus (σ/ε) is approximately 1.3 GN/m^2.

Therefore, using (2.23) and ignoring ($W/12$) on each side

$$70 \times 2^3 = 1.3 \times D^3$$

$$D = 7.6 \text{ mm}$$

Hence an acetal beam 7.6 mm deep and of the same width and length as the aluminium beam will perform in exactly the same way when the load of 200 N is applied.

However, as will be shown in Chapters 4 and 5, it would not be good design practice to use a plastic beam Section 7.6 mm thick. Such a section would take a long time to mould and would probably have sink marks on its surface. It would be preferable to use thinner wall sections and introduce ribs to provide the desired stiffness. This is illustrated below.

(b) Consider the flat beam to be replaced by a ribbed section as shown in Fig. 2.27. It is necessary for the second moment of area, I, to be the same for both sections. Clearly there are many permutations for the dimensions of the

Fig. 2.27   Flat beam and ribbed section, Example 2.11

ribbed section which would meet this requirement. One practical consideration is that the thickness of the rib should be less than the thickness of the top plate (*d*) in order that there will be no evidence (sink mark) to show the presence of the rib on the underside of the plate. A typical ratio (β) of rib thickness to plate thickness is 0.6.

In order to assist with the design of the ribbed plate, Fig. 2.28 shows a chart to enable permissible combinations of dimensions to be chosen. For example, suppose that we wish the plate thickness to be 2 mm and there are to be five ribs (= *N*) across the plate.

Fig. 2.28   Design of ribbed plates

Then

$$W/ND = \frac{50}{5 \times 7.6} = 0.66 \text{ and } h/D = \frac{2}{7.6} = 0.26$$

From Fig. 2.28, $d/D = 1.3$.

So the depth of the 5 ribs would be $1.3 \times 7.6 = 9.9$ mm and their width would be $0.6 \times 2 = 1.2$ mm.

Such a ribbed beam would perform in exactly the same way as the original aluminium beam or the flat acetal beam. However, an additional consideration is buckling of the ribs. If the ribs are thin and deep there is a possibility that they will buckle along the lower unsupported edge if they are loaded in compression (i.e. downward deflection of the cantilever in this Example).

Buckling is complex to analyse but if we consider the rib as a flat plate, clamped along one edge then Roark gives the formula for the critical buckling stress as

$$\sigma_c = \frac{1.2E}{(1 - v^2)} \left(\frac{\beta h}{d}\right)^2 \tag{2.24}$$

or

$$\left(\frac{h}{d}\right) = \sqrt{\left(\frac{1 - v^2}{1.2\beta^2}\right) \frac{\sigma_c}{E}}$$

If we take the critical stress as the yield stress then for many plastics, the ratio of σ_c/E is approximately 35×10^{-3}. Using Poisson's ratio, $v = 0.35$ and taking $\beta = 0.6$, as before, then

$$\frac{h}{d} = \frac{h}{D} \cdot \frac{D}{d} = 0.267$$

or

$$\frac{d}{D} = 3.75 \frac{h}{D}$$

This line may be superimposed on the rib design data as shown in Fig. 2.28. Combinations of dimensions above this line are likely to provide ribs which are too slender and so are liable to buckling. Combinations below the line are likely to be acceptable but do remember the assumptions made in the determination of the buckling line – in particular, the ratio of σ_y/E will increase with time due to creep and this will cause the buckling line to move downwards.

Returning to the Example, it is apparent that the dimensions chosen lie above the buckling line. It is necessary therefore, to choose other dimensions. For example, $h = 2.5$ mm gives

$$\frac{h}{D} = 0.33, \qquad \frac{W}{ND} = 0.66 \text{ (as before)}$$

So $d/D = 1.15$ which gives $d = 11.15(7.6) = 8.7$ mm.

These dimensions lie below the buckling line and so are acceptable. The solution would therefore be a ribbed beam with five ribs, plate thickness = 2.5 mm, rib thickness = 1.5 mm and rib depth = 8.7 mm.

It should be noted that ribbed sections play an extremely important part in the design of plastic products. Not only do they reduce manufacturing times (because they utilise thinner sections), but they also save material. It may easily be shown in this case that the volume of the ribbed beam is half the volume of the flat acetal beam and this will result in a substantial cost saving.

Fig. 2.29 shows a design chart for slightly thicker ribs ($\beta = 0.8$) so as to reduce the likelihood of rib buckling.

Fig. 2.29 Design of ribbed beams

2.10 Stiffening Mechanisms in Other Moulding Situations

In some moulding methods such as blow moulding, thermoforming and rotational moulding, it is difficult if not impossible to produce stiffening ribs. Hence other stiffening configurations such as corrugations have to be used. Fig. 2.30 illustrates a typical repeat unit in such a corrugation. The second moment of

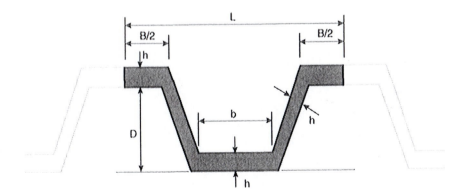

Fig. 2.30 Corrugation profile

area, I, will quantify the stiffening effect of the cross-section and if this is related to the I value for the flat sheet of the same length and thickness then it is possible to define a stiffness enhancement factor, q, as

$$q = \frac{1}{L} \left\{ mh + \frac{12m}{h} \left(\bar{y} - \frac{h}{2} \right)^2 + ph + \frac{12p}{h} \left(\bar{y} - \left(nh + \frac{h}{2} \right) \right)^2 \right.$$
$$\left. + \frac{2nh}{\sin \alpha} + \frac{24n}{h \sin \alpha} \left(\bar{y} - \frac{nh}{2} \right)^2 \right\} \tag{2.25}$$

where $\bar{y} =$ distance to centre of area from the base of the cross-section, i.e.

$$\bar{y} = h \left\{ \frac{0.5m + pn + 0.5p + \dfrac{n^2}{\sin \alpha}}{\left(p + m + \dfrac{2n}{\tan \alpha} \right)} \right\}$$

and

$$m = \frac{b}{h}, \qquad p = \frac{B}{h}, \qquad n = \frac{D}{h}$$

Clearly there are many permutations of D, b, h, α, etc and Fig. 2.31 shows how the stiffness enhancement factor, q, changes with various values of these parameters. In each case the angle α has been fixed at $85°$ and the corrugation dimensions have been expressed as a function of the wall thickness, h.

Fig. 2.31   Stiffness of corrugations as function of dimensions

The obvious question is 'Is there an optimum design for the corrugations?' Unfortunately the answer is 'No' because if one wishes to increase transverse stiffness then the obvious thing to do is to increase D up to the point where buckling problems start to be a concern. Usually this is when $D/h = 10$, for short-term loading and less than this for long term loading because of the decrease in modulus of viscoelastic materials.

Another approach is to recognise that initially for a flat sheet, the axial stiffness is high but the transverse stiffness is relatively low. As the corrugation depth increases then the transverse stiffness increases but at the expense of the axial stiffness. It is readily shown that the axial deflection per unit load for the corrugations for the new geometry compared with the flat sheet is given by

$$\text{Axial stiffness ratio} = \frac{4n^3h}{L\sin^2\alpha} \qquad (2.26)$$

If this is then divided into the previous enhancement ratio, q, it is possible to observe the way in which one stiffness increases at the expense of the other. Fig. 2.32 shows this transverse/axial stiffness ratio as a function of the depth of the corrugations. It may be seen that when the depth is less than four times the wall thickness then the axial stiffness ratio is better than the transverse stiffness ratio. However, when the depth is greater than four times the wall thickness then the transverse stiffness ratio dominates.

Fig. 2.32   Optimisation of corrugation depth

This would suggest that $D/h = 4$ offers the best balance of transverse and axial properties. In fact the 'optimum' ratio is essentially independent of the thickness, h, but it depends on B and b. Fig. 2.33 shows this relationship and reflects the combinations of B, b, D and h which will give the best design of corrugations. Generally it is desirable to have B and b approximately equal and it is also good design practice to keep D/h well below 10 (to avoid buckling). Hence values of B, h and D in the shaded area tend to be best. A combination of $D/h = 4$, $b/h = 5$ and $B/h = 4$ is known to give good results for rotationally moulded products where shape must be used very effectively to ensure good performance.

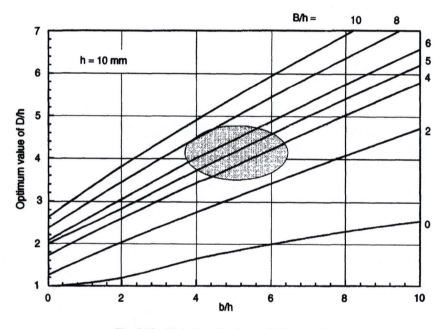

Fig. 2.33   Variation of optimum D/h with b/h

2.11 Mathematical Models of Viscoelastic Behaviour

Over the years there have been many attempts to simulate the behaviour of viscoelastic materials. This has been aimed at (i) facilitating analysis of the behaviour of plastic products, (ii) assisting with extrapolation and interpolation of experimental data and (iii) reducing the need for extensive, time-consuming creep tests. The most successful of the mathematical models have been based on spring and dashpot elements to represent, respectively, the elastic and viscous responses of plastic materials. Although there are no discrete molecular structures which behave like the individual elements of the models, nevertheless they do aid in the understanding and analysis of the behaviour of viscoelastic materials. Some of the more important models will now be considered.

(a) Maxwell Model

The Maxwell Model consists of a spring and dashpot in series at shown in Fig. 2.34. This model may be analysed as follows.

σ_1
ε_1
ξ

σ_2
ε_2
η

Stress σ

Fig. 2.34   The Maxwell model

Stress–Strain Relations

The spring is the elastic component of the response and obeys the relation

$$\sigma_1 = \xi \cdot \varepsilon_1 \qquad (2.27)$$

where σ_1 and ε_1 are the stress and strain respectively and ξ is a constant.

The dashpot is the viscous component of the response and in this case the stress σ_2 is proportional to the rate of strain $\dot{\varepsilon}_2$, ie

$$\sigma_2 = \eta \cdot \dot{\varepsilon}_2 \qquad (2.28)$$

where η is a material constant.

Equilibrium Equation

For equilibrium of forces, assuming constant area

$$\text{Applied Stress, } \sigma = \sigma_1 = \sigma_2 \qquad (2.29)$$

Geometry of Deformation Equation

The total strain, ε is equal to the sum of the strains in the two elements. So

$$\varepsilon = \varepsilon_1 + \varepsilon_2 \qquad (2.30)$$

From equations (2.27), (2.28) and (2.30)

$$\dot{\varepsilon} = \frac{1}{\xi}\dot{\sigma}_1 + \frac{1}{\eta}\sigma_2$$

$$\dot{\varepsilon} = \frac{1}{\xi} \cdot \dot{\sigma} + \frac{1}{\eta} \cdot \sigma \qquad (2.31)$$

This is the governing equation of the Maxwell Model. It is interesting to consider the response that this model predicts under three common time-dependent modes of deformation.

(i) Creep

If a constant stress, σ_o, is applied then equation (2.31) becomes

$$\dot{\varepsilon} = \frac{1}{\eta} \cdot \sigma_o \qquad (2.32)$$

which indicates a constant rate of increase of strain with time.

From Fig. 2.35 it may be seen that for the Maxwell model, the strain at any time, t, after the application of a constant stress, σ_o, is given by

$$\varepsilon(t) = \frac{\sigma_o}{\xi} + \frac{\sigma_o}{\eta}t$$

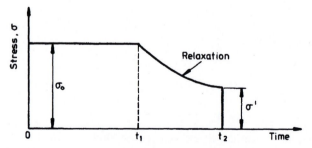

Fig. 2.35   Response of Maxwell model

Hence, the creep modulus, $E(t)$, is given by

$$E(t) = \frac{\sigma_o}{\varepsilon(t)} = \frac{\xi\eta}{\eta + \xi t} \tag{2.33}$$

(ii) Relaxation

If the strain is held constant then equation (2.31) becomes

$$0 = \frac{1}{\xi} \cdot \dot{\sigma} + \frac{1}{\eta} \cdot \sigma$$

Solving this differential equation (see Appendix B) with the initial condition $\sigma = \sigma_o$ at $t = t_o$ then,

$$\sigma(t) = \sigma_o e^{-\frac{\xi}{\eta}t} \tag{2.34}$$

$$\sigma(t) = \sigma_o e^{-t/T_R} \tag{2.35}$$

where $T_R = \eta/\xi$ is referred to as the *relaxation time*.

This indicates that the stress decays exponentially with a time constant of η/ξ (see Fig. 2.35).

(iii) Recovery

When the stress is removed there is an instantaneous recovery of the elastic strain, ε^1, and then, as shown by equation (2.31), the strain rate is zero so that there is no further recovery (see Fig. 2.35).

It can be seen therefore that although the relaxation behaviour of this model is acceptable as a first approximation to the actual materials response, it is inadequate in its prediction for creep and recovery behaviour.

(b) Kelvin or Voigt Model

In this model the spring and dashpot elements are connected in parallel as shown in Fig. 2.36.

Stress, σ

Fig. 2.36   The Kelvin or Voigt Model

Stress–Strain Relations

These are the same as the Maxwell Model and are given by equations (2.27) and (2.28).

Equilibrium Equation

For equilibrium of forces it can be seen that the applied load is supported jointly by the spring and the dashpot, so

$$\sigma = \sigma_1 + \sigma_2 \tag{2.36}$$

Geometry of Deformation Equation

In this case the total strain is equal to the strain in each of the elements, i.e.

$$\varepsilon = \varepsilon_1 = \varepsilon_2 \tag{2.37}$$

From equations (2.27), (2.28) and (2.36)

$$\sigma = \xi \cdot \varepsilon_1 + \eta \dot{\varepsilon}_2$$

or using equation (2.37)

$$\sigma = \xi \cdot \varepsilon + \eta \cdot \dot{\varepsilon} \tag{2.38}$$

This is the governing equation for the Kelvin (or Voigt) Model and it is interesting to consider its predictions for the common time dependent deformations.

(i) Creep

If a constant stress, σ_o, is applied then equation (2.38) becomes

$$\sigma_o = \xi \cdot \varepsilon + \eta \dot{\varepsilon}$$

and this differential equation may be solved for the total strain, ε, to give

$$\varepsilon(t) = \frac{\sigma_o}{\xi} \left[1 - e^{-\frac{\xi}{\eta}t} \right]$$

where the ratio η/ξ is referred to as the *retardation time*, T_R.

This indicates an exponential increase in strain from zero up to the value, σ_o/ξ, that the spring would have reached if the dashpot had not been present. This is shown in Fig. 2.37. As for the Maxwell Model, the creep modulus may be determined as

$$E(t) = \frac{\sigma_o}{\varepsilon(t)} = \xi \left[1 - e^{-\frac{t}{T_R}} \right]^{-1} \tag{2.39}$$

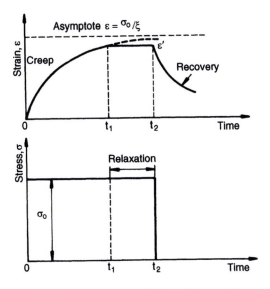

Fig. 2.37   Response of Kelvin/Voigt model

(ii) Relaxation

If the strain is held constant then equation (2.38) becomes

$$\sigma = \xi \cdot \varepsilon$$

That is, the stress is constant and supported by the spring element so that the predicted response is that of an elastic material, i.e. no relaxation (see Fig. 2.37)

(iii) Recovery

If the stress is removed, then equation (2.38) becomes

$$0 = \xi \cdot \varepsilon + \eta \dot{\varepsilon}$$

Solving this differential equation with the initial condition $\varepsilon = \varepsilon'$ at the time of stress removal, then

$$\varepsilon(t) = \varepsilon' e^{-\frac{\xi t}{\eta}} \tag{2.40}$$

This represents an exponential recovery of strain which is a reversal of the predicted creep.

(c) More Complex Models

It may be seen that the simple Kelvin model gives an acceptable first approximation to creep and recovery behaviour but does not account for relaxation. The Maxwell model can account for relaxation but was poor in relation to creep and recovery. It is clear therefore that some compromise may be achieved by combining the two models. Such a set-up is shown in Fig. 2.38. In this case the stress–strain relations are again given by equations (2.27) and (2.28). The geometry of deformation yields.

$$\text{Total strain, } \varepsilon = \varepsilon_1 + \varepsilon_2 + \varepsilon_k \tag{2.41}$$

Fig. 2.38   Maxwell and Kelvin models in series

where ε_k is the strain response of the Kelvin Model. From equations (2.27), (2.28) and (2.41).

$$\varepsilon(t) = \frac{\sigma_o}{\xi_1} + \frac{\sigma_o t}{\eta_1} + \frac{\sigma_o}{\xi_2}\left[1 - e^{-\frac{\xi_2 t}{\eta_2}}\right] \tag{2.42}$$

From this the strain rate may be obtained as

$$\dot{\varepsilon} = \frac{\sigma_o}{\eta_1} + \frac{\sigma_o}{\eta_2}e^{-\frac{\xi_2}{\eta_2}t} \tag{2.43}$$

The response of this model to creep, relaxation and recovery situations is the sum of the effects described for the previous two models and is illustrated in Fig. 2.39. It can be seen that although the exponential responses predicted in these models are not a true representation of the complex viscoelastic response of polymeric materials, the overall picture is, for many purposes, an acceptable approximation to the actual behaviour. As more and more elements are added to the model then the simulation becomes better but the mathematics become complex.

Example 2.12 An acrylic moulding material is to have its creep behaviour simulated by a four element model of the type shown in Fig. 2.38. If the creep curve for the acrylic at 14 MN/m^2 is as shown in Fig. 2.40, determine the values of the four constants in the model.

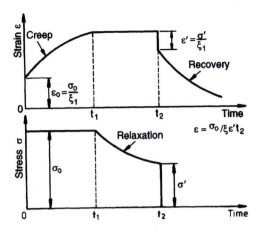

Fig. 2.39   Response of combined Maxwell and Kelvin models

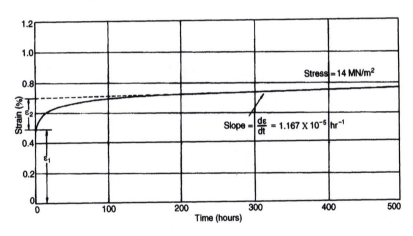

Fig. 2.40   Creep curve for acrylic at 20°C

Solution The spring element constant, ξ_1, for the Maxwell model may be obtained from the instantaneous strain, ε_1. Thus

$$\xi_1 = \frac{\sigma_o}{\varepsilon_1} = \frac{14}{0.005} = 2800 \text{ MN/m}^2$$

The dashpot constant, η_1, for the Maxwell element is obtained from the slope of the creep curve in the steady state region (see equation (2.32)).

$$\eta_1 = \frac{\sigma_o}{\dot{\varepsilon}} = \frac{14}{1.167 \times 10^{-6}} = 1.2 \times 10^7 \text{ MN.hr/m}^2$$

$$= 4.32 \times 10^{10} \text{ MN.s/m}^2$$

The spring constant, ξ_2, for the Kelvin–Voigt element is obtained from the maximum retarded strain, ε_2, in Fig. 2.40.

$$\xi_2 = \frac{\sigma_o}{\varepsilon_2} = \frac{14}{(0.7 - 0.5)10^{-2}} = 7000 \text{ MN/m}^2$$

The dashpot constant, η_2, for the Kelvin–Voigt element may be determined by selecting a time and corresponding strain from the creep curve in a region where the retarded elasticity dominates (i.e. the knee of the curve in Fig. 2.40) and substituting into equation (2.42). If this is done then $\eta_2 = 3.7 \times 10^8$ MN.s/m^2.

Having thus determined the constants for the model the strain may be predicted for any selected time or stress level assuming of course these are within the region where the model is applicable.

(d) Standard Linear Solid

Another model consisting of elements in series and parallel is that attributed to Zener. It is known as the Standard Linear Solid and is illustrated in Fig. 2.41. The governing equation may be derived as follows.

Fig. 2.41   The standard linear solid

Stress–Strain Relations

As shown earlier the stress–strain relations are

$$\sigma_1 = \xi_1 \varepsilon_1 \tag{2.44}$$

$$\sigma_2 = \xi_2 \varepsilon_2 \tag{2.45}$$

$$\sigma_3 = \eta_3 \dot{\varepsilon}_3 \tag{2.46}$$

Equilibrium Equation

In a similar manner to the previous models, equilibrium of forces yields.

$$\sigma_1 = \sigma_3$$

$$\sigma = \sigma_1 + \sigma_2 \tag{2.47}$$

Geometry of Deformation Equation

In this case the total deformation, ε, is given by

$$\varepsilon = \varepsilon_2 = \varepsilon_1 + \varepsilon_3 \qquad (2.48)$$

From equation (2.48)

$$\dot{\varepsilon} = \dot{\varepsilon}_1 + \dot{\varepsilon}_3$$

but from equation (2.47)

$$\dot{\sigma}_1 = \dot{\sigma} - \dot{\sigma}_2$$

and $\sigma_3 = \sigma - \sigma_2$

$$\dot{\varepsilon} = \frac{\dot{\sigma} - \xi_2 \dot{\varepsilon}}{\xi_1} + \frac{\sigma - \xi_2 \varepsilon}{\eta_3}$$

Rearranging gives

$$\eta_3 \dot{\sigma} + \xi_1 \sigma = \eta_3 (\xi_1 + \xi_2) \dot{\varepsilon} + \xi_2 \xi_1 \varepsilon \qquad (2.49)$$

This is the governing equation for this model.
The behaviour of this model can be examined as before

(i) Creep

If a constant stress, σ_o, is applied then the governing equation becomes

$$\dot{\varepsilon}\{\eta_3(\xi_1 + \xi_2)\} + \xi_1 \xi_2 \varepsilon - \xi_1 \sigma_o = 0$$

The solution of this differential equation may be obtained using the boundary condition $\varepsilon = \sigma_o/(\xi_1 + \xi_2)$ at $t = 0$. So

$$\text{strain, } \varepsilon(t) = \frac{\sigma_o}{\xi_2} - \frac{\sigma_o \xi_1}{(\xi_1 + \xi_2)\xi_2} e^{\frac{-\xi_1 \xi_2 t}{\eta_3(\xi_1 \xi_2)}} \qquad (2.50)$$

It may be seen in Fig. 2.42 that this predicts the initial strain when the stress is first applied as well as an exponential increase in strain subsequently.

(ii) Relaxation

If the strain is held constant at ε', then the governing equation becomes

$$\eta_3 \dot{\sigma} + \xi_1 \sigma - \xi_1 \xi_2 \varepsilon' = 0$$

This differential equation may be solved with the boundary condition that $\sigma = \sigma_o = \varepsilon'(\xi_1 + \xi_2)$ when the strain is first kept constant.

$$\text{Stress, } \sigma(t) = \frac{\sigma_o}{\xi_1 + \xi_2} \left\{ \xi_2 + \xi_1 e^{\frac{-\xi_1 t}{\eta_3}} \right\} \qquad (2.51)$$

This predicts an exponential decay of stress as shown in Fig. 2.42.

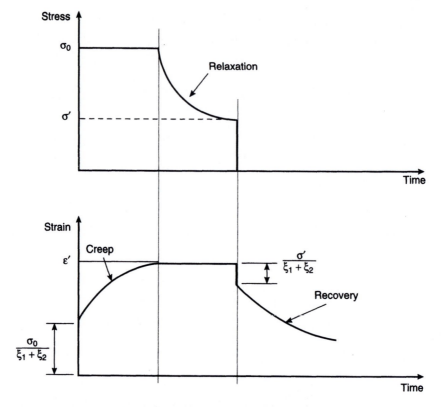

Fig. 2.42   Response of standard linear solid

(iii) Recovery

If the stress is at a value of σ' and then completely removed, the governing equation becomes

$$\eta_3(\xi_1 + \xi_2)\dot{\varepsilon} + \xi_1\xi_2\varepsilon = 0$$

The solution of this differential equation may be obtained using the boundary condition that when the stress is removed, the strain is given by

$$\varepsilon' = \sigma'/(\xi_1 + \xi_2)$$

$$\text{strain, } \varepsilon(t) = \left\{ \varepsilon' - \frac{\sigma'}{(\xi_1 + \xi_2)} \right\} e^{\frac{-\xi_1\xi_2 t}{\eta_3(\xi_1 + \xi_2)}} \tag{2.52}$$

This predicts an instantaneous recovery of strain followed by an exponential decay.

It may be observed that the governing equation of the standard linear solid has the form

$$a_1\dot{\sigma} + a_o\sigma = b_1\dot{\varepsilon} + b_o\varepsilon$$

where a_1, a_o, b_1 and b_o are all material constants. In the more modern theories of viscoelasticity this type of equation or the more general form given in equation (2.53) is favoured.

$$a_n \frac{\partial^n \sigma}{\partial t^n} + a_{n-1} \frac{\partial^{n-1} \sigma}{\partial t^n} + \cdots + a_o \sigma = b_m \frac{\partial^m \varepsilon}{\partial t^m} \cdots + b_o \varepsilon \qquad (2.53)$$

The models described earlier are special cases of this equation.

2.12 Intermittent Loading

The creep behaviour of plastics considered to date has assumed that the level of the applied stress is constant. However, in service the material may be subjected to a complex pattern of loading and unloading cycles. This can cause design problems in that clearly it would not be feasible to obtain experimental data to cover all possible loading situations and yet to design on the basis of constant loading at the maximum stress would not make efficient use of material or be economical. In these cases it is useful to have methods of predicting the extent of the recovered strain which occurs during the rest periods of conversely the accumulated strain after N cycles of load changes.

There are several approaches that can be used to tackle this problem and two of these will be considered now.

2.12.1 Superposition Principle

The simplest theoretical model proposed to predict the strain response to a complex stress history is the Boltzmann Superposition Principle. Basically this principle proposes that for a linear viscoelastic material, the strain response to a complex loading history is simply the algebraic sum of the strains due to each step in load. Implied in this principle is the idea that the behaviour of a plastic is a function of its entire loading history. There are two situations to consider.

(a) Step Changes of Stress

When a linear viscoelastic material is subjected to a constant stress, σ_1, at time, t_1, then the creep strain, $\varepsilon(t)$, at any subsequent time, t, may be expressed as

$$\varepsilon(t) = \frac{1}{E(t - t_1)} \cdot \sigma_1 \qquad (2.54)$$

where $E(t - t_1)$ is the time-dependent modulus for the elapsed time $(t - t_1)$.

Then suppose that instead of this stress σ_1, another stress, σ_2 is applied at some arbitrary time, t_2, then at any subsequent time, t, the stress will have been applied for a time $(t - t_2)$ so that the strain will be given by

$$\varepsilon(t) = \frac{1}{E(t - t_2)} \cdot \sigma_1$$

Now consider the situation in which the stress, σ_1, was applied at time, t_1, and an additional stress, σ_2, applied at time, t_2, then Boltzmanns' Superposition Principle states that the total strain at time, t, is the algebraic sum of the two independent responses.

$$\varepsilon(t) = \frac{1}{E(t - t_1)}\sigma_o + \frac{1}{E(t - t_2)} \cdot \sigma_1$$

This equation can then be generalised, for any series of N step changes of stress, to the form

$$\varepsilon(t) = \sum_{i=1}^{i=N} \sigma_i \left[\frac{1}{E(t - t_i)} \right] \tag{2.55}$$

where σ_i is the step change of stress which occurs at time, t_i.

To illustrate the use of this expression, consider the following example.

Example 2.13 A plastic which can have its creep behaviour described by a Maxwell model is to be subjected to the stress history shown in Fig. 2.43(a). If the spring and dashpot constants for this model are 20 GN/m² and 1000 GNs/m² respectively then predict the strains in the material after 150 seconds, 250 seconds, 350 seconds and 450 seconds.

Solution From Section 2.11 for the Maxwell model, the strain up to 100s is given by

$$\varepsilon(t) = \frac{\sigma}{\xi} + \frac{\sigma t}{\eta}$$

Also the time dependent modulus $E(t)$ is given by

$$E(t) = \frac{\sigma}{\varepsilon(t)} = \frac{\xi\eta}{\eta + \xi t} \tag{2.56}$$

Then using equation (2.54) the strains may be calculated as follows:
(i) at $t = 150$ seconds; $\sigma_i = 10$ MN/m² at $t_1 = 0$, $\sigma_2 = -10$ MN/m² at $t_2 = 100$ s

$$\varepsilon(150) = \sigma_1 \left[\frac{\eta + \xi \cdot (t - t_1)}{\xi\eta} \right] + (\sigma_2) \left[\frac{\eta + \xi \cdot (t - t_2)}{\xi\eta} \right]$$

$$= 0.002 - 0.001 = 0.1\%$$

(ii) at 250 seconds; σ_1, σ_2 as above, $\sigma_3 = 5$ MN/m² at $t_3 = 200$ s

$$\varepsilon(250) = 10 \left[\frac{\eta + \xi \cdot (250 - 0)}{\xi\eta} \right] + (-10) \left[\frac{\eta + \xi \cdot (250 - 100)}{\xi\eta} \right]$$

$$+ 5 \left[\frac{\eta + \xi \cdot (250 - 200)}{\xi\eta} \right]$$

$$= 0.003 - 0.002 + 0.0005 = 0.15\%$$

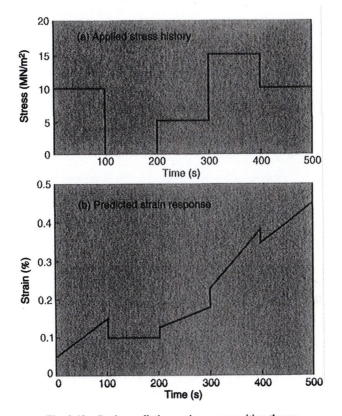

Fig. 2.43 Strain predictions using superposition theory

(iii) at 350 seconds; σ_1, σ_2, σ_3 as above, $\sigma_4 = 10$ MN/m^2 at $t_4 = 300$ s
So,

$$\varepsilon(350) = 0.003 = 0.3\%$$

(iv) and in the same way

$$\varepsilon(450) = 0.004 = 0.4\%$$

The predicted strain variation is shown in Fig. 2.43(b). The constant strain rates predicted in this diagram are a result of the Maxwell model used in this example to illustrate the use of the superposition principle. Of course superposition is not restricted to this simple model. It can be applied to any type of model or directly to the creep curves. The method also lends itself to a graphical solution as follows. If a stress σ_1 is applied at zero time, then the creep curve will be the time dependent strain response predicted by equation (2.54). When a second stress, σ_2 is added then the new creep curve will be obtained by adding the creep due to σ_2 to the anticipated creep if stress σ_1 had remained alone. This is illustrated in Fig. 2.44(a). Then if all the stress is removed this is equivalent to removing the creep strain due to σ_1 and σ_2 independently as shown in Fig. 2.44(b). The procedure is repeated in a similar way for any other stress changes.

Fig. 2.44(a)   Stress history

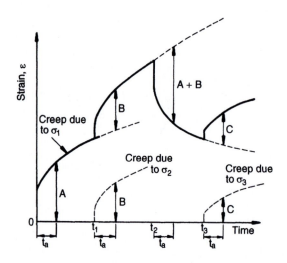

Fig. 2.44(b)   Predicted strain response using Boltzmann's superposition principle

(b) Continuous Changes of Stress

If the change in stress is continuous rather than a step function then equation (2.55) may be generalised further to take into account a continuous loading cycle. So

$$\varepsilon(t) = \int_{-\infty}^{t} \frac{1}{E(t - t_1)} \cdot \frac{d\sigma(t)}{dt} \cdot dt \tag{2.57}$$

where $\sigma(t)$ is the expression for the stress variation that begins at time, t_1. The lower limit is taken as minus infinity since it is a consequence of the

Superposition Principle that the entire stress history of the material contributes to the subsequent response.

It is worth noting that in exactly the same way, a material subjected to a continuous variation of strain may have its stress at any time predicted by

$$\sigma(t) = \int_{-\infty}^{t} E(t-t) \cdot \frac{d\varepsilon(t)}{dt} dt \qquad (2.58)$$

To illustrate the use of equation (2.57) consider the following Example.

Example 2.14 A plastic is subjected to the stress history shown in Fig. 2.45. The behaviour of the material may be assumed to be described by the Maxwell model in which the elastic component $\xi = 20$ GN/m² and the viscous component $\eta = 1000$ GNs/m². Determine the strain in the material (a) after u_1 seconds (b) after u_2 seconds and (c) after u_3 seconds.

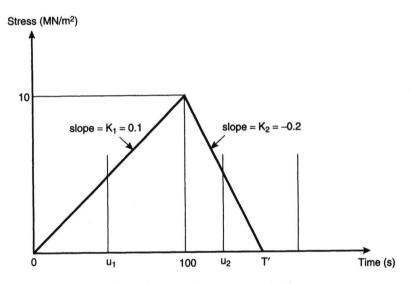

Fig. 2.45   Stress history to be analysed

Solution As shown in the previous Example, the modulus for a Maxwell element may be expressed as

$$E(t) = \frac{\xi\eta}{\eta + \xi t}$$

(a) The stress history can be defined as

$$-\infty < t < 0, \sigma(t) = 0 \rightarrow d\sigma(t)/dt = 0$$

$$0 < t < T, \sigma(t) = K_1 t \rightarrow d\sigma(t)/dt = K_1$$

Substituting into equation (2.57)

$$\varepsilon(u_1) = \int_{-\infty}^{0} \frac{1}{E(t)} \cdot 0 \cdot du + \int_{0}^{u_1} \frac{(\eta + \xi t - \xi t_1)}{\xi \eta} K_1 \cdot dt$$

$$\varepsilon(u_1) = \frac{K_1}{\xi \eta} \cdot \left[\eta t + \frac{\xi t^2}{2} - \xi t_1 t \right]_{0}^{u_1} \quad \text{where } t = 0$$

$$\varepsilon(u_1) = K_1 u_1 \left(\frac{1}{\xi} + \frac{u_1}{2\eta} \right)$$

then for $u_1 = 50$ seconds and $K_1 = 0.1$ MN/m^2 s

$$\varepsilon(50) = 0.1(50) \left(\frac{1}{20 \times 10^3} + \frac{50}{2 \times 10^6} \right) = 0.037\%$$

$$\varepsilon(100) = 0.1(100) \left(\frac{1}{20 \times 10^3} + \frac{100}{2 \times 10^6} \right) = 0.1\%$$

It is interesting to note that if K_1 was large (say $K_1 = 10$ in which case $T = 1$ second) then the strain predicted after application of the total stress (10 MN/m^2) would be $\varepsilon(1) = 0.0505\%$. This agrees with the result in the previous Example in which the application of stress was regarded as a step function. The reader may wish to check that if at time $T = 1$ second, the stress was held constant at 10 MN/m^2 then after 100 seconds the predicted strain using the integral expression would be $\varepsilon(100) = 0.1495\%$ which again agrees with the previous example.

(b) After the time T, the *change* in stress is given by

$$\text{Change in stress, } \sigma(t) = -K_1(t - T) + K_2(t - T)$$

$$\frac{d\sigma(t)}{dt} = -K_1 + K_2$$

Hence,

$$\varepsilon(u_2) = \int_{t_1}^{u_2} \frac{\eta + \xi t - \xi t_1}{\xi \eta} K_1 \cdot dt + \int_{t_2}^{u_2} \frac{\eta + \xi t - \xi t_2}{\xi \eta} (-K_1 + K_2) dt$$

where $t_1 = 0$ and $t_2 = T = 100$ s.

$$\varepsilon(u_2) = K_1 u_2 \left[\frac{1}{\xi} + \frac{u_2}{2\eta} \right] + (K_1 - K_2)T \left[\frac{1}{\xi} - \frac{T}{2\eta} \right]$$

$$- (K_1 - K_2)u_2 \left[\frac{1}{\xi} + \frac{u_2}{2\eta} - \frac{T}{\eta} \right]$$

then for $K_2 = -0.2$ MN/m^2 s, $T = 100$ seconds and $u_2 = 125$ seconds

$$\varepsilon(125) = 0.094\%$$

(c) After the time T^1, the change in stress is given by

$$\text{Change in stress, } \sigma(t) = K_2(t - T^1)$$

$$\frac{d\sigma(t)}{dt} = K_2$$

Hence,

$$\varepsilon(u_3) = \int_{t_1}^{u_3} \frac{\eta + \xi t - \xi t_1}{\xi \eta} \cdot K_1 dt + \int_{t_2}^{u_3} \frac{\eta + \xi t - \xi t_2}{\xi \eta} (-K_1 + K_2) dt$$

$$+ \int_{T'}^{u_3} \frac{\eta + \xi t - \xi T^1}{\xi \eta} K_2 \, dt$$

and from this, for $u_3 = 200$ seconds and $K_2 = -0.2$ MN/m^2 s

$$\varepsilon(200) = 0.075\%$$

This will in fact be constant for all values of u_3 because the Maxwell Model cannot predict changes in strain if there is no stress. The overall variation in strain is shown in Fig. 2.46.

Example 2.15 In the previous Example, what would be the strain after 125 seconds if (a) the stress remained constant at 10 MN/m^2 after 100 seconds and (b) the stress was reduced to zero after 100 seconds.

Solution
(a) If the stress was kept constant at 10 MN/m^2 after 100 seconds as shown in Fig. 2.47 then the effective change in stress would be given by

$$\text{change in stress, } \sigma(t) = -K_1(t - T)$$

$$\frac{d\sigma(t)}{dt} = -K_1$$

So

$$\varepsilon(u_2) = \int_{t_1}^{u_2} \frac{\eta + \xi t - \xi t_1}{\xi \eta} K_1 dt + \int_{t_2}^{u_2} \frac{\mu + \xi t - \xi t_2}{\xi \eta} (-K_1) dt$$

$$\varepsilon(u_2) = \frac{K_1 u_2 T}{\eta} + \frac{K_1 T}{\xi} - \frac{K_1 T^2}{2\eta}$$

$$\varepsilon(125) = 0.125\%$$

Fig. 2.46   Variation of strain with time

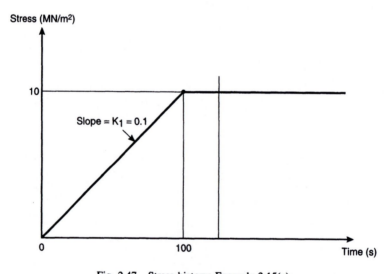

Fig. 2.47   Stress history: Example 2.15(a)

(b) If the stress was completely removed after 100 seconds as shown in Fig. 2.48 then the effective change in stress would be given by

$$\text{change in stress,} \quad \sigma(t) = -K_1(t - T) - \Delta\sigma$$

$$\frac{d\sigma(t)}{dt} = -K_1$$

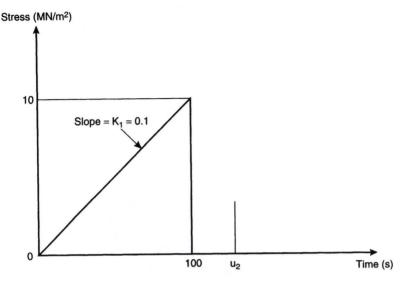

Fig. 2.48   Stress history: Example 2.15(b)

So

$$\varepsilon(u_3) = \int_{t_1}^{u_2} \frac{\eta + \xi t - \xi t_1}{\xi \eta} K_1 dt + \int_{t_2}^{u_2} \frac{\eta + \xi t - \xi t_2}{\xi \eta} (-K_1) dt - \frac{\Delta \sigma}{E(u_2 - T)}$$

$$\varepsilon(u_3) = \frac{K_1 T^2}{2\eta}$$

$$\varepsilon(125) = 0.05\%$$

It is apparent therefore that the Superposition Principle is a convenient method of analysing complex stress systems. However, it should not be forgotten that the principle is based on the assumption of linear viscoelasticity which is quite inapplicable at the higher stress levels and the accuracy of the predictions will reflect the accuracy with which the equation for modulus (equation (2.33)) fits the experimental creep data for the material. In Examples (2.13) and (2.14) a simple equation for modulus was selected in order to illustrate the method of solution. More accurate predictions could have been made if the modulus equation for the combined Maxwell/Kelvin model or the Standard Linear Solid had been used.

2.12.2 Empirical Approach

As mentioned earlier, it is not feasible to generate test data for all possible combinations of load variations. However, there have been a number of experimental investigations of the problem and these have resulted in some very useful design aids. From the experimental point of view the most straightforward situation to analyse and one that has considerable practical relevance is the load/no-load cycle. In this case a constant load is applied for a period and then completely removed for a period. The background to this approach is as follows.

For a linear viscoelastic material in which the strain recovery may be regarded as the reversal of creep then the material behaviour may be represented by Fig. 2.49. Thus the time-dependent residual strain, $\varepsilon_r(t)$, may be expressed as

$$\varepsilon_r(t) = \varepsilon_c(t) - \varepsilon_c(t - T) \qquad (2.59)$$

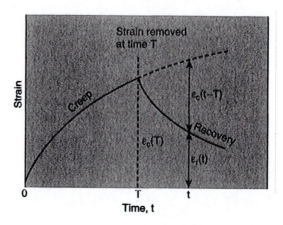

Fig. 2.49   Typical creep and recovery behaviour of a plastic

where ε_c is the creep strain during the specified period denoted by (t) or $(t - T)$.

Since there can be an infinite number of combinations of creep and recovery periods it has been found convenient to express this behaviour in terms of two dimensionless variables. The first is called the **Fractional Recovery**, defined as

$$\text{Fractional recovery, } F_r = \frac{\text{Strain recovered}}{\text{Max. creep strain}} = \frac{\varepsilon_c(T) - \varepsilon_r(t)}{\varepsilon_c(T)} \qquad (2.60)$$

where $\varepsilon_c(T)$ is the creep strain at the end of creep period and $\varepsilon_r(t)$ is the residual strain at any selected time during the recovery period.

The second dimensionless variable is called the **Reduced Time**, t_R, defined as

$$\text{Reduced time, } t_R = \frac{\text{Recovery time}}{\text{Creep time}} = \left(\frac{t - T}{T}\right) \qquad (2.61)$$

Extensive tests have shown that if the final creep strain is not large then a graph of Fractional Recovery against Reduced Time is a master curve which

Fig. 2.50   Typical recovery behaviour of a plastic

describes recovery behaviour with acceptable accuracy (see Fig. 2.50). The relationship between F_r and t_R may be derived in the following way.

When creep curves are plotted on logarithmic strain and time scales they are approximately straight lines so that the creep strain, $\varepsilon_c(t)$ may be expressed as

$$\varepsilon_c(t) = At^n \qquad (2.62)$$

and using this relationship in equation (2.59)

$$\varepsilon_r(t) = At^n - A(t - T)^n$$

and so the Fractional Recovery may be written as

$$F_r = \frac{AT^n - [At^n - A(t - T)^n]}{AT^n}$$

$$= 1 - \left(\frac{t}{T}\right)^n + \left(\frac{t}{T} - 1\right)^n$$

$$F_r = 1 + t_R^n - (t_R + 1)^n \qquad (2.63)$$

In practice this relationship is only approximately correct because most plastics are not linearly viscoelastic, nor do they obey completely the power law expressed by equation (2.62). However this does not detract from the considerable value of this simple relationship in expressing the approximate solution to a complex problem. For the purposes of engineering design the expression provides results which are sufficiently accurate for most purposes. In addition, equation (2.63) permits the problem of intermittent loading to be analysed in a relatively straightforward manner thus avoiding uneconomical overdesign which would result if the recovery during the rest periods was ignored.

From equation (2.63) and the definition of Fractional Recovery, F_r, the residual strain is given by

$$\varepsilon_r(t) = \varepsilon_c(T) - F_r \cdot \varepsilon_c(T)$$

$$= \varepsilon_c(T) \left[\left(\frac{t}{T} \right)^n - \left(\frac{t}{T} - 1 \right)^n \right]$$

If there have been N cycles of creep and recovery the accumulated residual strain would be

$$\varepsilon_r(t) = \varepsilon_c(T) \sum_{x=1}^{x=N} \left[\left(\frac{t_p x}{T} \right)^n - \left(\frac{t_p x}{T} - 1 \right)^n \right] \qquad (2.64)$$

where t_p is the period of each cycle and thus the time for which the total accumulated strain is being calculated is $t = t_p N$.

Note also that the total accumulated strain after the load application for the $(N + 1)$th time will be the creep strain for the load-on period ie $\varepsilon_c(T)$ plus the residual strain $\varepsilon_r(t)$.

ie $\qquad (\varepsilon_{N+1})_{max} = \varepsilon_c(T) \left\{ 1 + \sum_{x=1}^{x=N} \left[\left(\frac{t_p x}{T} \right)^n - \left(\frac{t_p x}{T} - 1 \right)^n \right] \right\} \qquad (2.65)$

Tests have shown that when total strain is plotted against the logarithm of the total creep time (ie NT or total experimental time minus the recovery time) there is a linear relationship. This straight line includes the strain at the end of the first creep period and thus one calculation, for say the 10th cycle allows the line to be drawn. The total creep strain under intermittent loading can then be estimated for any combinations of loading/unloading times.

In many design calculations it is necessary to have the creep modulus in order to estimate deflections etc from standard formulae. In the steady loading situation this is straightforward and the method is illustrated in the Examples (2.1)–(2.5). For the intermittent loading case the modulus of the material is effectively increased due to the apparent stiffening of the material caused by the recovery during the rest periods. For example, if a constant stress of 17.5 MN/m² was applied to acetal (see Fig. 2.51) for 9600 hours then the total creep strain after this time would be 2%. This would give a 2% secant modulus of $17.5/0.02 = 875$ MN/m². If, however, this stress was applied intermittently for 6 hours on and 18 hours off, then the total creep strain after 400 cycles (equivalent to a total time of 9600 hours) would only be 1.4%. This would be equivalent to a stress of 13 MN/m² being applied continuously for 9600 hours and so the *effective creep modulus* would be $13/0.014 = 929$ MN/m².

Fig. 2.51   Creep curves for acetal

The effective increase in modulus due to intermittent loading is shown in Fig. 2.52. This illustrates that considerable material savings can be made if intermittent loading of the plastic can be ensured.

Example 2.16 Analysis of the creep curves given in Fig. 2.51 shows that they can be represented by an equation of the form $\varepsilon(t) = A\sigma t^n$ where the constants $n = 0.083$ and $A = 0.0486$. A component made from this material is subjected to a loading pattern in which a stress of 10.5 MN/m^2 is applied for 100 hours and then completely removed. Estimate (a) the residual strain in the material 100 hours after the stress has been removed, (b) the total creep strain after the 5th loading cycle in which the stress has been applied for 100 hours and removed for 100 hours in each cycle and (c) the residual strain after 1000 cycles of the type described in (b).

Solution (a) The strain $\varepsilon_c(T)$ after 100 hours at 10.5 MN/m^2 may be obtained from the creep curves in Fig. 2.23 or by calculation from the equation $\varepsilon_c(T) = A\sigma t^n AT^n$.

So

$$\varepsilon_c(100) = 0.0486(10.5)(100)^{0.083}$$
$$= 0.51(100)^{0.083} = 0.747\%$$

Fig. 2.52 Variation of modulus for continuous and intermittent loading

Now from equation (2.60) the Fractional Recovery is given by

$$F_r = \frac{\varepsilon_c(100) - \varepsilon_r(t)}{\varepsilon_c(100)} = 1 - \frac{\varepsilon_r(t)}{0.747}.$$

Also, the Reduced Time is given by

$$t_R = \frac{t - T}{T} = \frac{200 - 100}{100} = 1$$

Then from equation (2.63)

$$F_r = 1 + t_R^n - (t_R + 1)^n$$

$$1 - \frac{\varepsilon_r(t)}{0.747} = 1 + 1 - (2)^{0.083}$$

$$\varepsilon_r(t) = 0.044\%$$

Alternatively, since recovery may be regarded as the reverse of creep this part of the problem may be solved as follows.

The projected creep strain after 200 hours at 10.5 MN/m² is

$$\varepsilon_c(200) = 0.51(200)^{0.083} = 0.792\%$$

(see Fig. 2.51 to confirm this value of strain).

However, since the stress was removed after 100 hours the recovered strain after a further 100 hours will be the same as $\varepsilon_c(T)$, i.e. 0.747%. Thus the residual strain after the recovery period may be determined by superposition.

$$\varepsilon_r(t) = 0.792 - 0.747 = 0.045\%$$

This is simpler than the first solution but this approach is only convenient for the simple loading sequence of stress on–stress off. If this sequence is repeated many times then this superposition approach becomes rather complex. In these cases the analytical solution shown below is recommended but it should be remembered that the equations used were derived on the basis of the superposition approach illustrated above.

(b) If the stress is applied for 100 hours and removed for 100 hours then in equation (2.64), $T = 100$ hours and $t_p = 200$ hours. Therefore after four cycles i.e. $t = 800$ hours.

$$\varepsilon_r(800) = \varepsilon_c(100) \sum_{x=1}^{x=4} \left[\left(\frac{t_p x}{T} \right)^n - \left(\frac{t_p x}{T} - 1 \right)^n \right]$$

$$= 0.747 \sum_{x=1}^{x=4} \left[\left(\frac{200x}{100} \right)^{0.083} - \left(\frac{200x}{100} - 1 \right)^{0.083} \right]$$

$$= 0.747(0.059 + 0.0265 + 0.0174 + 0.0131)$$

$$= 0.0868\%$$

Thus the total creep strain after the 5th load application for 100 hours would be $0.0868 + 0.747 = 0.834\%$.

(c) The residual strain after the 1000th cycle may be calculated as shown in (b) with the limits $x = 1$ to $x = 1000$. Clearly this repetitive calculation is particularly suitable for a computer and indeed there are many inexpensive desk-top programmable calculators which could give the solution quickly and easily. Thus for a time, t, of 1000×200 hours.

$$\varepsilon_r(2 \times 10^5) = \varepsilon_c(100) \left\{ \sum_{x=1}^{x=1000} \left[\left(\frac{200x}{100} \right)^{0.083} - \left(\frac{200x}{100} - 1 \right)^{0.083} \right] \right\}$$

$$= 0.342\%.$$

However, in the absence of a programmable calculator or computer the problem may be solved as follows. If the residual strain is calculated for, say 10 cycles then the value obtained is

$$\varepsilon(10^3) = 0.121\%$$

The total creep strain after the stress of 10.5 MN/m^2 has been applied for the 11th time would be $0.121 + 0.747 = 0.868\%$. Now tests have shown that a plot of total creep strain plotted against the logarithm of the total creep time (i.e. ignoring the recovery times) is a straight line which includes the point $\varepsilon_c(T)$.

Therefore after the 11th cycle the total creep time is $11 \times 100 = 1.1 \times 10^3$ hours. If the total strain at this time is plotted on Fig. 2.51 then a straight line can be drawn through this point and the point $\varepsilon_c(T)$, and this line may be extrapolated to any desired number of cycles. For the case in question the line must be extrapolated to (1001×100) hours at which point the total strain may be obtained as 1.09%. Thus the accumulated residual strain after 1000 cycles would be $1.09 - 0.747 = 0.343\%$ as calculated on the computer.

Of course it should always be remembered that the solutions obtained in this way are only approximate since the assumptions regarding linearity of relationships in the derivation of equation (2.64) are inapplicable as the stress levels increase. Also in most cases recovery occurs more quickly than is predicted by assuming it is a reversal of creep. Nevertheless this approach does give a useful approximation to the strains resulting from complex stress systems and as stated earlier the results are sufficiently accurate for most practical purposes.

2.13 Dynamic Loading of Plastics

So far the deformation behaviour of plastics has been considered for situations where (i) the stress (or strain) is constant (ii) the stress (or strain) is changing relatively slowly or (iii) the stress (or strain) is changed intermittently but the frequency of the changes is small. In practice it is possible to have situations where the stress and strain are changing quite rapidly in a regular manner. This can lead to fatigue fracture which will be discussed in detail in later. However, it is also interesting to consider the stress/strain relationships when polymers are subjected to dynamic loading.

The simplest dynamic system to analyse is one in which the stress and strain are changing in a sinusoidal fashion. Fortunately this is probably the most common type of loading which occurs in practice and it is also the basic deformation mode used in dynamic mechanical testing of plastics.

When a sinusoidally varying stress is applied to a material it can be represented by a rotating vector as shown in Fig. 2.53. Thus the stress at any moment in time is given by

$$\sigma = \sigma_0 \sin \omega t \tag{2.66}$$

where ω is the angular velocity of the vector $(= 2\pi f = 2\pi/T$, where f is the cyclic frequency in hertz (H_z) and T is the period of the sinusoidal oscillations).

If the material being subjected to the sinusoidal stress is elastic then there will be a sinusoidal variation of strain which is in phase with the stress, i.e.

$$\varepsilon = \varepsilon_0 \sin \omega t$$

However, for a viscoelastic material the strain will lag behind the stress. The strain is thus given by

$$\varepsilon = \varepsilon_0 \sin(\omega t - \delta) \tag{2.67}$$

where δ is the phase lag.

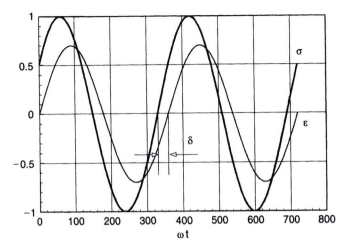

Fig. 2.53   Sinusoidal variation of stress and strain in viscoelastic material

It is more usual to write these equations in a form which shows the stress leading the strain, i.e.

$$\varepsilon = \varepsilon_0 \sin \omega t \qquad (2.68)$$

$$\sigma = \sigma_0 \sin(\omega t + \delta) \qquad (2.69)$$

The latter equation may be expanded to give

$$\sigma = \sigma_0 \sin \omega t \cos \delta + \sigma_0 \cos \omega t \sin \delta \qquad (2.70)$$

Thus the stress can be considered to have two components:

(i) $\sigma_0 \cos \delta$ which is in phase with the strain and
(ii) $\sigma_0 \sin \delta$ which is 90° out of phase with the strain.

This leads to the definition of two dynamic moduli, E_1 and E_2:

(i) $E_1 = (\sigma_0 \cos \delta)/\varepsilon_0$ in phase with the strain
(ii) $E_2 = (\sigma_0 \sin \delta)/\varepsilon_0$ 90° out of phase with strain

We could thus represent these two moduli on a phasor diagram as shown in Fig. 2.54. E_1 leads E_2 by 90° ($\pi/2$ radians) and from this diagram it is possible to define a *complex modulus*, E^* where

$$E^* = \frac{\sigma}{\varepsilon} = \sqrt{E_1^2 + E_2^2} = E_1 + iE_2 \qquad (2.71)$$

where $i = \sqrt{-1}$.

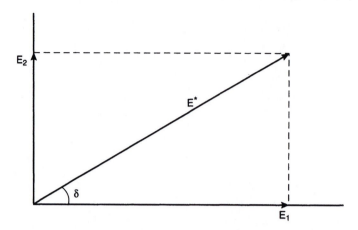

Fig. 2.54   Phasor diagram showing complex modulus (E^*) relative to loss (E_2) and storage (E_1) moduli

This presentation format leads to the terminology

E_1 = real modulus or storage modulus

E_2 = imaginary modulus or loss modulus.

Also, from Fig. 2.54 it is possible to relate E_1 and E_2 to the lag or phase angle, δ

$$\tan \delta = \frac{E_2}{E_1} \qquad (2.72)$$

In dynamic mechanical analysis of plastics, the material is subjected to a sinusoidal variation of stress and the strain is recorded so that E_1, E_2 and δ can be determined. The classical variation of these parameters is illustrated in Fig. 2.55.

At very low frequencies the polymer is rubber-like and has a low modulus (E_1 typically 0.1 MN/m^2). At very high frequencies, the material behaves like a stiff, glassy solid with a high modulus (typically about 10^3 MN/m^2). It may be seen that this E_1 plot is the inverse of that shown in Fig. 2.9 since $\omega \propto 1/t$. The loss modulus, E_2, will be zero at low and high frequencies but will rise to a maximum close to the region where E_1 is increasing most rapidly. The loss factor, $\tan \delta (= E_2/E_1)$, will also have a maximum in the viscoelastic region but this will occur at a lower frequency than the peak in the E_2 curve.

Example 2.17 Establish and plot the variation with frequency of the storage and loss moduli for materials which can have their viscoelastic behaviour described by the following models

  (i) A Maxwell model with $\xi = 1$ GN/m^2 and $\eta = 0.1$ GNs/m^2.
 (ii) A Standard Linear Solid with $\xi_1 = 2$ GN/m^2, $\xi_2 = 0.1$ GN/m^2, $\eta = 1$ GNs/m^2.

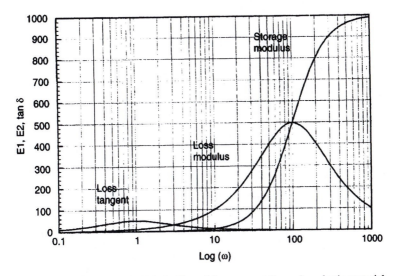

Fig. 2.55 Typical variation of E_1, E_2 and loss tangent for a viscoelastic material

Solution As illustrated above, the stress and strain will be given by

$$\sigma = \sigma_0 \sin(\omega t + \delta)$$

$$\varepsilon = \varepsilon_0 \sin \omega t$$

However, in this type of question it is preferable to utilise complex numbers and so the transformations will be used so that

$$\sigma = \sigma_0 e^{i(\omega t + \delta)} \tag{2.73}$$

$$\varepsilon = \varepsilon_0 e^{i\omega t} \tag{2.74}$$

(i) As shown earlier in this chapter, the governing equation for the Maxwell model is given by

$$\dot{\varepsilon} = \frac{1}{\xi}\dot{\sigma} + \frac{1}{\eta}\sigma$$

Then from equations (2.73) and (2.74)

$$\varepsilon_0 i\omega e^{i\omega t} = \frac{1}{\xi}\sigma_0 i\omega e^{i(\omega t + \delta)} + \frac{1}{\eta}\sigma_0 e^{i(\omega t + \delta)}$$

$$i\omega = \frac{i\omega}{\xi} \cdot \frac{\sigma_0 e^{i(\omega t + \delta)}}{\varepsilon_0 e^{i\omega t}} + \frac{1}{\eta} \cdot \frac{\sigma_0 e^{i(\omega t + \delta)}}{\varepsilon_0 e^{i\omega t}}$$

$$i\omega = \frac{i\omega}{\xi} \cdot E^* + \frac{1}{\eta}E^*$$

where E^* is the complex modulus. Rearranging gives

$$E^* = \frac{i\omega\eta\xi}{\xi + \omega\eta i}$$

Using the fact that $(a + bi)(a - bi) = a^2 + b^2$

$$E^* = \frac{i\omega\eta\xi}{(\xi + \omega\eta i)} \cdot \frac{(\xi - \omega\eta i)}{(\xi - \omega\eta i)} = \frac{\omega^2\eta^2\xi + \omega\eta\xi^2 i}{\omega^2\eta^2 + \xi^2}$$

and since $E^* = E_1 + iE_2$

then storage modulus,

$$E_1 = \frac{\omega^2\eta^2\xi}{\omega^2\eta^2 + \xi^2}$$

and loss modulus,

$$E_2 = \frac{\omega\eta\xi^2}{\omega^2\eta^2 + \xi^2}$$

$$\tan\delta = \frac{E_2}{E_1} = \frac{\xi}{\omega\eta}$$

The variations of E_1, E_2 and $\tan\delta$ with ω are shown in Fig. 2.56. It may be seen that the variations of E_1 and E_2 are of the general shape that would be expected. The $\tan\delta$ curve does not go through a maximum and so this is a further shortcoming of this simple model.

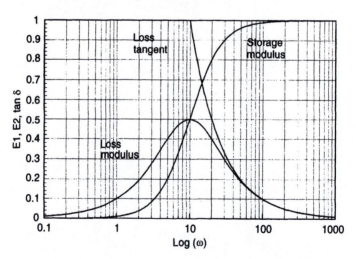

Fig. 2.56   Variation of E_1, E_2 and loss tangent for Maxwell model

(ii) For the Standard Linear Solid, the governing equation was derived earlier as

$$\eta\dot{\sigma} + \xi_1\sigma = \eta(\xi_1 + \xi_2)\dot{\varepsilon} + \xi_2\xi_1\varepsilon$$

substituting from equations (2.73) and (2.74) once again

$$\eta\sigma_0 i\omega e^{i(\omega t+\delta)} + \xi_1\sigma_0 e^{i(\omega t+\delta)} = \eta(\xi_1 + \xi_2)i\omega\varepsilon_0 e^{i\omega t} + \xi_1\xi_2\varepsilon_0 e^{i\omega t}$$

Dividing across by $\varepsilon_0 e^{i\omega t}$ and letting $E^* = \sigma_0 e^{i(\omega t+\delta)}/\varepsilon_0 e^{i\omega t}$

then,
$$E^* = \frac{\eta(\xi_1 + \xi_2)i\omega + \xi_1\xi_2}{\xi_1 + \eta\omega i}$$

Multiplying top and bottom by the conjugate of the denominator, we get

$$E^* = \frac{\{\xi_1^2\xi_2 + \eta^2(\xi_1 + \xi_2)\omega^2\} + \{\xi^2\eta\omega\}i}{\xi_1^2 + \omega^2\eta^2}$$

and since $E^* = E_1 + iE_2$,

storage modulus, $\quad E_1 = \dfrac{\xi_1^2\xi_2 + \eta^2(\xi_1 + \xi_2)\omega^2}{\xi_1^2 + \omega^2\eta^2}$

loss modulus, $\quad E_2 = \dfrac{\xi_1^2\eta\omega}{\xi_1^2 + \omega^2\eta^2}$

$$\tan\delta = \frac{E_2}{E_1} = \frac{\xi_1^2\eta\omega}{\xi_1^2\xi_2 + \eta^2(\xi_1 + \xi_2)\omega^2}$$

It may be seen in Fig. 2.57 that the variations of E_1, E_2 and $\tan\delta$ follow the classical pattern referred to earlier in this section.

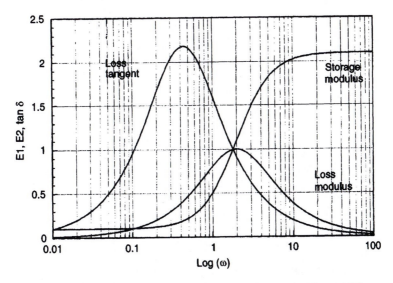

Fig. 2.57   Variation of E_1, E_2 and loss tangent for standard linear solid

2.14 Time–Temperature Superposition

It has been shown throughout this chapter that the properties of plastics are dependent on time. In Chapter 1 the dependence of properties on temperature was also highlighted. The latter is more important for plastics than it would be for metals because even modest temperature changes below 100°C can have a significant effect on properties. Clearly it is not reasonable to expect creep curves and other physical property data to be available at all temperatures. If information is available over an appropriate range of temperatures then it may be possible to attempt some type of interpolation. For example, if creep curves are available at 20°C and 60°C whereas the service temperature is 40°C then a linear interpolation would provide acceptable design data.

If creep curves are available at only one temperature then the situation is a little more difficult. It is known that properties such as modulus will decrease with temperature, but by how much? Fortunately it is possible to use a time–temperature superposition approach as follows:

It was shown earlier that the variation of creep or relaxation moduli with time are as illustrated in Fig. 2.9. If we now introduce temperature as a variable then a series of such curves will be obtained as shown in Fig. 2.58. In general the relaxed and unrelaxed modulus terms are independent of temperature. The remainder of the moduli curves are essentially parallel and so this led to the thought that a shift factor, a_T, could be applied to move from one curve to another.

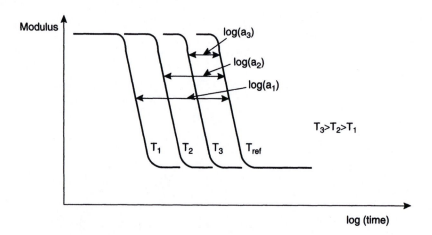

Fig. 2.58   Modulus–time curves for a range of temperatures

It may be seen from Fig. 2.59 that the two modulus curves for temperatures T_1 and T_2 are separated by a uniform distance $(\log a_T)$. Thus, if the material behaviour is known at T_1, in order to get the modulus at time, t, and temperature

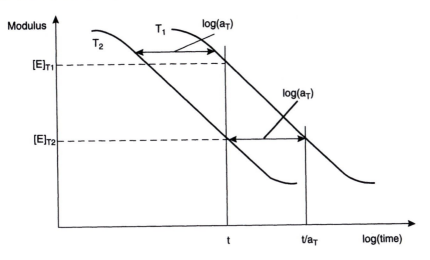

Fig. 2.59   Modulus curves at temperatures T_1 and T_2

T_2, it would be necessary to use a time (t/a_T) as shown in Fig. 2.59, in relation to the T_1 data. This means that

$$[E(t)]_{T_2} = \left[E\left(\frac{t}{a_T} \right) \right]_{T_1} \qquad (2.75)$$

where $T_2 > T_1$. Log (a_T) is negative and so $a_T < 1$.

Williams, Landel and Ferry developed an empirical relationship for this type of shift factor. This has the form

$$\log a_T = \frac{C_1(T - T_{ref})}{C_2 + (T - T_{ref})} \qquad (2.76)$$

where C_1 and C_2 are constants and T_{ref} is a reference temperature.

For many polymers it has been found that C_1 and C_2 are constants and T_{ref} is taken as T_g, the glass transition temperature for the polymer (values are given in Table 1.8). The WLF equation then takes the form

$$\log a_T = \frac{-17.4(T - T_g)}{51.6 + (T - T_g)} \qquad (2.77)$$

Thus all the different temperature related data in Fig. 2.58 could be shifted to a single master curve at the reference temperature (T_g). Alternatively if the properties are known at T_{ref} then it is possible to determine the property at any desired temperature. It is important to note that the shift factor cannot be applied to a single value of modulus. This is because the shift factor is on the horizontal time-scale, not the vertical, modulus scale. If a single value of modulus E_{T_1} is known as well as the shift factor a_T it is not possible to determine E_{T_2} without knowing how the modulus varies with time. This is illustrated in Fig. 2.60 where it is readily seen that a knowledge of E_{T_1} and a_T do not lead to a single value of E_{T_2} unless the $E(t)$ relationship is known.

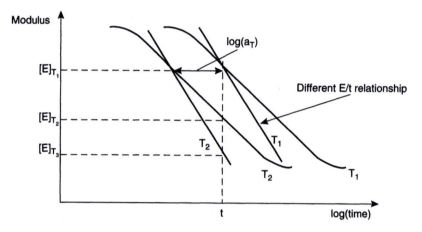

Fig. 2.60   Various modulus–time curves

Example 2.18 A particular grade of polypropylene can have its relaxation modulus described by the equation

$$E(t) = 1.8t^{-0.1}$$

where $E(t)$ is in GN/m^2 when 't' is in seconds. The temperature of the material is 20°C. Use the WLF equation to determine the 1 year modulus of the material at 60°C. The glass transition temperature for the polypropylene is −10°C.

Solution To use equation (2.76) it would be necessary to know the properties at −10°C. In this example, the properties are known at 20°C which becomes the reference temperature (T_1). The approach taken will be to get the shift factor at $T_2(= 60°C)$ and the shift factor at $T_1(= 20°C)$ and then subtract these to get the shift factor from T_1 to T_2.

$$\log a_T = \frac{17.4(T_1 - T_g)}{51.6 + (T_1 - T_g)} - \frac{17.4(T_2 - T_g)}{51.6 + (T_2 - T_g)} \tag{2.78}$$

$$= \frac{17.4(20 + 10)}{51.6 + (20 + 10)} - \frac{17.4(60 + 10)}{51.6 + (60 + 10)}$$

$$= -3.62$$

$$a_T = 2.4 \times 10^{-4}$$

Therefore, using equation (2.75)

$$[E(t)]_{T_2} = \left[E\left(\frac{t}{a_T}\right)\right]_{T_1}$$

$$= 1.8\left(\frac{t}{2.4 \times 10^{-4}}\right)^{-0.1}$$

$$[E(t)]_{T_2} = 0.78t^{-0.1}$$

The modulus at $T_2(= 60°C)$ can now be calculated at any desired time. For example, at 1 year ($t = 3.15 \times 10^6$ s)

$$[E(t)]_{60} = 0.78(3.15 \times 10^6)^{-0.1}$$

$$= 0.14 \text{ GN/m}^2$$

The two modulus curves at 20°C and 60°C are illustrated in Fig. 2.61 along with the shift factor.

Fig. 2.61   Variation of modulus with time and temperature for polypropylene

2.15 Fracture Behaviour of Unreinforced Plastics

If a plastic moulding fails in the performance of its normal function it is usually caused by one of two factors – excessive deformation or fracture. In the previous sections it was pointed out that, for plastics, more often than not it will be excessive creep deformation which is the limiting factor. However, fracture, if it occurs, can have more catastrophic results. Therefore it is essential that designers recognise the factors which are likely to initiate fracture in plastics so that steps can be taken to avoid this.

Fractures are usually classified as brittle or ductile. Although any type of fracture is serious, brittle fractures are potentially more dangerous because there is no observable deformation of the material prior to or during breakage. When a material fails in a ductile fashion, large non-recoverable deformations are evident and these serve as a warning that all is not well. In polymeric materials, fracture may be ductile or brittle depending on such variables as the nature of the additives, the processing conditions, the strain rate, the temperature and the stress system. The principal external causes of fracture are the application of a stress in a very short period of time (impact), the prolonged action of a steady stress (creep rupture) or the continuous application of a cyclically varying stress (fatigue). In all cases the fracture processes will be accelerated if the plastic is in an aggressive environment.

Basically there are two approaches to the fracture of a material. These are usually described as the microscopic and the continuum approaches. The former approach utilises the fact that the macroscopic fracture of the material must involve the rupture of atomic or molecular bonds. A study of the forces necessary to break these bonds should, therefore, lead to an estimate of the fracture strength of the material. In fact such an estimate is usually many times greater than the measured strength of the material. This is because any real solid contains multitudes of very small inherent flaws and microcracks which give rise to local stresses far in excess of the average stress on the material. Therefore although the stress calculated on the basis of the cross-sectional area might appear quite modest, in fact the localised stress at particular defects in the material could quite possibly have reached the fracture stress level. When this occurs the failure process will be initiated and cracks will propagate through the material. As there is no way of knowing the value of the localised stress, the strength is quoted as the average stress on the section and this is often surprisingly small in comparison with the theoretical strength.

The second approach to fracture is different in that it treats the material as a continuum rather than as an assembly of molecules. In this case it is recognised that failure initiates at microscopic defects and the strength predictions are then made on the basis of the stress system and the energy release processes around developing cracks. From the measured strength values it is possible to estimate the size of the inherent flaws which would have caused failure at this stress level. In some cases the flaw size prediction is unrealistically large but in many cases the predicted value agrees well with the size of the defects observed, or suspected to exist in the material.

In this chapter the various approaches to the fracture of plastics are described and specific causes such as impact loading, creep and fatigue are described in detail.

2.16 The Concept of Stress Concentration

Any material which contains a geometrical discontinuity will experience an increase in stress in the vicinity of the discontinuity. This stress concentration effect is caused by the re-distribution of the lines of force transmission through the material when they encounter the discontinuity. Causes of stress concentration include holes, notches, keyways, corners, etc as illustrated in Fig. 2.62.

(a)   (b)

Fig. 2.62   Stress concentration

The classical equation for calculating the magnitude of the stress concentration at a defect of the type shown in Fig. 2.62(b) is

$$\sigma_c = \sigma \left(1 + 2\sqrt{a/r}\right) \tag{2.79}$$

where σ_c is the local stress, σ is the nominal stress on the material, $2a$ is the defect size and r is the radius of the defect at the area in question.

The parameter $(1 + 2\sqrt{a/r})$ is commonly termed the *stress concentration factor* (K_t) and for a hole where $a = r$ then $K_t = 3$, i.e. the stresses around the periphery of the hole are three times as great as the nominal stress in the material.

It should be noted, however, that for a crack-like defect in which $r \rightarrow 0$ then $K_t \rightarrow \infty$. Obviously this does not occur in practice. It would mean that a material containing a crack could not withstand any stress applied to it. Therefore it is apparent that the stress concentration approach is not suitable for allowing for the effects of cracks. This has given rise to the use of **Fracture Mechanics** to deal with this type of situation.

2.17 Energy Approach to Fracture

When a force is applied to a material there is work done in the sense that a force moves through a distance (the deformation of the material). This work is converted to elastic (recoverable) energy absorbed in the material and surface energy absorbed in the creation of new surfaces at cracks in the material. The original work on Fracture Mechanics was done by Griffith and he proposed that unstable crack growth (fracture) would occur if the incremental change in the net energy (work done − elastic energy) exceeded the energy which could be absorbed in the creation of the new surface. In mathematical terms this may be expressed as

$$\frac{\partial}{\partial a}(W - U) > \gamma \frac{\partial A}{\partial a} \qquad (2.80)$$

where γ is the surface energy per unit area.

Note that for a situation where the applied force does no work (i.e. there is no overall change in length of the material) then $W = 0$ and equation (2.80) becomes

$$\frac{\partial U}{\partial a} > \gamma \frac{\partial A}{\partial a} \qquad (2.81)$$

Now, for a through crack propagating in a sheet of material of thickness, B, we may write

$$\partial A = 2B\partial a$$

So equation (2.80) becomes

$$\frac{\partial}{\partial a}(W - U) > 2\gamma B \qquad (2.83)$$

In the context of fracture mechanics the term 2γ is replaced by the G_c so that the condition for fracture is written as

$$\frac{1}{B}\frac{\partial}{\partial a}(W - U) > G_c \qquad (2.84)$$

G_c is a material property which is referred to as the **toughness, critical strain energy release rate or crack extension force**. It is effectively the energy required to increase the crack length by unit length in a piece of material of unit width. It has units of J/m^2.

Equation (2.84) may be converted into a more practical form as follows. Consider a piece of material of thickness, B, subjected to a force, F, as shown in Fig. 2.63(a). The load-deflection graph is shown as line (i) in Fig. 2.63(b). From this the elastic stored energy, U_1, may be expressed as

$$U_1 = \tfrac{1}{2}F\delta \qquad (2.85)$$

If the crack extends by a small amount ∂a then the stiffness of the material changes and there will be a small change in both load, ∂F, and deflection, $\partial\delta$. This is shown as line (ii) in Fig. 2.63(b). The elastic stored energy would then be

$$U_2 = \tfrac{1}{2}(F + \partial F)(\delta + \partial\delta) \qquad (2.86)$$

From equations (2.85) and (2.86) the change in stored energy as a result of the change in crack length ∂a would be given by

$$\partial U = U_2 - U_1 = \tfrac{1}{2}(F\partial\delta + \delta\partial F + \partial F\partial\delta) \qquad (2.87)$$

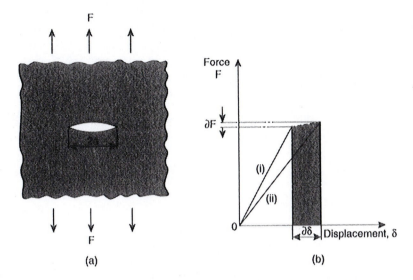

Fig. 2.63 Force displacement behaviour of an elastic cracked plate

The work done, ∂W, as a result of the change in crack length ∂a is shown shaded in Fig. 2.63(b). This will be given by

$$\partial W = F\partial\delta + \tfrac{1}{2}\partial F\partial\delta \qquad (2.88)$$

Now using equations (2.87) and (2.88) in equation (2.84)

$$\frac{1}{2B}\left\{\frac{F\partial\delta}{\partial a} - \frac{\delta\partial F}{\partial a}\right\} = G_c \qquad (2.89)$$

This equation may be simplified very conveniently if we consider the *Compliance, C,* of the material. This is the reciprocal of stiffness and is given by

$$C = \frac{\delta}{F} \qquad (2.90)$$

From this equation we may write

$$\partial\delta = F\partial C + C\partial F$$

and using this in equation (2.89) we get

$$G_c = \frac{F_c^2}{2B}\frac{\partial C}{\partial a} \qquad (2.91)$$

where F_c is the applied force at fracture.

This is a very important relationship in that it permits the fundamental material property G_c to be calculated from the fracture force, F_c, and the variation of compliance with crack length.

Example 2.19 During tensile tests on 4 mm thick acrylic sheets of the type shown in Fig. 2.63(a), the force-displacement characteristics shown in Fig. 2.64(a) were recorded when the crack lengths were as indicated. If the sheet containing a 12 mm long crack fractured at a force of 330 N, determine the fracture toughness of the acrylic and calculate the applied force necessary to fracture the sheets containing the other crack sizes.

Solution The compliance (δ/F) of each sheet may be determined from the slope of the graph in Fig. 2.64(a). A plot of compliance, C, against crack dimension, a, is shown in Fig. 2.64(b) and from this the parameter dC/da may be obtained. This is also shown plotted on Fig. 2.64(b). Using this, for a = 6 mm, it may be seen that $dC/da = 115 \times 10^{-6}$ N^{-1}.

Thus, from equation (2.91)

$$G_c = \frac{F_c^2}{2B} \frac{\partial C}{\partial a} = \frac{330^2(115 \times 10^{-6})}{2(0.004)} = 1.56 \text{ kJ/m}^2.$$

As this is a material constant, it may be used to calculate F_c for the other crack sizes. For example, for a = 2 mm, $dC/da = 7 \times 10^{-6}$ N^{-1} so

$$F_c = \sqrt{\frac{2(0.004)1.56 \times 10^3}{7 \times 10^{-6}}} = 1.34 \text{ kN}.$$

Similarly for $a = 3, 4, 5$ and 5.5 mm the fracture loads would be 1.15 kN, 0.98 kN, 0.71 kN and 0.5 kN respectively.

Fig. 2.64(a)   Force-displacement characteristics for cracked plate

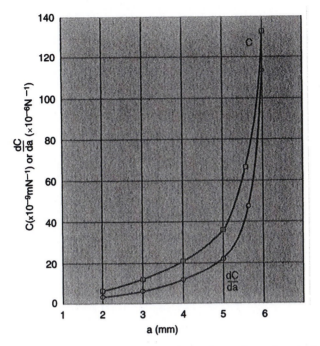

Fig. 2.64(b)   Compliance and rate of change of compliance for various crack lengths

An alternative energy approach to the fracture of polymers has also been developed on the basis of non-linear elasticity. This assumes that a material without any cracks will have a uniform strain energy density (strain energy per unit volume). Let this be U_0. When there is a crack in the material this strain energy density will reduce to zero over an area as shown shaded in Fig. 2.65. This area will be given by ka^2 where k is a proportionality constant. Thus the loss of elastic energy due to the presence of the crack is given by

$$-U = ka^2 B U_0 \qquad (2.92)$$

and

$$-\left(\frac{\partial U}{\partial a}\right) = 2kaBU_0 \qquad (2.93)$$

Comparing this with equation (2.84) and assuming that the external work is zero then it is apparent that

$$G_c = 2kaU_c \qquad (2.94)$$

where U_c is the value of strain energy density at which fracture occurs.

Now, for the special case of a *linear* elastic material this is readily expressed in terms of the stress, σ_c, on the material and its modulus, E.

$$U_c = \tfrac{1}{2}\sigma_c \, \epsilon_c = \tfrac{1}{2}\frac{\sigma_c^2}{E} \qquad (2.95)$$

Fig. 2.65   Loading of cracked plate

So, combining equations (2.94) and (2.95)

$$G_c = \frac{k\sigma^2 a}{E}$$

and in practice it is found that $k \simeq \pi$, so

$$G_c = \frac{\pi\sigma_c^2 a}{E} \qquad (2.96)$$

This is an alternative form of equation (2.91) and expresses the fundamental material parameter G_c in terms the applied stress and crack size. From a knowledge of G_c it is therefore possible to specify the maximum permissible applied stress for a given crack size, or vice versa. It should be noted that, strictly speaking, equation (2.96) only applies for the situation of plane stress. For plane strain it may be shown that material toughness is related to the stress system by the following equation.

$$G_{1c} = \frac{\pi\sigma_c^2 a}{E}(1 - v^2) \qquad (2.97)$$

where v is the lateral contraction ratio (Poissons ratio).

Note that the symbol G_{1c} is used for the plane strain condition and since this represents the least value of toughness in the material, it is this value which is usually quoted. Table 2.2 gives values for G_{1c} for a range of plastics.

Table 2.2

Typical fracture toughness parameters for a range of materials (at 20°C)

Material	G_{1c} (kJ/m^2)	K_{1c} (MN/m$^{3/2}$)	$\left(\dfrac{K_{1c}}{\sigma_y}\right)$	Ductility Factor (in mm) $\left(\dfrac{K_{1c}}{\sigma_y}\right)^{1/2}$
ABS	5	2–4	0.13	17
Acetal	1.2–2	4	0.08	6
Acrylic	0.35–1.6	0.9–1.6	0.014–0.023	0.2–0.5
Epoxy	0.1–0.3	0.3–0.5	0.005–0.008	0.02–0.06
Glass reinforced polyester	5–7	5–7	0.12	14
LDPE	6.5	1	0.125	16
MDPE/HDPE	3.5–6.5	0.5–5	0.025–0.25	5–100
Nylon 66	0.25–4	3	0.06	3.6
Polycarbonate	0.4–5	1–2.6	0.02–0.5	0.4–2.7
Polypropylene copolymer	8	3–4.5	0.15–0.2	22–40
Polystyrene	0.3–0.8	0.7–1.1	0.02	0.4
uPVC	1.3–1.4	1–4	0.03–0.13	1.1–18
Glass	0.01–0.02	0.75	0.01	0.1
Mild Steel	100	140	0.5	250

2.18 Stress Intensity Factor Approach to Fracture

Although Griffith put forward the original concept of linear elastic fracture mechanics (LEFM), it was Irwin who developed the technique for engineering materials. He examined the equations that had been developed for the stresses in the vicinity of an elliptical crack in a large plate as illustrated in Fig. 2.66. The equations for the elastic stress distribution at the crack tip are as follows.

$$\left.\begin{aligned}
\sigma_x &= \frac{K}{(2\pi r)^{1/2}} \cos\left(\frac{\theta}{2}\right)\left\{1 - \sin\left(\frac{\theta}{2}\right)\sin\left(\frac{3\theta}{2}\right)\right\} \\
\sigma_y &= \frac{K}{(2\pi r)^{1/2}} \cos\left(\frac{\theta}{2}\right)\left\{1 + \sin\left(\frac{\theta}{2}\right)\sin\left(\frac{3\theta}{2}\right)\right\} \\
\tau_{xy} &= \frac{K}{(2\pi r)^{1/2}} \sin\left(\frac{\theta}{2}\right)\cos\left(\frac{\theta}{2}\right)\cos\left(\frac{3\theta}{2}\right)
\end{aligned}\right\} \qquad (2.98)$$

and for plane strain

$$\sigma_z = \frac{2K}{(2\pi r)^{1/2}} \cos\left(\frac{\theta}{2}\right)$$

or for plane stress, $\sigma_z = 0$.

Irwin observed that the stresses are proportional to $(\pi a)^{1/2}$ where 'a' is the half length of the crack. On this basis, a **Stress Intensity Factor, K,** was

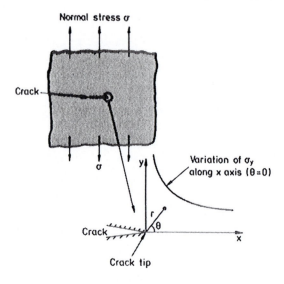

Fig. 2.66   Stress distribution in vicinity of a crack

defined as

$$K = \sigma(\pi a)^{1/2} \tag{2.99}$$

The stress intensity factor is a means of characterising the elastic stress distribution near the crack tip but in itself has no physical reality. It has units of MN m$^{-3/2}$ and should not be confused with the elastic stress concentration factor (K_t) referred to earlier.

In order to extend the applicability of LEFM beyond the case of a central crack in an infinite plate, K is usually expressed in the more general form

$$K = Y\sigma(\pi a)^{1/2} \tag{2.100}$$

where Y is a geometry factor and 'a' is the half length of a central crack or the full length of an edge crack.

Fig. 2.67 shows some crack configurations of practical interest and expressions for K are as follows.

(a) Central crack of length $2a$ in a sheet of finite width

$$K = \sigma(\pi a)^{1/2} \left\{ \frac{W}{\pi a} \cdot \tan\left(\frac{\pi a}{W}\right) \right\}^{1/2} \tag{2.101}$$

(b) Edge cracks in a plate of finite width

$$K = \sigma(\pi a)^{1/2} \left\{ \frac{W}{\pi a} \tan\left(\frac{\pi a}{W}\right) + \frac{0.2W}{\pi a} \sin\left(\frac{\pi a}{W}\right) \right\}^{1/2} \tag{2.102}$$

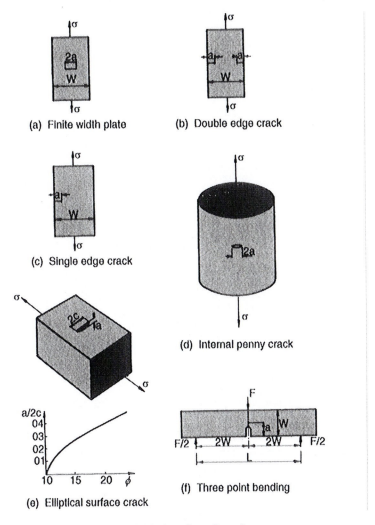

(a) Finite width plate

(b) Double edge crack

(c) Single edge crack

(d) Internal penny crack

(e) Elliptical surface crack

(f) Three point bending

Fig. 2.67 Typical crack configurations

(c) Single edge cracks in a plate of finite width

$$K = \sigma(\pi a)^{1/2} \left\{ 1.12 - 0.23 \left(\frac{a}{W} \right) + 10.6 \left(\frac{a}{W} \right)^2 - 21.7 \left(\frac{a}{W} \right)^3 \right.$$

$$\left. + 30.4 = \left(\frac{a}{W} \right)^4 \right\} \tag{2.103}$$

Note: in most cases $\left(\dfrac{a}{W} \right)$ is very small so $Y = 1.12$.

(d) Penny shaped internal crack

$$K = \sigma(\pi a)^{1/2} \left(\frac{2}{\pi}\right) \tag{2.104}$$

assuming $a \ll D$

(e) Semi-elliptical surface flaw

$$K = \sigma(\pi a)^{1/2} \left(\frac{1.12}{\phi^{1/2}}\right) \tag{2.105}$$

(f) Three point bending

$$K = \frac{3FL}{2BW^{3/2}} \left\{ 1.93 \left(\frac{a}{W}\right)^{1/2} - 3.07 \left(\frac{a}{W}\right)^{3/2} + 14.53 \left(\frac{a}{W}\right)^{5/2} \right.$$
$$\left. - 25.11 \left(\frac{a}{W}\right)^{7/2} + 25.8 \left(\frac{a}{W}\right)^{9/2} \right\} \tag{2.106}$$

or

$$K = \frac{F}{BW^{1/2}} \cdot f_1 \left(\frac{a}{W}\right) \tag{2.107}$$

Thus the basis of the LEFM design approach is that
(a) all materials contain cracks or flaws
(b) The stress intensity value, K, may be calculated for the particular loading and crack configuration
(c) failure is predicted if K exceeds the critical value for the material.

The **critical stress intensity factor** is sometimes referred to as the **fracture toughness** and will be designated K_c. By comparing equations (2.96) and (2.99) it may be seen that K_c is related to G_c by the following equation

$$(EG_c)^{1/2} = K_c \tag{2.108}$$

This is for plane stress and so for the plane strain situation

$$\left(\frac{EG_{1c}}{1 - v^2}\right)^{1/2} = K_{1c} \tag{2.109}$$

Table 2.2 gives typical values of K_{1c} for a range of plastics

Example 2.20 A cylindrical vessel with an outside radius of 20 mm and an inside radius of 12 mm has a radial crack 3.5 mm deep on the outside surface. If the vessel is made from polystyrene which has a critical stress intensity factor of 1.0 MN m$^{-3/2}$ calculate the maximum permissible pressure in this vessel.

Solution The stress intensity factor for this configuration is

$$K = 3.05 \left(\frac{PR_1^2 (\pi a)^{1/2}}{(R_2 - R_1)(R_2 + R_1)} \right)$$

The information given in the question may be substituted directly into this equation to give the bursting pressure, P_B, as

$$P_B = \frac{1.0(8 \times 10^{-3})(32 \times 10^{-3})}{3.05(12 \times 10^{-3})^2 (\pi \times 3.5 \times 10^{-3})^{1/2}} = 5.6 \text{ MN/m}^2$$

2.19 General Fracture Behaviour of Plastics

If the defect or crack in the plastic is very blunt then the stress intensification effect will be small and although failure will originate from the crack, the failure stress based on the net section will correspond to the failure stress in the uncracked material. If the stress on the material is based on the gross area then what will be observed is a reduction in the failure stress which is directly proportional to the size of the crack. This is shown as line A in Fig. 2.68.

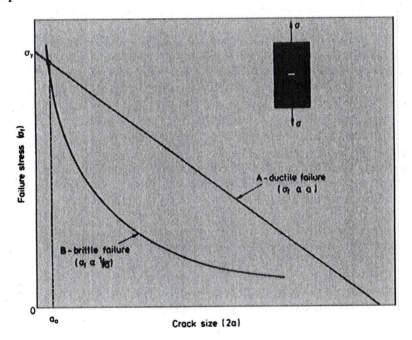

Fig. 2.68 Brittle and ductile failure characteristics for plastics

If, however, the defect or crack is sharp then the picture can change significantly. Although ABS and MDPE are special cases, where the materials are insensitive to notch condition, other thermoplastics will exhibit brittle failure if they contain sharp cracks of significant dimensions.

Polycarbonate is perhaps the most notoriously notch-sensitive of all thermoplastics, although nylons are also susceptible to ductile/brittle transitions in failure behaviour caused by notch sharpening. Other plastics such as acrylic, polystyrene and thermosets are always brittle – whatever the crack condition.

For brittle failures we may use the fracture mechanics analysis introduced in the previous sections. From equations (2.96) and (2.99) we may write

$$G_c = \frac{K_c^2}{E} = \frac{\pi \sigma_f^2 a}{E} = \text{constant} \tag{2.110}$$

From this therefore it is evident that the failure stress, σ_f, is proportional to $a^{-1/2}$. This relationship is plotted as line B on Fig. 2.68. This diagram is now very useful because it illustrates the type of ductile/brittle transitions which may be observed in plastics. According to line B, as the flaw size decreases the failure stress tends towards infinity. Clearly this is not the case and in practice what happens is that at some defect size (a_0) the material fails by yielding (line A) rather than brittle fracture.

This diagram also helps to illustrate why the inherent fracture toughness of a material is not the whole story in relation to brittle fracture. For example, Table 2.2 shows that polystyrene, which is known to be a brittle material, has a K value of about 1 MN m$^{-3/2}$. However, LDPE which has a very high resistance to crack growth also has a K value of about 1 MN m$^{-3/2}$. The explanation is that polyethylene resists crack growth not because it is tough but because it has a low yield strength. If a material has a low yield stress then its yield locus (line A in Fig. 2.68) will be pulled down, possibly below the brittle locus as happens for polyethylene. Fig. 2.69 illustrates some of the variations which are possible in order to alter the ductile/brittle characteristics of plastics. The brittle failure line can be shifted by changes in chemical structure, use of alloying techniques, changes in processing conditions, etc. The yield locus line can be shifted by the use of additives or changes in the ambient temperature or straining rate.

It is apparent therefore that a materials resistance to crack growth is defined not just by its inherent toughness but by its ratio of toughness to yield stress. Some typical values of K_{1c}/σ_y are given in Table 2.2.

Another approach to the question of resistance to crack growth is to consider the extent to which yielding occurs prior to fracture. In a ductile material it has been found that yielding occurs at the crack tip and this has the effect of blunting the crack. The extent of the plastic zone (see Fig. 2.70) is given by

$$r_p = \frac{1}{2\pi} \left(\frac{K}{\sigma_y} \right)^2 \tag{2.111}$$

for plane stress. The plane strain value is about one third of this.

Fig. 2.69 Effect of varying stress field on flaw size for ductile/brittle transition (K = constant)

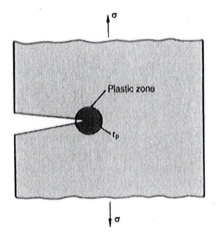

Fig. 2.70 Extent of plastic zone at crack tip

The size of the plastic zone can be a useful parameter in assessing toughness and so the ratio $(K_{1c}/\sigma_y)^2$ has been defined as a *ductility factor*. Table 2.2 gives typical values of this for a range of plastics. Note that although the ratio used in the ductility factor is conceptually related to plastic zone size, it utilises K_{1c}. This is to simplify the definition and to remove any ambiguity in relation to the stress field conditions when related to the plastic zone size. It is important that consistent strain rates are used to determine K_{1c} and σ_y, particularly when materials are being compared. For this reason the values in Table 2.2. should not be regarded as definitive. They are given simply to illustrate typical orders of magnitude.

2.20 Creep Failure of Plastics

When a constant stress is applied to a plastic it will gradually change in size due to the creep effect which was described earlier. Clearly the material cannot continue indefinitely to get larger and eventually fracture will occur. This behaviour is referred to as **Creep Rupture** although occasionally the less acceptable (to engineers) term of **Static Fatigue** is used. The time taken for the material to fracture will depend on the stress level, the ambient temperature, the type of environment, the component geometry, the molecular structure, the fabrication method, etc. At some stresses the creep rate may be sufficiently low that for most practical purposes the endurance of the material may be regarded as infinite. On the other hand, at high stresses the material is likely to fail shortly after the stress is applied.

The mechanism of time-dependent failure in polymeric materials is not completely understood and is the subject of much current research. In the simplest terms it may be considered that as the material creeps, the stress at some point in the material becomes sufficiently high to cause a micro-crack to develop but not propagate catastrophically. The stress in the remaining unbroken section of the material will than be incremented by a small amount. This causes a further stable growth of the microcrack so that over a period of time the combined effects of creep and stable crack growth cause a build up of true stress in the material. Eventually a stage is reached when the localised stress at the crack reaches a value which the remaining cross-section of the material is unable to sustain. At this point the crack propagates rapidly across the whole cross-section of the material.

Creep rupture data is usually presented as applied static stress, σ, against the logarithm of time to fracture, t, as shown in Fig. 2.71. If fracture is preceded by phenomena such as crazing (see Section 2.20.2), whitening and/or necking, then it is usual to indicate on the creep rupture characteristics the stage at which these were first observed. It may be seen from Fig. 2.71 that the appearance of crazing or whitening is not necessarily a sign the fracture is imminent. In many cases the material can continue to sustain the applied load for weeks, months or even years after these phenomena are observed. However, there is no doubt that when a load bearing component starts to craze or whiten, it can be disconcerting and so it is very likely that it would be taken out of service at this point. For this reason it is sometimes preferable to use the term **Creep Failure** rather than creep rupture because the material may have been deemed to have failed before it fractures.

Isometric data from the creep curves may also be superimposed on the creep rupture data in order to give an indication of the magnitudes of the strains involved. Most plastics behave in a ductile manner under the action of a steady load. The most notable exceptions are polystyrene, injection moulding grade acrylic and glass-filled nylon. However, even those materials which are ductile at short times tend to become embrittled at long times. This can cause

Fig. 2.71   Typical creep rupture behaviour of plastics

difficulties in the extrapolation of short-term tests, as shown in Fig. 2.71. This problem has come to the fore in recent years with the unexpected brittle fracture of polyethylene pipes after many years of being subjected to moderate pressures. On this basis the British Standards Institution (Code of Practice 312) has given the following stresses as the design values for long term usage of plastics.

Plastic	Safe working stresses
LDPE	2.1 MN/m^2
HDPE	5.0 MN/m^2
PP	5.0 MN/m^2
ABS	6.3 MN/m^2
uPVC	10.0–12.0 MN/m^2

Other factors which promote brittleness are geometrical discontinuities (stress concentrations) and aggressive environments which are likely to cause ESC (see Section 1.4.2). The absorption of fluids into plastics (e.g. water into nylon) can also affect their creep rupture characteristics, so advice should be sought where it is envisaged that this may occur.

It may be seen from Fig. 2.71 that in most cases where the failure is ductile the isometric curves are approximately parallel to the fracture curve, suggesting that this type of failure is primarily strain dominated. However, the brittle fracture line cuts across the isometric lines. It may also be seen that whitening or crazing occur at lower strains when the stress is low.

Many attempts have been made to obtain mathematical expressions which describe the time dependence of the strength of plastics. Since for many plastics a plot of stress, σ, against the logarithm of time to failure, t_f, is approximately a straight line, one of the most common expressions used is of the form

$$t_f = Ae^{-B\sigma} \qquad (2.112)$$

where A and B are nominally constants although in reality they depend on such things as the structure of the material and on the temperature. Some typical values for A and B at 20°C are given below. It is recommended that the material manufacturers should be consulted to obtain values for particular grades of their materials.

	Acrylic		Polypropylene	
	Sheet	Moulded	Homopolymer	Copolymer
A(s)	1.33×10^{14}	4.73×10^{12}	1.4×10^{13}	1.03×10^{14}
B(m²/MN)	0.404	0.42	0.88	1.19

It is recommended that the material manufacturers should be consulted to obtain values for particular grades of their materials.

One of the most successful attempts to include the effects of temperature in a relatively simple expression similar to the one above, has been made by Zhurkov and Bueche using an equation of the form

$$t_f = t_0 e^{\{U_0^{\gamma\sigma}/RT\}} \qquad (2.113)$$

where t_0 is a constant which is approximately 10^{-12} s for most plastics

U_0 is the activation energy of the fracture process

γ is a coefficient which depends on the structure of the material

R is the molar gas constant (= 8.314 J/mol° K)

and T is the absolute temperature.

If the values for U_0 and γ for the material are not known then a series of creep rupture tests at a fixed temperature would permit these values to be determined from the above expression. The times to failure at other stresses and temperatures could then be predicted.

2.20.1 Fracture Mechanics Approach to Creep Fracture

Fracture mechanics has also been used to predict failure under static stresses. The basis of this is that observed crack growth rates have been found to be related to the stress intensity factor K by the following equation

$$\frac{da}{dt} = C_1 K^m \tag{2.114}$$

where C_1 and m are material constants.

Now using equation (2.100) we may write

$$\frac{da}{dt} = C_1 (Y\sigma)^m (\pi a)^{m/2}$$

If the material contains defects of size $(2a_i)$ and failure occurs when these reach a size $(2a_c)$ then the time to failure, t_f, may be obtained by integrating the above equation.

$$t_f = \frac{2}{C_1 (Y\sigma)^m \pi^{m/2} (m-2)} \left[a_i^{1-m/2} - a_c^{1-m/2} \right] \tag{2.115}$$

Although equations (2.112), (2.113) and (2.115) can be useful they must not be used indiscriminately. For example, they are seldom accurate at short times but this is not a major worry since such short-time failures are usually not of practical interest. At long times there can also be inaccurate due to the embrittlement problem referred to earlier. In practice therefore it is generally advisable to use the equations in combination with safety factors as recommended by the appropriate National Standard.

2.20.2 Crazing in Plastics

When a tensile stress is applied to an amorphous (glassy) plastic, such as polystyrene, crazes may be observed to occur before fracture. Crazes are like cracks in the sense that they are wedge shaped and form perpendicular to the applied stress. However, they may be differentiated from cracks by the fact that they contain polymeric material which is stretched in a highly oriented manner perpendicular to the plane of the craze, i.e. parallel to the applied stress direction. Another major distinguishing feature is that unlike cracks, they are able to bear stress. Under static loading, the strain at which crazes start to form, decreases as the applied stress decreases. In constant strain rate testing the crazes always start to form at a well defined stress level. Of course, as with all aspects of the behaviour of plastics other factors such as temperature will influence the levels of stress and strain involved. Even a relatively low stress may induce crazing after a period of time, although in some glassy plastics there is a lower stress limit below which crazes will never occur. This is clearly an important stress for design considerations. However, the presence of certain liquids (organic solvents) can initiate crazing at stresses far below this lower stress limit. This phenomenon of solvent crazing has been the cause of many service failures because it is usually impossible to foresee every environment in which a plastic article will be used.

There is considerable evidence to show that there is a close connection between crazing and crack formation in amorphous plastics. At certain stress levels, crazes will form and studies have shown that cracks can nucleate in the crazes and then propagate through the preformed craze matter. In polystyrene, crazes are known to form at relatively low stresses and this has a significant effect on crack growth mechanisms in the material. In particular, during fracture toughness testing, unless great care is taken the material can appear to have a greater toughness than acrylic to which it is known to be inferior in practice. The reason is that the polystyrene can very easily form bundles of crazes at the crack tip and these tend to blunt the crack.

If a plastic article has been machined then it is likely that crazes will form at the surface. In moulded components, internal nucleation is common due to the presence of localised residual stresses.

2.21 Fatigue of Plastics

The failure of a material under the action of a fluctuating load, namely **fatigue**, has been recognised as one of the major causes of fracture in metals. Although plastics are susceptible to a wider range of failure mechanisms it is likely that fatigue still has an important part to play. For metals the fatigue process is generally well understood, being attributed to stable crack propagation from existing crack-like defects or crack initiation and propagation from structural microflaws known as dislocations. The cyclic action of the load causes the crack to grow until it is so large that the remainder of the cross-section cannot support the load. At this stage there is a catastrophic propagation of the crack across the material in a single cycle. Fatigue failures in metals are always brittle and are particularly serious because there is no visual warning that failure is imminent. The knowledge of dislocations in metals stems from a thorough understanding of crystal structure, and dislocation theory for metals is at an advanced stage. Unfortunately the same cannot be said for polymer fatigue. In this case the completely different molecular structure means that there is unlikely to be a similar type of crack initiation process although it is possible that once a crack has been initiated, the subsequent propagation phase may be similar.

If a plastic article has been machined then it is likely that this will introduce surface flaws capable of propagation, and the initiation phase of failure will be negligible. If the article has been moulded this tends to produce a protective skin layer which inhibits fatigue crack initiation/propagation. In such cases it is more probable that fatigue cracks will develop from within the bulk of the material. In this case the initiation of cracks capable of propagation may occur through slip of molecules if the polymer is crystalline. There is also evidence to suggest that the boundaries of spherulites are areas of weakness which may develop cracks during straining as well as acting as a crack propagation path.

In amorphous polymers it is possible that cracks may develop in the voids which are formed during viscous flow.

Moulded plastics will also have crack initiation sites created by moulding defects such as weld lines, gates, etc and by filler particles such as pigments, stabilisers, etc. And, of course, stress concentrations caused by sharp geometrical discontinuities will be a major source of fatigue cracks. Fig. 2.72 shows a typical fatigue fracture in which the crack has propagated from a surface flaw.

Free surface

Fig. 2.72   Typical fatigue fracture surface

There are a number of additional features which make polymer fatigue a complex subject and not one which lends itself to simple analysis. The very nature of the loading means that stress, strain and time are all varying simultaneously. The viscoelastic behaviour of the material means that strain rate (or frequency) is an important factor. There are also special variables peculiar to this type of testing such as the type of control (whether controlled load or controlled deformation), the level of the mean load or mean deformation and the shape of the cyclic waveform. To add to the complexity, the inherent damping and low thermal conductivity of plastics causes a temperature rise during fatigue. This may bring about a deterioration in the mechanical properties of the material or cause it to soften so much that it becomes useless in any load bearing application.

Another important aspect of the fatigue of all materials is the statistical nature of the failure process and the scatter which this can cause in the results. In a particular sample of plastic there is a random distribution of microcracks, internal flaws and localised residual stresses. These defects may arise due to structural imperfections (for example, molecular weight variations) or as a result of the fabrication method used for the material. There is no doubt that failure processes initiate at these defects and so the development and propagation of a crack will depend on a series of random events. Since the distribution and size of the flaws are likely to be quite different, even in outwardly identical samples, then the breaking strength of the plastic is a function of the probability of a sufficiently large defect being correctly oriented in a highly stressed region of the material. Since there is a greater probability of a suitable defect existing in a large piece of material there may be a size effect. The most important point to be realised is that the breaking strength of a material is not a unique value which can be reproduced at will. At best there may be a narrow distribution of strength values but in all cases it is essential to satisfy oneself about the statistical significance of a single data point. The design procedures which are most successful at avoiding fracture usually involve the selection of a factor of safety which will reduce the probability of failure to an acceptably low value.

2.21.1 Effect of Cyclic Frequency

Consider a sample of plastic which is subjected to a fixed cyclic stress amplitude of $\pm\sigma_1$. The high damping and low thermal conductivity of the material means that some of the input energy will be dissipated in each cycle and will appear as heat. The temperature of the material will rise therefore, as shown in Fig. 2.73. Eventually a stage will be reached when the heat transfer to the surroundings equals the energy dissipation. At this point the temperature of the material stabilises until a conventional brittle fatigue failure occurs. This failure may be plotted on a graph of stress amplitude against the logarithm of the number of cycles to fracture as shown in Fig. 2.74. If, in the next test, the stress amplitude is increased to σ_2 then the material temperature will rise again and stabilise at a higher value as shown in Fig. 2.73. Continued cycling then leads to a fatigue

Fig. 2.73   Temperature rise during cyclic loading

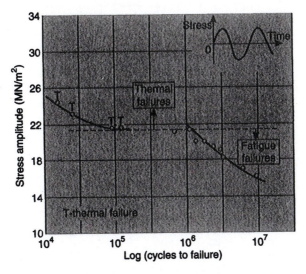

Fig. 2.74   Typical fatigue behaviour of acetal at 5 Hz

failure as shown in Fig. 2.74. Higher stress amplitudes in subsequent tests will repeat this pattern until a point is reached when the temperature rise no longer stabilises. Instead the temperature continues to rise and results in a short term thermal softening failure in the material. Stress amplitudes above this cross-over stress level will cause thermal failures in an even shorter time. The nett result of this is that the fatigue curve in Fig. 2.74 has two distinct regimes. One for the relatively short-term thermal failures and one for the long-term conventional fatigue failures.

If the frequency of cycling is reduced then stress amplitudes which would have produced thermal softening failures at the previous frequency, now result in stable temperature rises and eventually fatigue failures. Normally it is found that these fatigue failures fall on the extrapolated curve from the fatigue failures at the previous frequency. Even at the lower frequency, however, thermal soft-ening failures will occur at high stress levels. If fatigue failures are to occur at these high stresses, then the frequency must be reduced still further. The overall picture which develops therefore is shown in Fig. 2.75. In some plastics the fatigue failure curve becomes almost horizontal at large values of N. The stress level at which this occurs is clearly important for design purposes and is known as the *fatigue limit*. For plastics in which fatigue failures continue to occur even at relatively low stress levels it is necessary to define an *endurance limit* i.e. the stress level which would not cause fatigue failure until an acceptably large number of stress cycles.

The occurrence of thermal failures in a plastic depends not only on the cyclic frequency and applied stress level but also on the thermal and damping characteristics of the material. For example, polycarbonate has very little

Fig. 2.75   Typical fatigue behaviour of acetal at several frequencies

tendency towards thermal failures whereas with polypropylene there is a marked propensity in this direction. Thermosets, of course, are very thermally stable and only exhibit brittle fatigue failures.

2.21.2 Effect of Waveform

Assuming that the cyclic waveform used in the previous section was sinusoidal then the effect of using a square wave is to reduce, at any frequency, the level of stress amplitude at which thermal softening failures start to occur. This is because there is a greater energy dissipation per cycle when a square wave is used. If a ramp waveform is applied, then there is less energy dissipation per cycle and so higher stresses are possible before thermal runaway occurs.

2.21.3 Effect of Testing Control Mode

During cyclic loading of a material the energy dissipated is proportional to the product of the stress and strain. If the loading on a plastic is such that the stress amplitude is maintained constant, then any temperature rise in the material will lead to an increase in strain since the modulus decreases with temperature. The increase in strain means that more energy is dissipated, leading to a further drop in modulus and a further increase in strain. It is this type of chain reaction which leads to the thermal softening failures if the heat transfer to the surroundings is insufficient.

The alternative mode of testing is to control the strain amplitude. In this case an increase in temperature again causes a drop in modulus but this leads to a drop in stress amplitude. There is therefore a drop in energy dissipation and hence temperature. In this case it is found that this self stabilising mechanism prevents the occurrence of thermal softening failures. The nett result is that under this mode of control the temperature rise always stabilises and only fatigue type failures are observed.

2.21.4 Effect of Mean Stress

For convenience, in the previous sections it has been arranged so that the mean stress is zero. However, in many cases of practical interest the fluctuating stresses may be always in tension (or at least biased towards tension) so that the mean stress is not zero. The result is that the stress system is effectively a constant mean stress, σ_m superimposed on a fluctuating stress σ_a. Since the plastic will creep under the action of the steady mean stress, this adds to the complexity because if the mean stress is large then a creep rupture failure may occur before any fatigue failure. The interaction of mean stress and stress amplitude is usually presented as a graph of $(\sigma_a V \sigma_m)$ as shown in Fig. 2.76. This represents the locus of all the combinations of σ_a and σ_m which cause fatigue failure in a particular number of cycles, N. For plastics the picture is slightly different from that observed in metals. Over the region WX the behaviour is similar in that as the mean stress increases, the stress amplitude must be decreased to cause failure in the same number of cycles. Over the region YZ, however, the mean stress is so large that creep rupture failures are dominant. Point Z may be obtained from creep rupture data at a time equal to that necessary to give N cycles at the test frequency. It should be realised that, depending on the level of mean stress, different phenomena may be the cause of failure.

The level of mean stress also has an effect on the occurrence of thermal failures. Typically, for any particular stress amplitude the stable temperature rise will increase as the mean stress increases. This may be to the extent that a stress amplitude which causes a stable temperature rise when the mean stress is zero, can result in a thermal runaway failure if a mean stress is superimposed.

For design purposes it is useful to have a relationship between σ_a and σ_m, similar to those used for metals (e.g. the Soderberg and Goodman relationships). It is suggested that the equation of a straight line joining points W and Z in Fig. 2.76 would be best because it is simple and will give suitably conservative estimates for the permissible combinations of σ_a and σ_m to produce failure in a pre-selected number of cycles. Such an equation would have the form

$$\sigma_a = \sigma_f \left(1 - \frac{\sigma_m}{\sigma_c} \right) \qquad (2.116)$$

where σ_f is the fatigue endurance at N cycles

σ_c is the creep rupture strength at a time equivalent to N cycles

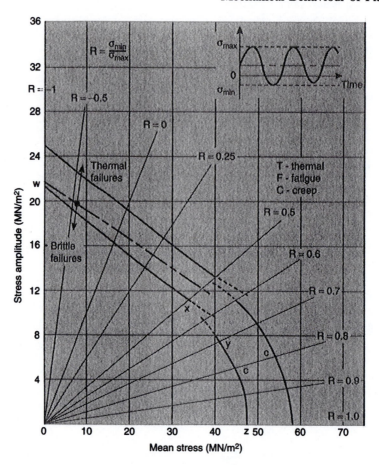

Fig. 2.76   Relationships between stress amplitude and mean stress

Example 2.21 A rod of plastic is subjected to a steady axial pull of 50 N and superimposed on this is an alternating axial load of ±100 N. If the fatigue limit for the material is 13 MN/m^2 and the creep rupture strength at the equivalent time is 40 MN/m^2, estimate a suitable diameter for the rod. Thermal effects may be ignored and a fatigue strength reduction factor of 1.5 with a safety factor of 2.5 should be used.

Solution The alternating stress, σ_a is given by

$$\sigma_a = \frac{\text{Alternating load}}{\text{area}} = \frac{4 \times 100}{\pi d^2} \ (\text{MN/m}^2)$$

Also the mean stress, σ_m, is given by

$$\sigma_m = \frac{\text{Steady load}}{\text{area}} = \frac{4 \times 50}{\pi d^2} \ (\text{MN/m}^2)$$

Then using equation (2.116)

$$\sigma_a = \sigma_f \left(1 - \frac{\sigma_m}{\sigma_c}\right)$$

So applying the fatigue strength reduction factor and the factor of safety

$$\frac{2.5 \times 4 \times 100}{\pi d^2} = \frac{13}{1.5} \left(1 - \frac{4 \times 50 \times 2.5}{\pi d^2 \times 40}\right)$$

This may be solved to give $d = 6.4$ mm.

2.21.5 Effect of Stress System

In the previous sections the stress system has been assumed to be cyclic uniaxial loading since this is the simplest to analyse. If, however, the material is subjected to bending, then this will alter the stress system and hence the fatigue behaviour. In general it is found that a sample subjected to fluctuating bending stresses will have a longer fatigue endurance than a sample of the same material subjected to a cyclic uniaxial stress. This is because of the stress gradient across the material in the bending situation. Fatigue cracks are initiated by the high stress at the surface but the rate of crack propagation is reduced due to the lower stresses in the bulk of the material. In addition, the crack initiation phase may have to be lengthened. This is because mouldings have a characteristic skin which appears to resist the formation of fatigue cracks. Under uniaxial loading the whole cross-section is subjected to the same stress and cracks will be initiated at the weakest region.

The stress gradient also means that the occurrence of thermal softening failures is delayed. At any particular frequency of stressing, thermal softening failures will not occur until higher stresses if the stress system is bending rather than uniaxial.

2.21.6 Fracture Mechanics Approach to Fatigue

During fatigue the stress amplitude usually remains constant and brittle failure occurs as a result of crack growth from a sub-critical to a critical size. Clearly the rate at which these cracks grow is the determining factor in the life of the component. It has been shown quite conclusively for many polymeric materials that the rate at which cracks grow is related to the stress intensity factor by a relation of the form

$$\frac{da}{dN} = C_2(\Delta K)^n \tag{2.117}$$

where $\dfrac{da}{dN}$ is the crack growth rate

ΔK is the alternating stress intensity factor corresponding to the stress range $\Delta\sigma$ (i.e. $\Delta K = K_{max} - K_{min}$) and C_2 and n are constants.

Hence a graph of $\log(da/dN)$ against $\log(\Delta K)$ will be a straight line of slope n as shown Fig. 2.77. Now, in Section 3.4 it was shown that the range of stress intensity factor could be represented by a general equation of the form

$$K = Y\sigma(\pi a)^{1/2} \quad \text{or} \quad \Delta K = Y(\Delta\sigma)(\pi a)^{1/2} \qquad (2.118)$$

where Y is a geometry function.

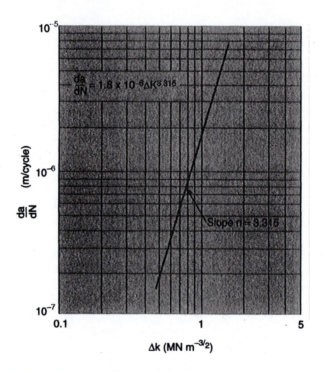

Fig. 2.77 Crack growth rate as a function of stress intensity factor

Thus, combining equations (2.117) and (2.118) gives

$$\frac{da}{dN} = C_2\{Y(\Delta\sigma)(\pi a)^{1/2}\}^n$$

Assuming that the geometry function, Y, does not change as the crack grows then this equation may be integrated to give the number of cycles, N_f, which are necessary for the crack to grow from its initial size ($2a_i$) to its critical size at fracture ($2a_c$).

$$\int_{a_i}^{a_c} \frac{1}{YC_2(\Delta\sigma)n\pi^{n/2}} \cdot \frac{da}{a^{n/2}} = \int_0^{N_f} dN$$

Assuming $n \neq 2$

$$\frac{2}{Y C_2 (\Delta\sigma)^n \pi^{n/2} (2 - n)} \left\{ a_c^{(1-n/2)} - a_i^{(1-n/2)} \right\} = N_f \qquad (2.119)$$

The way in which this sort of approach may be used to design articles subjected to fatigue loading is illustrated in the following example.

Example 2.22 A certain grade of acrylic has a K_c value of 1.6 MN m$^{-3/2}$ and the fatigue crack growth data as shown in Fig. 2.77. If a moulding in this material is subjected to a stress cycle which varies from 0 to 15 MN/m^2, estimate the maximum internal flaw size which can be tolerated if the fatigue endurance is to be at least 10^5 cycles.

Solution The first step is to calculate the critical flaw size which will cause brittle failure to occur in one cycle. This may be obtained from equation (2.100) assuming $Y = 1$.

$$K = \sigma (\pi a)^{1/2}$$

or

$$a_c = \left(\frac{K_c}{\sigma} \right)^2 \frac{1}{\pi} = \left(\frac{1.6}{15} \right)^2 \frac{1}{\pi} = 3.62 \times 10^{-3} \text{ m}$$

During cyclic loading, any cracks in the material will propagate until they reach this critical size. If the article is to have an endurance of at least 10^5 cycles then equation (2.119) may be used to determine the size of the smallest flaw which can be present in the material before cycling commences.

$$a_c^{(2-n)/2} - a_i^{(2-n)/2} = \frac{C_2}{2} N (\Delta\sigma)^n \pi^{n/2} (2 - n)$$

Using $C_2 = 1.8 \times 10^{-6}$ and $n = 3.315$ from Fig. 2.77 then $a_i = 1.67$ μm. Therefore the inspection procedures must ensure that there are no defects larger than $(2 \times 1.67) = 3.34$ μm in the material before the cyclic stress is applied.

2.22 Impact Behaviour of Plastics

The resistance to impact is one of the key properties of materials. The ability of a material to withstand accidental knocks can decide its success or failure in a particular application. It is ironical therefore that for plastics this is one of the least well defined properties. Although impact test data is widely quoted in the literature, most of it is of little value because impact strength is not an inherent material property and hence it is not possible to specify a unique universal value for the impact strength of any plastic. Impact strength depends on a range of variables including temperature, straining rate, stress system, anisotropy, geometry of the article, fabrication conditions, environment and so on. As a result of this there is often a poor correlation (a) between laboratory test data and product performance and (b) between test results from different laboratories. The first of these problems is the more serious because it can raise doubts in the mind of the designer about the use of plastics.

Fortunately the situation in practice is not quite as complex as it might seem. In general, very acceptable designs are achieved by using impact data obtained under conditions which relate as closely as possible to the service conditions. Impact strength values available in the literature may be used for the initial selection of a material on the basis of a desired level of toughness. Then, wherever possible this should be backed up by tests on the plastic article, or a specimen cut from it, to ensure that the material, as moulded, is in a satisfactory state to perform its function.

As always, of course, to alleviate fracture problems it is essential to avoid the factors which are likely to cause brittleness. These include stress concentrations and low temperatures and the effects of these will be considered in the following sections.

2.22.1 Effect of Stress Concentrations

During service the impact behaviour of a plastic article will be influenced by the combined effects of the applied stress system and the geometry of the article. Although the applied stress system may appear simple (for example, uniaxial) it may become triaxial in local areas due to a geometrical discontinuity. Fig. 2.78

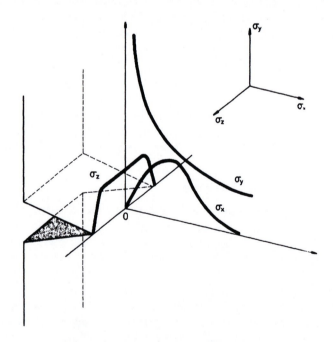

Fig. 2.78   Triaxial stress distribution at a notch

shows the triaxial stresses which exist at the tip of a notch. It is this triaxiality which promotes brittleness in the material. Therefore, in practice one should avoid abrupt changes in section, holes, notches, keyways etc at critical, highly stressed areas in a moulding.

In the laboratory the impact behaviour of a material could be examined by testing plain samples, but since brittle failures are of particular interest it is more useful to ensure that the stress system is triaxial. This may be achieved most conveniently by means of a notch in the sample. The choice of notch depth and tip radius will affect the impact strengths observed. A sharp notch is usually taken as 0.25 mm radius and a blunt notch as 2 mm radius.

Fig. 2.79 shows the typical variation of impact strength with notch tip radius for several thermoplastics. The first important fact to be noted from this graph is that the use of a sharp notch will rank the plastics materials in a different order to that obtained using a blunt notch. This may be explained by considering the total impact strength as consisting of both crack initiation and crack propagation

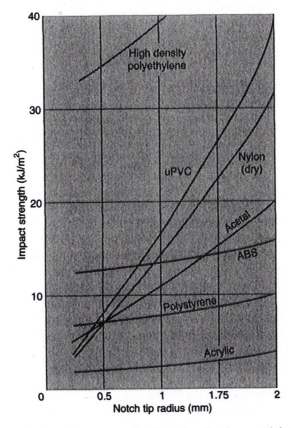

Fig. 2.79   Variation of impact strength with notch radius for several thermoplastics

energy. When the very sharp notch (0.25 mm radius) is used it may be assumed that the energy necessary to initiate the crack is small and the main contribution to the impact strength is propagation energy. On this basis, Fig. 2.79 shows that high density polyethylene and ABS have relatively high crack propagation energies whereas materials such as PVC, nylon and acrylic have low values. The significant improvement in impact strength for PVC and nylon when a blunt notch is used would suggest that their crack initiation energies are high. However, the smaller improvement in the impact behaviour of ABS with a blunt notch indicates that the contribution from crack initiation energy is low.

Graphs such as Fig. 2.79 also give a convenient representation of the notch sensitivity of materials. For example it may be seen that sharp notches are clearly detrimental to all the materials tested and should be avoided in any good design. However, it is also apparent that the benefit derived from using generously rounded corners is much less for ABS than it is for materials such as nylon or PVC.

Impact strength also increases as the notch depth is decreased. The variation of impact strength with notch depth and radius may be rationalised for some materials by use of the linear elastic stress concentration expression.

$$K_t = 1 + 2(a/r)^{1/2} \tag{2.120}$$

where 'r' is the notch radius and 'a' is the notch depth.

It has been shown that for acrylic, glass-filled nylon and methyl pentene there is reasonable correlation between the reciprocal of the stress concentration factor, K_t, and impact strength. However, for PVC good correlation could only be achieved if the finite dimensions of the sample were taken into account in the calculation of stress concentration factor.

2.22.2 Effect of Temperature

In most cases thermoplastic components are designed for use at room temperature. It might appear, therefore, that data on the impact properties at this temperature (approximately 20°C) would provide sufficient information for design. However, this approach would be rather naive since even indoors, temperatures may vary by an amount which can have a significant effect on impact behaviour. For components used outdoors of course, the situation can be much worse with conditions varying from sub-zero to tropical. In common with metals, many plastics exhibit a transition from ductile behaviour to brittle as the temperature is reduced.

Fig. 2.80 is typical of the effects which may be observed with several common plastics materials. Quite apart from the changes in impact strength with temperature an important lesson which should be learned from this diagram is that the ranking of the materials is once again influenced by the test conditions. For example, at 20°C polypropylene is superior to acetal whereas at −20°C it

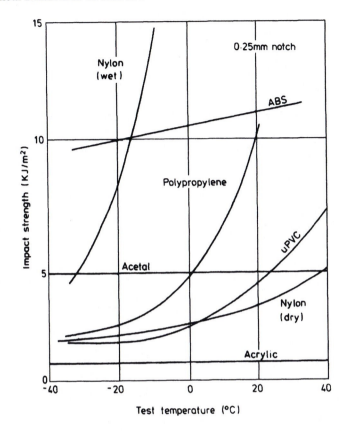

Fig. 2.80   Variation of impact strength with temperature for several thermoplastics

exhibits a considerable drop in impact strength to give it a poorer performance than acetal.

It may be seen from Fig. 2.80 that some plastics experience the change from ductile to brittle behaviour over a relatively narrow temperature range. This permits a tough/brittle transition temperature to be quoted. In other plastics this transition is much more gradual so that it is not possible to attribute it to a single value of temperature. In these circumstances it is common to quote a Brittleness Temperature, $T_B(1/4)$. This temperature is defined as the value at which the impact strength of the material with a sharp notch (1/4 mm tip radius) equals 10 kJ/m². This temperature, when quoted, gives an indication of the temperature above which there should be no problems with impact failures. It does not mean that the material should never be used below $T_B(1/4)$ because by definition it refers only to the sharp notch case. When the material has a blunt notch or is un-notched its behaviour may still be satisfactory well below $T_B(1/4)$.

2.22.3 Miscellaneous Factors Affecting Impact

Other factors which can affect impact behaviour are fabrication defects such as internal voids, inclusions and additives such as pigments, all of which can cause stress concentrations within the material. In addition, internal welds caused by the fusion of partially cooled melt fronts usually turn out to be areas of weakness. The environment may also affect impact behaviour. Plastics exposed to sunlight and weathering for prolonged periods tend to become embrittled due to degradation. Alternatively if the plastic is in the vicinity of a fluid which attacks it, then the crack initiation energy may be reduced. Some plastics are affected by very simple fluids e.g. domestic heating oils act as plasticisers for polyethylene. The effect which water can have on the impact behaviour of nylon is also spectacular as illustrated in Fig. 2.80.

The surface finish of the specimen may also affect impact behaviour. Machined surfaces usually have tool marks which act as stress concentrations whereas moulded surfaces have a characteristic skin which can offer some protection against crack initiation. If the moulded surface is scratched, then this protection no longer exists. In addition, mouldings occasionally have an embossed surface for decorative effect and tests have shown that this can cause a considerable reduction in impact strength compared to a plain surface.

2.22.4 Impact Test Methods

The main causes of brittleness in materials are known to be

(1) triaxiality of stress
(2) high strain rates, and
(3) low temperatures.

In order to provide information on the impact behaviour of materials, metallurgists developed tests methods which involved striking a notched bar with a pendulum. This conveniently subjected the material to triaxiality of stress (at the notch tip) and a high strain rate so as to encourage brittle failures. The standard test methods are the Izod and Charpy tests which use the test procedures illustrated in Fig. 2.81(a) and (b). The specimens have a standard notch machined in them and the impact energy absorbed in breaking the specimen is recorded. With the ever-increasing use of plastics in engineering applications it seemed appropriate that these well established test methods should be adopted. However, even the metallurgists recognised that the tests do have certain shortcomings. The main problem is that the test conditions are arbitrary. The speed of impact, method of stressing and specimen geometry are fixed and experience has shown that it was too much to expect the results to be representative of material behaviour under different conditions.

In particular, standard specimens contain a sharp notch so that it is propagation energy rather than initiation energy which is the dominant factor. In general the standard tests are useful for quality control and specification purposes but not

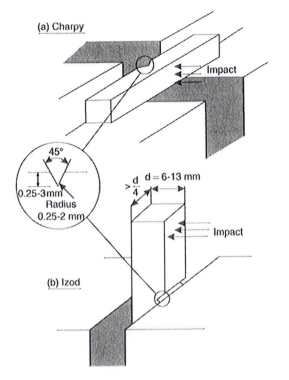

(a) Charpy

Impact

45°

0.25-3mm
Radius
0.25-2 mm

$> \dfrac{d}{4}$ d = 6-13 mm

Impact

(b) Izod

Fig. 2.81   Pendulum impact tests

for the prediction of end-product performance. The complex interaction of the variables does not permit component designs to be based on the data. A material which appears bad in the standard tests will not necessarily be bad in service.

Although the Izod and Charpy tests are widely used for plastics, other types of test are also popular. These include tensile impact tests and flexural plate (falling weight) tests. The latter is particularly useful in situations where the effects of flow anisotropy are being assessed. In addition, arbitrary end-product tests are widely used to provide reassurance that unforseen factors have not emerged to reduce the impact performance of the product.

The results of impact tests are often scattered even with the most careful test procedures. In these circumstances it is normal practice to quote the median strength rather than the average. This is because the median will be more representative of the bulk of the results if there are odd very high or very low results. A non-broken sample can also be allowed for in median analysis but not when the average is used.

Impact strength are normally quoted as

$$\text{Impact Strength} = \frac{\text{Energy to break}}{\text{area at notch section}} \quad (\text{J/m}^2)$$

Occasionally the less satisfactory term of energy to break per unit width may be quoted in units of J/m.

In some applications impact performance may not be critical and only a general knowledge of materials behaviour is needed. In these circumstances it would be unrealistic to expect the designer to sift through all the combination of multi-point data. Therefore diagrams such as Fig. 2.82 can be useful for providing an overall indication of the general impact performance to be expected from different plastics. However, this type of general guide should be used with caution because it oversimplifies in at least two important respects. It ignores the plane stress/plane strain toughness transition which causes the order of merit to depend on the material thickness. Also it ignores the effect of molecular orientation except insofar as this particular diagram refers to specimens cut from one sort of moulding.

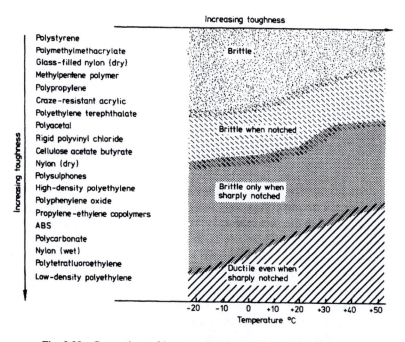

Fig. 2.82   Comparison of impact strengths as measured by Charpy test

2.22.5 Fracture Mechanics Approach to Impact

In recent years impact testing of plastics has been rationalised to a certain extent by the use of fracture mechanics. The most successful results have been achieved by assuming that LEFM assumptions (bulk linear elastic behaviour and presence of sharp notch) apply during the Izod and Charpy testing of a plastic.

During these types of test it is the energy absorbed at fracture, U_c, which is recorded. In terms of the applied force, F_c, and sample deformation, δ, this will be given by

$$U_c = \tfrac{1}{2}F_c\delta \qquad (2.121)$$

or expressing this in terms of the compliance, from equation (2.90)

$$U_c = \tfrac{1}{2}F_c^2 C \qquad (2.122)$$

Now, from equation (2.91) we have the expression for the toughness, G_c, of the material

$$G_c = \frac{F_c^2}{2B}\frac{\partial C}{\partial a}$$

So using equation (2.122) and introducing the material width, D

$$G_c = \frac{U_c}{BD\emptyset} \qquad (2.123)$$

where $\emptyset = ((1/C)(\partial C/\partial a))^{-1}$. This is a geometrical function which can be evaluated for any geometry (usually by finite element analysis). Fig. 2.83 shows the preferred test geometry for a Charpy-type test and Table 2.3 gives the values of $\emptyset$ for this test configuration. Other values of $\emptyset$ may be determined by interpolation.

Recommended specimen sizes are

S/D	D	B	L	S
4	10	10	55	40
6	6·7	6·7	55	40

Fig. 2.83   Charpy test piece

It is apparent from equation (2.123) that a graph of $BD\emptyset$ against fracture energy U_c (using different crack depths to vary $\emptyset$) will be a straight line, the slope of which is the material toughness, G_c.

Table 2.3

Charpy calibration factor (Ø)

a/D	Ø Values		
	S/D = 4	S/D = 6	S/D = 8
0.06	1.183	1.715	2.220
0.10	0.781	1.112	1.423
0.20	0.468	0.631	0.781
0.30	0.354	0.450	0.538
0.40	0.287	0.345	0.398
0.50	0.233	0.267	0.298
0.60	0.187	0.205	0.222

Fig. 2.84   Plot of U_c against $BDØ$

Example 2.23 A series of Charpy impact tests on uPVC specimens with a range of crack depths gave the following results

Crack length (mm)	1	2	3	4	5
Fracture Energy (mJ)	100	62	46.5	37	31

If the sample section is 10 mm × 10 mm and the support width is 40 mm, calculate the fracture toughness of the uPVC. The modulus of the uPVC is 2 GN/m^2.

Solution Since $B = D = 10$ mm and using the values of $\emptyset$ from Table 2.3 we may obtain the following information.

a(mm)	a/D	$\emptyset$	$BD\emptyset$	$U(mJ)$
1	0.1	0.781	78.1×10^{-6}	100
2	0.2	0.468	46.8×10^{-6}	62
3	0.3	0.354	35.4×10^{-6}	46.5
4	0.4	0.287	28.7×10^{-6}	37
5	0.5	0.233	23.3×10^{-6}	31

A graph of U against $BD\emptyset$ is given in Fig. 2.84. The slope of this gives $G_c = 1.33$ kJ/m^2.
Then from equation (2.108) the fracture toughness is given by

$$K_c = \sqrt{EG_c} = \sqrt{2 \times 10^9 \times 1.33 \times 10^3} = 1.63 \text{ MN m}^{-3/2}$$

Bibliography

Throne, J.L. Mechanical properties of thermoplastic structural foams, in Wendle, B.C. (ed.) *Engineering Guide to Structural Foams*, Technomic, Lancaster, PA, USA (1976) pp. 91–114.

Marshall, G.P. *Design for toughness in polymers – Fracture Mechanics*, Plastics and Rubber Proc. and Appl. 2(1982) p 169–182.

Moore, D.R., Hooley, C.J. and Whale, M. *Ductility factors for thermoplastics*, Plastics and Rubber Proc. and Appl. 1(1981) p 121–127.

Kinloch, A.J. and Young R.J. *Fracture Behaviour of Polymers*, Applied Science, London (1983)

Williams, J.G. *Fracture Mechanics of Polymers*, Ellis Horwood, Chicester (1984).

Andrews, E.H. *Developments in Polymer Fracture – 1*, Applied Science, London (1979).

Kausch, H.H. *Polymer Fracture*, Springer-Verlag, Berlin (1978).

Hertzberg, R.W. and Manson, J.A., *Fatigue of Engineering Plastics*, Academic Press, New York (1980).

Ogorkiewicz, R.M. *Engineering Properties of Plastics*, Wiley Interscience, London (1970).

Powell, P.C. *Engineering with Polymers*, Chapman and Hall, London (1983).

Levy, S. and Dubois, J.H. *Plastics Product Design Engineering Handbook*, Van Nostrand, New York (1977).

Crawford R.J. *Plastics and Rubber – Engineering Design and Applications*, MEP Ltd, London (1985).

Moore, D.R., Couzens, K.H. and Iremonger, M.J., *J. Cellular Plastics*, May/June (1974) pp. 135–139.

Benham, P.P., Crawford, R.J., and Armstrong, C.G., *Mechanics of Engineering Materials*, 2nd Edition, Longmans (1996).

Sterrett, T. and Miller, E., *J. Elastomers and Plastics*, **20** (1988) p. 346.

Young, W.C. *Roark's Formulas for Stress and Strain* 5th Edition, McGraw-Hill (1975).

Williams, M.L., Landel, R.F. and Ferry, J.D., *J. Amer. Chem. Soc.*, **77** (1955) p 3701.
Thomas, D.A. Uniaxial compressive creep studies, *Plastics and Polymers*, Oct(1969) p. 485.

Questions

Where appropriate the creep curves in Fig. 2.5 should be used.

2.1 A plastic beam is to be subjected to load for a period of 1500 hours. Use the 1500 hour modulus values given below and the data in Table 1.5 to decide which of the materials listed would provide the most cost effective design (on a stiffness basis).

Material	PP	uPVC	ABS	Nylon 66	Polycarb.	Acetal	Poly-sulphone
1500 hr modulus (GN/m^2)	0.3	2.1	1.2	1.2	2.0	1.0	2.1

2.2 An extruded T-section beam in polypropylene has a cross-sectional area of 225 mm and a second moment of area, I, of 12.3×10^3 mm^4. If it is to be built-in at both ends and its maximum deflection is not to exceed 4 mm after 1 week, estimate a suitable length for the beam. The central deflection, δ, is given by

$$\delta = WL^3/384EI$$

where W is the weight of the beam. Use a limiting strain of 2%.

2.3 In the previous question the use of the 2% limiting strain will produce a conservative estimate for the beam length because the actual strain in the beam will be less than 2%. If the T-section is 25 mm wide and 25 mm deep with a general wall thickness of 5 mm, what is the % error incurred by using the 2% modulus?. Calculate the likely beam deflection after 1 week. The central bending moment on the beam is given by $WL/24$.

2.4 A polypropylene pipe with an outside diameter of 80 mm is required to withstand a constant pressure of 0.5 MN/m^2 for at least 3 years. If the density of the material is 909 kg/m^3 and the maximum allowable strain is 1.5% estimate a suitable value for the wall thickness of the pipe. If a lower density grade of polypropylene ($\rho = 905$ kg/m^3) was used under the same design conditions, would there be any weight saving per unit length of pipe?

2.5 Show that for a simply supported beam of length, L, subjected to a point load, W, at mid-span the maximum strain, ε, in the material is given by

$$\varepsilon = \frac{6\delta d}{L^2}$$

where d is the beam depth and δ is the central deflection.

2.6 A piece of thin wall polypropylene pipe with a diameter of 300 mm is rotated about its longitudinal axis at a speed of 3000 rev/min. Calculate how long it would take for the diameter of the pipe to increase by 1.2 mm. The density of the polypropylene is 909 kg/m^3.

2.7 The maximum strain in a vertical liquid storage tank is given by $\varepsilon = \rho g H R/2Eh$ where H is the level above the base of the liquid of density ρ, and h is the wall thickness of the tank.

If a polypropylene tank of radius, $R = 0.625$ m, and height 3 m is to be filled with water for a period of one year, calculate the thickness of the tank material so that the change in its diameter will not exceed 12.5 mm. The density of the polypropylene is 904 kg/m^3.

2.8 The value of the external pressure, P_c, which would cause a thin wall sphere of radius, R, to collapse is given by

$$P_c = 0.365E \ (h/R)^2$$

where h is the wall thickness.

An underground polypropylene storage tank is a sphere of diameter 1.4 m. If it is to be designed to resist an external pressure of 20 kN/m² for at least 3 years, estimate a suitable value for the wall thickness. Tensile creep data may be used and the density of the polypropylene is 904 kg/m³.

2.9 A polypropylene bar with a square section (10 mm × 10 mm) is 225 mm long. It is pinned at both ends and an axial compressive load of 140 N is applied. How long would it be before buckling would occur. The relationship between the buckling load, F_c, and the bar geometry is

$$F_c = \pi^2 EI/L^2$$

where L is the length of the bar and I is the second moment of area of the cross-section.

2.10 Show that a ratio of depth to thickness equal to 10 is the normal limit if buckling is to be avoided during short-term loading of plastics. What is likely to happen to this ratio for long-term loading? You should consider the situation of buckling of a strut fixed at both ends for which the critical buckling load is given by

$$P_c = \frac{4\pi^2 EI}{L^2}$$

2.11 Show that the critical buckling strain in a strut with pinned ends is dependent only on the geometry of the strut.

A polypropylene rod, 150 mm long is to be designed so that it will buckle at a critical strain of 0.5%. Calculate a suitable diameter for the rod and the compressive load which it could transmit for at least one year.

2.12 A circular polypropylene plate, 150 mm in diameter is simply supported around its edge and is subjected to a uniform pressure of 40 kN/m². If the stress in the material is not to exceed 6 MN/m², estimate a suitable thickness for the plate and the deflection, δ, after one year. The stress in the plate is given by

$$\sigma = 3(1 + v)PR^2/8h^2$$

and

$$\delta = [3(1 - v)(5 + v)PR^4]/16Eh^3$$

2.13 A cylindrical polypropylene bottle is used to store a liquid under pressure. It is designed with a 4 mm skirt around the base so that it will continue to stand upright when the base bulges under pressure. If the diameter of the bottle is 64 mm and it has a uniform wall thickness of 2.5 mm, estimate the maximum internal pressure which can be used if the container must not rock on its base after one year. Calculate also the diameter change which would occur in the bottle after one year under pressure.

2.14 A rectangular section polypropylene beam has a length, L of 200 mm and a width of 12 mm. It is subjected to a load, W, of 150 N uniformly distributed over its length, L, and it is simply supported at each end. If the maximum deflection of the beam is not to exceed 6 mm after a period of 1 year estimate a suitable depth for the beam. The central deflection of the beam is given by

$$\delta = 5WL/384EI$$

2.15 In a particular application a 1 m length of 80 mm diameter polypropylene pipe is subjected to two dimetrically opposite point loads. If the wall thickness of the pipe is 3 mm, what is the maximum value of the load which can be applied if the change in diameter between the loads is not to exceed 3 mm in one year.

The deflection of the pipe under the load is given by

$$\delta = \frac{W}{Eh}[0.48(L/R)^{0.5}(R/h)^{1.22}]$$

and the stress is given by $\sigma = 2.4 \ W/h^2$ where W is the applied load and h is the wall thickness of the pipe.

2.16 The stiffness of a closed coil spring is given by the expressions:

$$\text{Stiffness} = Gd^4/64R^3N$$

where d is the diameter of the spring material, R is the radius of the coils and N is the number of coils.

In a small mechanism, a polypropylene spring is subjected to a fixed extension of 10 mm. What is the initial force in the spring and what will it exert after one week. The length of the spring is 30 mm, its diameter is 10 mm and there are 10 coils. The design strain and creep contraction ratio for the polypropylene may be taken as 2% and 0.4 respectively.

2.17 A closed coil spring made from polypropylene is to have a steady force, W, of 3 N applied to it for 1 day. If there are 10 coils and the spring diameter is 15 mm, estimate the minimum diameter for the spring material if it is to recover completely when the force is released.

If the spring is subjected to a 50% overload for 1 day, estimate the percentage increase in the extension over the normal 1 day extension. The shear stress in the material is given by 16 WR/d^3. Use the creep curves supplied and assume a value of 0.4 for the lateral contraction ratio.

2.18 A rod of polypropylene, 10 mm in diameter, is clamped between two rigid fixed supports so that there is no stress in the rod at 20°C. If the assembly is then heated quickly to 60°C estimate the initial force on the supports and the force after 1 year. The tensile creep curves should be used and the effect of temperature may be allowed for by making a 56% shift in the creep curves at short times and a 40% shift at long times. The coefficient of thermal expansion for polypropylene is $1.35 \times 10^{-4}\text{°C}^{-1}$ in this temperature range.

2.19 When a pipe fitting is tightened up to a 12 mm diameter polypropylene pipe at 20°C the diameter of the pipe is reduced by 0.05 mm. Calculate the stress in the wall of the pipe after 1 year and if the inside diameter of the pipe is 9 mm, comment on whether or not you would expect the pipe to leak after this time. State the minimum temperature at which the fitting could be used. Use the tensile creep curves and take the coefficient of thermal expansion of the polypropylene to be $9.0 \times 10^{-5}\text{°C}^{-1}$.

2.20 A polypropylene pipe of inside diameter 10 mm and outside diameter 12 mm is pushed on to a rigid metal tube of outside diameter 10.16 mm. If the polypropylene pipe is in contact with the metal tube over a distance of 15 mm, calculate the axial force necessary to separate the two pipes (a) immediately after they are connected (b) 1 year after connection. The coefficient of friction between the two materials is 0.3 and the creep data in Fig. 2.5 may be used.

2.21 A nylon bush is to be inserted into a metal housing as illustrated in Fig. 2.85 The housing has a diameter of 40 mm and the inside diameter of the bush is 35 mm. If the length of the bush is 10 mm and the initial extraction force is to be 1.2 kN, calculate (a) the necessary interference on radius between the bush and the housing (b) the temperature to which the bush must be cooled to facilitate easy assembly (c) the internal diameter of the bush when it is in the housing and (d) the long term extraction force for the bush. The short term modulus of the nylon is 2 GN/m², its coefficient of friction is 0.24 and its coefficient of thermal expansion is $100 \times 10^{-6}\text{°C}^{-1}$. Poissons Ratio for the Nylon is 0.4 and its long term modulus may be taken as 1 GN/m².

2.22 If the bobbin illustrated in Example 2.6 (Fig. 2.16) is cooled from 20°C to −40°C, estimate the maximum hoop stress set up in the acetal. The modulus of the acetal at −40°C is 3 GN/m² and Poisson's ratio is 0.33. The coefficients of thermal expansion for acetal and steel are $80 \times 10^{-6}\text{°C}^{-1}$ and $11 \times 10^{-6}\text{°C}^{-1}$, respectively.

2.23 From the creep curves for a particular plastic the following values of creep rate at various stress levels were recorded for times between 10^6 and 10^7 seconds:

Stress (MN/m²)	1.5	3.0	4.5	6.0	7.5	9.0	12.0
strain rate (s)	4.1×10^{-11}	7×10^{-11}	9.5×10^{-11}	1.2×10^{-10}	1.4×10^{-10}	1.6×10^{-10}	2×10^{-10}

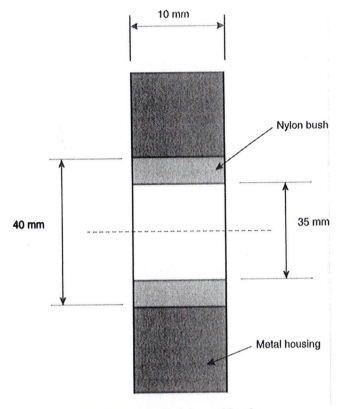

Fig. 2.85   Nylon bush in metal housing

Confirm whether or not this data obeys a law of the form

$$\dot{\varepsilon} = A\sigma^n$$

and if so, determine the constants A and n. When a stress of 5 MN/m² is applied to this material the strain after 10^6 seconds is 0.95%. Predict the value of the strain after 9×10^6 seconds at this stress.

2.24 For the grade of polypropylene whose creep curves are given in Fig. 2.5, confirm that the strain may be predicted by a relation of the form

$$\varepsilon(t) = At^n$$

where A and n are constants for any particular stress level. A small component made from this material is subjected to a constant stress of 5.6 MN/m² for 3 days at which time the stress is completely removed. Estimate the strain in the material after a further 3 days.

2.25 A small beam with a cross-section 15 mm square is foam moulded in polypropylene. The skin has a thickness of 2.25 mm and the length of the beam is 250 mm. It is to be built in at both ends and subjected to a uniformly distributed load, w, over its entire length. Estimate the dimensions of a square section solid polypropylene beam which would have the same stiffness when loaded in this way and calculate the percentage weight saving by using the foam moulding. (Density of skin = 909 kg/m³, density of core = 450 kg/m³).

2.26 If the stress in the composite beam in the previous question is not to exceed 7 MN/m^2 estimate the maximum uniformly distributed load which it could carry over its whole length. Calculate also the central deflection after 1 week under this load. The bending moment at the centre of the beam is $WL/24$.

2.27 A rectangular section beam of solid polypropylene is 12 mm wide, 8 mm deep and 300 mm long. If a foamed core polypropylene beam, with a 2 mm solid skin on the upper and lower surfaces only, is to be made the same width, length and weight estimate the depth of the composite beam and state the ratio of the stiffness of the two beams. ($\rho = 909$ kg/m^3, $\rho = 500$ kg/m^3).

2.28 Compare the flexural stiffness to weight ratios for the following three plastic beams. (a) a solid beam of depth 12 mm, (b) a beam of foamed material 12 mm thick and (c) a composite beam consisting of an 8 mm thick foamed core sandwiched between two solid skin layers 2 mm thick. The ratio of densities of the solid and foamed material is 1.5. (hint: consider unit width and unit length of beam).

2.29 For a sandwich beam with solid skins and a foamed core, show that (a) the weight of the core should be twice the weight of the skin if the beam is to be designed for maximum stiffness at minimum overall weight and (b) the weight of the core should equal the weight of the skin if the beam is to be designed to provide maximum strength for minimum weight.

2.30 The viscoelastic behaviour of a certain plastic is to be represented by spring and dashpot elements having constants of 2 GN/m^2 and 90 GNs/m^2 respectively. If a stress of 12 MN/m^2 is applied for 100 seconds and then completely removed, compare the values of strain predicted by the Maxwell and Kelvin–Voigt models after (a) 50 seconds (b) 150 seconds.

2.31 Maxwell and Kelvin–Voigt models are to be set up to simulate the creep behaviour of a plastic. The elastic and viscous constants for the Kelvin–Voigt models are 2 GN/m^2 and 100 GNs/m^2 respectively and the viscous constant for the Maxwell model is 200 GNs/m^2. Estimate a suitable value for the elastic constant for the Maxwell model if both models are to predict the same creep strain after 50 seconds.

2.32 During a test on a polymer which is to have its viscoelastic behaviour described by the Kelvin model the following creep data was obtained when a stress of 2 MN/m^2 was applied to it.

Time(s) 0	0.5×10^3	1×10^3	3×10^3	5×10^3	7×10^3	10×10^4	15×10^4
Strain 0	3.1×10^{-3}	5.2×10^{-3}	8.9×10^{-3}	9.75×10^{-3}	9.94×10^{-3}	9.99×10^{-3}	9.99×10^{-3}

Use this information to predict the strain after 1500 seconds at a stress of 4.5 MN/m^2. State the relaxation time for the polymer.

2.33 A *Standard Model* for the viscoelastic behaviour of plastics consists of a spring element in series with a Voigt model as shown in Fig. 2.86. Derive the governing equation for this model and from this obtain the expression for creep strain. Show that the Unrelaxed Modulus for this model is ξ_1 and the Relaxed Modulus is $\xi_1\xi_2/(\xi_1 + \xi_2)$.

2.34 The grade of polypropylene whose creep curves are given in Fig. 2.5 is to have its viscoelastic behaviour fitted to a Maxwell model for stresses up to 6 MN/m^2 and times up to 1000 seconds. Determine the two constants for the model and use these to determine the stress in the material after 900 seconds if the material is subjected to a constant strain of 0.4% throughout the 900 seconds.

2.35 The creep curve for polypropylene at 4.2 MN/m^2 (Fig. 2.5) is to be represented for times up to 2×10^6 s by a 4-element model consisting of a Maxwell unit and a Kelvin–Voigt unit in series. Determine the constants for each of the elements and use the model to predict the strain in this material after a stress of 5.6 MN/m^2 has been applied for 3×10^5 seconds.

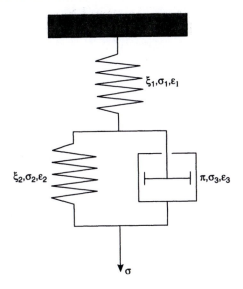

Fig. 2.86   Standard model for viscoelastic material

2.36 Show that for a viscoelastic material in which the modulus is given by $E(t) = At^{-n}$, there will be a non-linear strain response to a linear increase in stress with time.

2.37 In a tensile test on a plastic, the material is subjected to a constant strain rate of 10^{-5} s. If this material may have its behaviour modelled by a Maxwell element with the elastic component $\xi = 20$ GN/m^2 and the viscous element $\eta = 1000$ GNs/m^2, then derive an expression for the stress in the material at any instant. Plot the stress–strain curve which would be predicted by this equation for strains up to 0.1% and calculate the initial tangent modulus and 0.1% secant modulus from this graph.

2.38 A plastic is stressed at a constant rate up to 30 MN/m^2 in 60 seconds and the stress then decreases to zero at a linear rate in a further 30 seconds. If the time dependent creep modulus for the plastic can be expressed in the form

$$E(t) = \frac{\xi\eta}{\eta + \xi t}$$

use Boltzmann's Superposition Principle to calculate the strain in the material after (i) 40 seconds (ii) 70 seconds and (iii) 120 seconds. The elastic component of modulus in 3 GN/m^2 and the viscous component is 45×10^9 Ns/m^2.

2.39 A plastic with a time dependent creep modulus as in the previous example is stressed at a linear rate to 40 MN/m^2 in 100 seconds. At this time the stress in reduced to 30 MN/m^2 and kept constant at this level. If the elastic and viscous components of the modulus are 3.5 GN/m^2 and 50×10^9 Ns/m^2, use Boltzmann's Superposition Principle to calculate the strain after (a) 60 seconds and (b) 130 seconds.

2.40 A plastic has a time-dependent modulus given by

$$E(t) = 10^3 t^{-0.05}$$

where $E(t)$ is in MN/m^2 when 't' is in seconds. If this material is subjected to a stress which increases steadily from 0 to 20 MN/m^2 in 800 seconds and is then kept constant, calculate the strain in the material after (a) 500 seconds and (b) 1000 seconds.

2.41 A plastic which behaves like a Kelvin–Voigt model is subjected to the stress history shown in Fig. 2.87. Use the Boltzmanns Superposition Principle to calculate the strain in the material after (a) 90 seconds (b) 150 seconds. The spring constant is 12 GN/m^2 and the dashpot constant is 360 GNs/m^2.

Fig. 2.87

2.42 A plastic component was subjected to a series of step changes in stress as follows. An initial constant stress of 10 MN/m^2 was applied for 1000 seconds at which time the stress level was increased to a constant level of 20 MN/m^2. After a further 1000 seconds the stress level was decreased to 5 MN/m^2 which was maintained for 1000 seconds before the stress was increased to 25 MN/m^2 for 1000 seconds after which the stress was completely removed. If the material may be represented by a Maxwell model in which the elastic constant $\xi = 1$ GN/m^2 and the viscous constant $\eta = 4000$ GNs/m^2, calculate the strain 4500 seconds after the first stress was applied.

2.43 In tests on a particular plastic it is found that when a stress of 10 MN/m^2 is applied for 100 seconds and then completely removed, the strain at the instant of stress removal is 0.8% and 100 seconds later it is 0.058%. In a subsequent tests on the same material the stress of 10 MN/m^2 is applied for 2400 seconds and completely removed for 7200 seconds and this sequence is repeated 10 times. Assuming that the creep curves for this material may be represented by an equation of the form $\varepsilon(t) = At^n$ where A and n are constants then determine the total accumulated residual strain in the material at the end of the 10th cycle.

2.44 In a small polypropylene component a tensile stress of 5.6 MN/m^2 is applied for 1000 seconds and removed for 500 seconds. Estimate how many of these stress cycles could be permitted before the component reached a limiting strain of 1%. What is the equivalent modulus of the material at his number of cycles? The creep curves in Fig. 2.5 may be used.

2.45 A cylindrical polypropylene pressure vessel of 150 mm outside diameter is to be pressurised to 0.5 MN/m^2 for 6 hours each day for a projected service life of 1 year. If the material can be described by an equation of the form $\varepsilon(t) = At^n$ where A and n are constants and the maximum strain in the material is not to exceed 1.5% estimate a suitable wall thickness for the vessel on the assumption that it is loaded for 6 hours and unloaded for 18 hours each day. Estimate the material saved compared with a design in which it is assumed that the pressure is constant at 0.5 MN/m^2 throughout the service life. The creep curves in Fig. 2.5 may be used.

2.46 For the type of Standard Linear Solid described in Q 2.33, derive equations for the storage modulus, the loss modulus and tan δ when the material is subjected to a sinusoidally varying stress. Confirm that for $\eta = 1$ GNs/m^2, $\xi_1 = 2$ GN/m^2 and $\xi_2 = 0.1$ GN/m^2, your equations predict the classical variation of E_1, E_2 and tan δ for values of ω in the range 0.01 to 100 s^{-1}.

2.47 Creep rupture tests on a particular grade of uPVC at 20°C gave the following results for applied stress, σ, and time to failure, t.

Stress (MN/m^2)	60	55	52	48	45	43
time(s)	800	7×10^3	3.25×10^4	2.15×10^5	8.9×10^6	2.4×10^6

Confirm that this data obeys a law of the form

$$t = Ae^{-B\sigma}$$

and determine the values of the constants A and B.

2.48 For the material in the previous question, use the Zhurkov-Beuche equation to calculate the time to failure under a steady stress of 44 MN/m^2 if the material temperature is 40°C. The activation energy, U_0, may be taken as 150 kJ/mol.

2.49 A 200 mm diameter plastic pipe is to be subjected to an internal pressure of 0.5 MN/m^2 for 3 years. If the creep rupture behaviour of the material is as shown in Fig. 3.10, calculate a suitable wall thickness for the pipe. You should use a safety factor of 1.5.

2.50 Fracture Mechanics tests on a grade of ABS indicate that its K value is 2 MN m$^{-3/2}$ and that under static loading its growth rate is described by the equation.

$$da/dt = 3 \times 10^{-11}K^{3.2}$$

where K has units MN m$^{-3/2}$. If, in service, the material is subjected to a steady stress of 20 MN/m^2 estimate the maximum defect size which could be tolerated in the material if it is to last for at least 1 year.

2.51 Use the data in Table 2.2 to compare crack tip plastic zone sizes in acrylic, ABS and polypropylene.

2.52 In a tensile test on an un-notched sample of acrylic the fracture stress is recorded as 57 MN/m^2. Estimate the likely size of the intrinsic defects in the material.

2.53 In a small timing mechanism an acetal copolymer beam is loaded as shown in Fig. 2.88. The end load varies from 0 to F at a frequency of 5 Hz. If the beam is required to withstand at least 10 million cycles, calculate the permissible value of F assuming a fatigue strength reduction factor of 2. The surface stress (in MN/m^2) in the beam at the support is given by$\frac{1}{60}FL$ where F is in Newtons and L is the beam length in mm. Fatigue and creep fracture data for the acetal copolymer are given in Figs 2.89 and 2.90.

Fig. 2.88   Beam in timing mechanism

Fig. 2.89   Fatigue behaviour of acetal

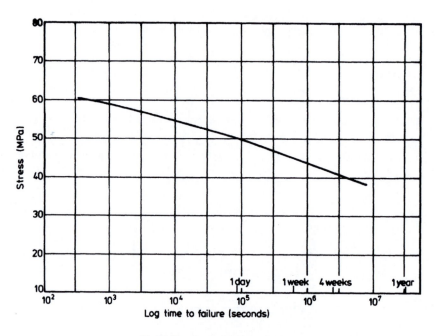

Fig. 2.90   Creep fracture of acetal

2.54 A plastic shaft of circular cross-section is subjected to a steady bending moment of 1 Nm and simultaneously to an alternating bending moment of 0.75 Nm. Calculate the necessary shaft diameter so as to avoid fatigue failure (the factor of safety is to be 2.5). The fatigue limit for the material in reversed bending is 25 MN/m^2 and the creep rupture strength at the equivalent time may be taken as 35 MN/m^2. Calculate also the shaft diameter if the fatigue strength reduction factor is to be taken as 2.

2.55 A 10 mm diameter uPVC shaft is subjected to a steady tensile load of 500 N. If the fatigue strength reduction factor is 1.8 and the factor of safety is to be 2 calculate the largest alternating bending moment which could be applied at a frequency of 5 Hz if fatigue failure is not to occur inside 10^7 cycles. The creep rupture characteristic for the material is given in question 3.1 and the reversed bending fatigue behaviour is described by the equation $\sigma = (43.4 - 3.8 \log N)$ (where N is the number of cycles to failure and σ is the stress in MN/m^2). It may be assumed that at 5 Hz, thermal softening will not occur.

2.56 A uPVC rod of diameter 12 mm is subjected to an eccentric axial force at a distance of 3 mm from the centre of the cross-section. If the force varies sinusoidally from $-F$ to F at a frequency of 10 Hz, calculate the value of F so that fatigue failure will not occur in 10 cycles. Assume a safety factor of 2.5 and use the creep rupture and fatigue characteristics described in the previous question. Thermal softening effects may be ignored at the stress levels involved.

2.57 For the purposes of performing an impact test on a material it is proposed to use an elastic stress concentration factor of 3.5. If the notch tip radius is to be 0.25 mm estimate a suitable notch depth.

2.58 On an impact testing machine for plastics the weight of the pendulum is 4.5 kgf. When the pendulum is raised to a height of 0.3 m and allowed to swing (a) with no specimen in position and (b) with a plain sample (4 × 12 mm cross-section) in position, the pendulum swings to heights of 0.29 and 0.2 m respectively. Estimate (i) the friction and windage losses in the machine (ii) the impact energy of the specimen (iii) the height the pendulum will swing to if it is released from a height of 0.25 m and breaks a sample of exactly the same impact strength as in (ii). (Assume that the losses remain the same and that the impact strength is independent of strike velocity).

2.59 A sheet of polystyrene 100 mm wide, 5 mm thick and 200 mm long contains a sharp single edge crack 10 mm long, 100 mm from one end. If the critical stress intensity factor is 1.75 MN m$^{-3/2}$, what is the maximum axial force which could be applied without causing brittle fracture.

2.60 A certain grade of PMMA has a K value of 1.6 MN m$^{-3/2}$ and it is known that under cyclic stresses, cracks grow at a rate given by $(2 \times 10^{-6} \Delta K^{3.32})$. If the intrinsic defects in the material are 50 mm long, how many hours will the material last if it is subjected to a stress cycle of 0 to 10 MN/m^2 at a frequency of 1 Hz.

2.61 A series of fatigue crack growth tests on a moulding grade of polymethyl methacrylate gave the following results

da/dN (m/cycle)	2.25×10^{-7}	4.0×10^{-7}	6.2×10^{-7}	11×10^{-7}	17×10^{-7}	29×10^{-7}
ΔK(MN m$^{-3/2}$)	0.42	0.53	0.63	0.79	0.94	1.17

If the material has a critical stress intensity factor of 1.8 MN m$^{-3/2}$ and it is known that the moulding process produces defects 40 m long, estimate the maximum repeated tensile stress which could be applied to this material for at least 10^6 cycles without causing fatigue failure.

2.62 A series of uniaxial fatigue tests on unnotched plastic sheets show that the fatigue limit for the material is 10 MN/m^2. If a pressure vessel with a diameter of 120 mm and a wall thickness of 4 mm is to be made from this material, estimate the maximum value of fluctuating internal pressure which would be recommended. The stress intensity factor for the pressure vessel is given by $K = 2\sigma_\theta (2a)^{1/2}$ where σ_θ is the hoop stress and 'a' is the half length of an internal defect.

CHAPTER 3 – Mechanical Behaviour of Composites

3.1 Deformation Behaviour of Reinforced Plastics

It was mentioned earlier that the stiffness and strength of plastics can be increased significantly by the addition of a reinforcing filler. A reinforced plastic consists of two main components; a matrix which may be either thermoplastic or thermosetting and a reinforcing filler which usually takes the form of fibres. A wide variety of combinations are possible as shown in Fig. 3.1. In general, the matrix has a low strength in comparison to the reinforcement which is also stiffer and brittle. To gain maximum benefit from the reinforcement, the fibres should bear as much as possible of the applied stress. The function of the matrix is to support the fibres and transmit the external loading to them by shear at the fibre/matrix interface. Since the fibre and matrix are quite different in structure and properties it is convenient to consider them separately.

3.2 Types of Reinforcement

The reinforcing filler usually takes the form of fibres but particles (for example glass spheres) are also used. A wide range of amorphous and crystalline materials can be used as reinforcing fibres. These include glass, carbon, boron, and silica. In recent years, fibres have been produced from synthetic polymers–for example, **Kevlar** fibres (from aromatic polyamides) and PET fibres. The stress–strain behaviour of some typical fibres is shown in Fig. 3.2.

Glass in the form of fibres is relatively inexpensive and is the principal form of reinforcement used in plastics. The fibres are produced by drawing off continuous strands of glass from an orifice in the base of an electrically heated platinum crucible which contains the molten glass. The earliest successful glass reinforcement had a calcium-alumina borosilicate composition developed

168

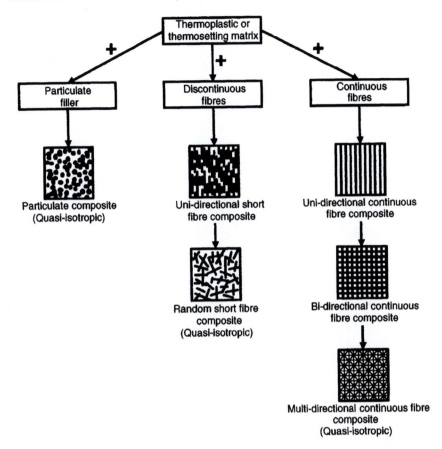

Fig. 3.1   Various types of composite (Adapted from Daniel and Ishai)

specifically for electrical insulation systems (E glass). Although other glasses were subsequently developed for applications where electrical properties are not critical, no commercial composition better than that of E-glass has been found. Certain special glasses for extra high strength or modulus have been produced in small quantities for special applications e.g. aerospace technology.

During production the fibres are treated with a fluid which performs several functions.

(a) it facilitates the production of strands from individual fibres
(b) it reduces damage to fibres during mechanical handling and
(c) it acts as a process aid during moulding.

This treatment is known as *sizing*. As mentioned earlier, the joint between the matrix and the fibre is critical if the reinforcement is to be effective and so the surface film on the glass ensures that the adhesion will be good.

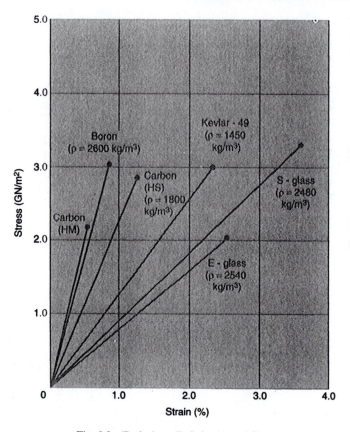

Fig. 3.2   Typical tensile behaviour of fibres

3.3 Types of Matrix

The matrix in a reinforced plastic may be either thermosetting or thermoplastic.

(a) Thermosets

In the early days nearly all thermosetting moulding materials were composites in that they contained fillers such as woodflour, mica, cellulose, etc to increase their strength. However, these were not generally regarded as reinforced materials in the sense that they did not contain fibres.

Nowadays the major thermosetting resins used in conjunction with glass fibre reinforcement are unsaturated polyester resins and to a lesser extent epoxy resins. The most important advantages which these materials can offer are that they do not liberate volatiles during cross-linking and they can be moulded using low pressures at room temperature. Table 3.1 shows typical properties of fibre reinforced epoxy.

Table 3.1

Typical properties of bi-directional fibre composites

Material	Volume fraction (V_f)	Density (kg/m^3)	Tensile strength (GN/m^2)	Tensile modulus (GN/m^2)
Epoxy	–	1200	0.07	6
Epoxy/E-Glass	0.57	1970	0.57	22
Epoxy/Kevlar	0.60	1400	0.65	40
Epoxy/Carbon	0.58	1540	0.38	80
Epoxy/Boron	0.60	2000	0.38	106

(b) Thermoplastics

A wide variety of thermoplastics have been used as the base for reinforced plastics. These include polypropylene, nylon, styrene-based materials, thermoplastic polyesters, acetal, polycarbonate, polysulphone, etc. The choice of a reinforced thermoplastic depends on a wide range of factors which includes the nature of the application, the service environment and costs. In many cases conventional thermoplastic processing techniques can be used to produce moulded articles (see Chapter 4). Some typical properties of fibre reinforced nylon are given in Table 3.2.

Table 3.2

Typical properties of fibre reinforced nylon 66

Material	Weight fraction (W_f)	Density (kg/m^3)	Tensile strength (GN/m^2)	Flexural modulus (GN/m^2)
Nylon 66	–	1140	0.07	2.8
Nylon 66/glass	0.40	1460	0.2	11.2
Nylon 66/carbon	0.40	1340	0.28	24.0
Nylon 66/glass/carbon	0.20C/0.20G	1400	0.24	20.0
Nylon 66/glass beads	0.40	1440	0.09	5.6

3.4 Forms of Fibre Reinforcement in Composites

Reinforcing fibres have diameters varying from 7 μm to 100 μm. They may be continuous or in the form of chopped strands (lengths 3 mm–50 mm). When chopped strands are used, the length to diameter ratio is called the **Aspect Ratio**. The properties of a short-fibre composite are very dependent on the aspect ratio – the greater the aspect ratio the greater will be the strength and stiffness of the composite.

The amount of fibres in a composite is often expressed in terms of the **volume fraction**, V_f. This is the ratio of the volume of the fibres, v_f, to the volume of the composite, v_c. The weight fraction of fibres, W_f, may be related to the volume fraction as follows.

$$W_f = \frac{W_f}{W_c} = \frac{\rho_f v_f}{\rho_c v_c} = \frac{\rho_f}{\rho_c} V_f \tag{3.1}$$

where ρ is the density and the subscripts f and c refer to fibres and composite respectively.

Table 3.3 indicates the extent to which the properties of plastics are influenced by the level of fibre content. Full details of the forms in which reinforcing fibres are available for inclusion in plastics are given in Chapter 4.

Table 3.3

Effect of fibre content on properties of glass reinforced nylon 66

Property	Weight fraction, W_f						
	0	0.10	0.20	0.30	0.40	0.50	0.60
Density	1140	1210	1280	1370	1460	1570	1700
Tensile strength (GN/m²)	0.07	0.09	0.13	0.18	0.21	0.23	0.24
% elongation at break	60	3.5	3.5	3.0	2.5	2.5	1.5
Flexural modulus (GN/m²)	2.8	4.2	6.3	9.1	11.2	15.4	19.6
Thermal expansion μm/m/°C	90	37	32	30	29	25	22
Water absorption (24 hr)	1.6	1.1	0.9	0.9	0.6	0.5	0.4

3.5 Analysis of Continuous Fibre Composites

The greatest improvement in the strength and stiffness of a plastic is achieved when it is reinforced with uni-directional continuous fibres. The analysis of such systems is relatively straightforward.

(i) Longitudinal Properties

Consider a composite with continuous aligned fibres as shown in Fig. 3.3. If the moduli of the matrix and fibres are E_m and E_f respectively then the modulus of the composite may be determined as follows.

Equilibrium Equation

The applied force on the composite will be shared by the fibres and the matrix. Hence

$$F_1 = F_f + F_m \tag{3.2}$$

Geometry of Deformation Equation

The strain, ε, is the same in the fibres and matrix and is equal to the strain in the composite.

$$\varepsilon_1 = \varepsilon_f = \varepsilon_m \tag{3.3}$$

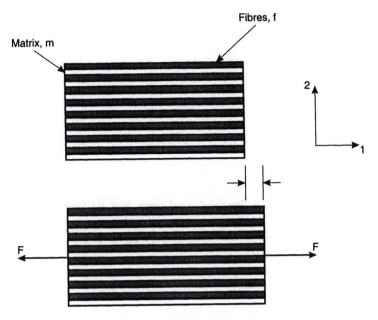

Fig. 3.3   Loading parallel to fibres

Stress–Strain Relationships

$$\sigma_1 = E_1 \varepsilon_{cL}, \qquad \sigma_f = E_f \varepsilon_f, \qquad \sigma_m = E_m \varepsilon_m \qquad (3.4)$$

Combining equations (3.3) and (3.4)

$$E_1 \varepsilon_{cL} A_c = E_f \varepsilon_f A_f + E_m \varepsilon_m A_m$$

and using equation (3.2)

$$E_1 = E_f \left(\frac{A_f}{A_c} \right) + E_m \left(\frac{A_m}{A_c} \right)$$

If the fibres have a uniform cross-section, then the area fraction will equal the volume fraction, so

$$E_1 = E_f V_f + E_m V_m \qquad (3.5)$$

This is an important relationship. It states that the modulus of a unidirectional fibre composite is proportional to the volume fractions of the materials in the composite. This is known as the **Rule of Mixtures**. It may also be used to determine the density of a composite as well as other properties such as the Poisson's Ratio, strength, thermal conductivity and electrical conductivity in the fibre direction.

Example 3.1 The density of a composite made from unidirectional glass fibres in an epoxy matrix is 1950 kg/m^3. If the densities of the glass and epoxy are known to be 2540 kg/m^3 and 1300 kg/m^3, calculate the weight fraction of fibres in the composite.

Solution From the rule of mixtures

$$\rho_c = \rho_f V_f + \rho_m V_m = \rho_f V_f + \rho_m(1 - V_f)$$

$$1950 = 2540 V_f + 1300(1 - V_f)$$

$$V_f = 0.52$$

Using equation (3.1). $W_f = \dfrac{\rho_f}{\rho_c} V_f = \dfrac{2540}{1950}(0.52) = 0.68$

Example 3.2 PEEK is to be reinforced with 30% by volume of unidirectional carbon fibres and the properties of the individual materials are given below. Calculate the density, modulus and strength of the composite in the fibre direction.

Material	Density (kg/m^3)	Tensile strength (GN/m^2)	Modulus (GN/m^2)
PEEK	1300	0.058[†]	3.8
Carbon fibre (HM)	1800	2.1	400

[†] Note that as shown below, this must be the matrix stress at the fibre fracture strain.

Solution From the rule of mixtures

$$\rho_c = \rho_f V_f + \rho_m V_m = 0.13(1800) + 0.7(1300) = 1450 \text{ kg/m}^3$$

$$\sigma_{cu} = \sigma_{fu} V_f + \sigma_m V_m = 0.3(2.1) + 0.7(0.058) = 0.67 \text{ GN/m}^2$$

$$E_{cL} = E_f V_f + E_m V_m = 0.3(400) + 0.7(3.8) = 122.7 \text{ GN/m}^2$$

Example 3.3 Calculate the fraction of the applied force which will be taken by the fibres in the composite referred to in Example 3.2

Solution From equations (3.2), (3.3) and (3.4), the force in the fibres is given by

$$F_f = \sigma_f A_f = E_f \varepsilon_f V_f v_c$$

Similarly the force in the composite is given by

$$F_1 = E_f \varepsilon_f V_f v_c + E_m \varepsilon_m V_m v_c$$

Hence,

$$\frac{F_f}{F_1} = \frac{E_f V_f}{E_f V_f + E_m V_m} = 0.978$$

It may be seen that a very large percentage of the applied force is carried by the fibres. Note also that the ratio of stresses in the fibre may be determined in a similar way.

From the rule of mixtures, the stresses are related as follows.

$$\sigma_1 = \sigma_f V_f + \sigma_m V_m$$

$$\frac{\sigma_1}{\sigma_f} = V_f + \frac{\sigma_m}{\sigma_f}(1 - V_f)$$

As the strains are the same in the matrix and fibres $\dfrac{\sigma_m}{\sigma_f} = \dfrac{E_m}{E_f}$

$$\frac{\sigma_1}{\sigma_f} = V_f + \frac{E_m}{E_f}(1 - V_f) = 0.3 + \frac{3.8}{400}(0.7) = 0.307$$

Thus,

$$\frac{\sigma_f}{\sigma_1} = 3.26$$

As shown in Fig. 3.4 stress–strain tests on uniaxially aligned fibre composites show that their behaviour lies somewhere between that of the fibres and that of the matrix. In regard to the strength of the composite, σ_{cu}, the rule of mixtures has to be modified to relate to the matrix stress, σ'_m at the fracture strain of the fibres rather than the ultimate tensile strength, σ_{mu} for the matrix.

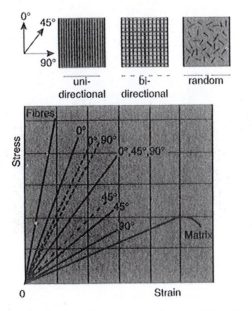

Fig. 3.4   Stress–strain behaviour for several types of fibre reinforcement

This is because, with brittle fibres, failure of the composite will occur when the fibres reach their fracture strain. At this point the matrix is subjected to the full applied load, which it is unable to sustain.

Thus, as shown earlier, the ultimate strength of the composite may be predicted by the rule of mixtures.

$$\sigma_{cu} = \sigma_{fu}V_f + \sigma'_m V_m \tag{3.6}$$

However, as shown in Fig. 3.5, this equation only applies when the volume fraction is greater than a certain critical value, V_{crit}. From Fig. 3.5. this will be defined by

$$\sigma_{mu}(1 - V_{crit}) = \sigma_{fu}V_{crit} + \sigma'_m(1 - V_{crit})$$

$$V_{crit} = \frac{\sigma_{mu} - \sigma'_m}{\sigma_{fu} + \sigma_{mu} - \sigma'_m} \tag{3.7}$$

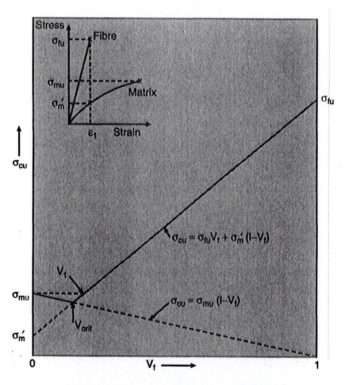

Fig. 3.5   Effect of volume fraction on strength

In addition it may be seen that the strengthening effect of the fibres is only observed (i.e. $\sigma_{cu} > \sigma_{mu}$) when the volume fraction is greater than a certain value V_1. From Fig. 3.5 the value of V_1 is obtained from

$$\sigma_{mu} = \sigma_{fu}V_1 + \sigma'_m(1 - V_1)$$

$$V_1 = \frac{\sigma_{mu} - \sigma'_m}{\sigma_{fu} - \sigma_m} \qquad (3.8)$$

In practice the maximum volume fraction, V_{max}, which can be achieved in unidirectional fibre composites is about 0.8. Designers must therefore arrange for volume fractions to be in the range $V_1 \rightarrow V_{max}$. It should also be noted that in commercial production it is not always possible to achieve the high standards of manufacture necessary to obtain full benefit from the fibres. It is generally found that although the stiffness is predicted quite accurately by equation (3.5), the strength is usually only about 65% of the value calculated by the rule of mixtures. For fibre reinforcement systems other than unidirectional fibres, these values can be reduced even more. To allow for this a constant 'k' is sometimes included in the fibre contribution to equation (3.6).

Example 3.4 For the PEEK/carbon fibre composite referred to in Example 3.2 calculate the values of V_1 and V_{crit} if it is known that the ultimate tensile strength of PEEK is 62 MN/m^2.

Solution From equation (3.7)

$$V_{crit} = \frac{\sigma_{mu} - \sigma'_m}{\sigma_{fu} + \sigma_{mu} - \sigma'_m} = \frac{0.062 - 0.058}{2.1 + 0.062 - 0.058} = 0.19\%$$

and from equation (2.46)

$$V_1 = \frac{\sigma_{mu} - \sigma'_m}{\sigma_{fu} - \sigma'_m} = \frac{0.062 - 0.058}{2.1 - 0.058} = 0.2\%$$

(ii) Properties Perpendicular to Longitudinal Axis

The properties of a unidirectional fibre will not be nearly so good in the transverse direction compared with the longitudinal direction. As a material in service is likely to be subjected to stresses and strains in all directions it is important to be aware of the properties in all directions. The transverse direction will, of course, be the weakest direction and so it is necessary to pay particular attention to this.

The transverse modulus, E_{cT}, may be determined in a manner similar to that described earlier for the longitudinal modulus. Consider a unidirectional fibre composite subjected to a transverse force, F_{cT}, in the direction perpendicular to the fibre axis.

Equilibrium Condition

Referring to Fig. 3.6. The force in the fibres will be the same as the force in the matrix:

$$F_2 = F_f = F_m$$

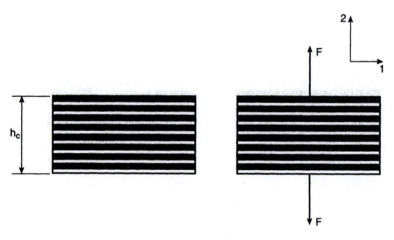

Fig. 3.6   Loading perpendicular to fibres

Assuming that the width and thickness (into the page) is the same for both the matrix and the fibres, then

$$\sigma_2 = \sigma_f = \sigma_m \qquad (3.9)$$

Geometry of Deformation Equation

The total transverse deformation will be the sum of the deformations in the matrix and the fibres

$$\delta_2 = \delta_f + \delta_m$$

$$\varepsilon_2 h_c = \varepsilon_f h_f + \varepsilon_m h_m \qquad (3.10)$$

where h is the thickness of the relevant component.

Stress–Strain Relations

$$\sigma_2 = E_{cT}\varepsilon_{cT}, \qquad \sigma_f = E_f \varepsilon_f, \qquad \sigma_m = E_m \varepsilon_m \qquad (3.11)$$

Then, from (3.10) and (3.11) we may write

$$\frac{\sigma_2}{E_2} = \frac{\sigma_f}{E_f}\frac{h_f}{h_c} + \frac{\sigma_m}{E_m}\frac{h_m}{h_c}$$

Using equation (3.9) and the fact that the thickness ratios will be equal to the corresponding volume fractions

$$\frac{1}{E_2} = \frac{V_f}{E_f} + \frac{V_m}{E_m} \tag{3.12}$$

$$E_2 = \frac{E_f E_m}{V_f E_m + V_m E_f} \tag{3.13}$$

Fig. 3.7 shows how the longitudinal and transverse moduli vary with volume fraction for a unidirectional fibre composite.

Fig. 3.7   Variation of longitudinal and transverse modulus in unidirectional composites

In practical terms the above analysis is too simplistic, particularly in regard to the assumption that the stresses in the fibre and matrix are equal. Generally the fibres are dispersed at random on any cross-section of the composite (see Fig. 3.8) and so the applied force will be shared by the fibres and matrix but not necessarily equally. Other inaccuracies also arise due to the mis-match of the Poisson's ratios for the fibres and matrix. Several other empirical equations have been suggested to take these factors into account. One of these is the Halpin–Tsai equation which has the following form

$$E_2 = E_m \left(\frac{1 + 2\beta V_f}{1 - \beta V_f} \right) \tag{3.14}$$

Fig. 3.8   Dispersion of fibres in cross-section of unidirectional composite

where

$$\beta = \frac{(E_f/E_m) - 1}{(E_f/E_m) + 2}$$

The predictions from this equation are shown in Fig. 3.7.

Another alternative is the Brintrup equation which gives E_2 as

$$E_2 = \frac{E'_m E_f}{E_f(1 - V_f) + V_f E'_m} \tag{3.15}$$

where

$$E'_m = E_m/(1 - \nu_m^2)$$

A simplistic mechanics of materials analysis of the shear modulus G_{12} gives

$$\frac{1}{G_{12}} = \frac{V_f}{G_f} + \frac{V_m}{G_m}$$

or

$$G_{12} = \frac{G_f G_m}{V_f G_m + V_m G_f}$$

where the shear modulus of the fibres is the G_{12f} but it has been written as G_f for simplicity.

As in the case of the transverse tensile modulus, E_2, the above analysis tends to underestimate the in-plane shear modulus. Therefore, once again it is common to resort to empirical relationships and the most popular is the Halpin–Tsai equation

$$G_{12} = G_m \left\{ \frac{(G_f + G_m) + V_f(G_f - G_m)}{(G_f + G_m) - V_f(G_f - G_m)} \right\} \tag{3.16}$$

Example 3.5 Calculate the transverse modulus of the PEEK/carbon fibre composite referred to in Example 3.2, using both the simplified solid mechanics approach and the empirical approach. For PEEK $v_m = 0.36$.

Solution From equation (3.13)

$$E_2 = \frac{3.8(400)}{0.3(3.8) + 0.7(400)} = 5.4 \text{ GN/m}^2$$

Using the Halpin–Tsai equation

$$\beta = \frac{(E_f/E_m) - 1}{(E_f/E_m) + 2} = \frac{(400/3.8) - 1}{(400/3.8) + 2} = 0.97$$

$$E_2 = 3.8 \left[\frac{1 + 2(0.3)(0.97)}{1 - (0.3)(0.97)} \right] = 8.5 \text{ GN/m}^2$$

Using the Brintrup equation

$$E'_m = 3.8/(1 - 0.36^2) = 4.37 \text{ GN/m}^2$$

$$E_2 = \frac{(4.37)400}{400(1 - 0.3) + 0.3(4.37)} = 6.2 \text{ GN/m}^2$$

Some typical elastic properties for unidirectional fibre composites are given in Table 3.4.

Table 3.4

Typical elastic properties of unidirectional fibre reinforced plastics

Material	Fibre volume fraction, V_f	E_1 (GN/m^2)	E_2 (GN/m^2)	G_{12} (GN/m^2)	v_{12}
GFRP (E glass/epoxy)	0.6	40	9	4	0.31
GFRP (E glass/polyester)	0.42	32	7	3.1	0.35
KFRP (Kevlar 49/epoxy)	0.6	79	4.1	1.5	0.43
CFRP (carbon/epoxy)	0.6	125	9	4.4	0.34
CFRP (carbon HM/epoxy)	0.62	497	5.3	5.6	0.31

GFRP – Glass Fibre Reinforced Plastic
KFRP – Kevlar Fibre Reinforced Plastic
CFRP – Carbon Fibre Reinforced Plastic

3.6 Deformation Behaviour of a Single Ply or Lamina

The previous section illustrated how to obtain the elastic properties of a uni-directional lamina. In practice considerably more information may be required about the behavioural characteristics of a single lamina. To obtain details of the stresses and strains at various orientations in a single ply the following type of analysis is required.

(i) Loading on Fibre Axis

Consider the situation of a thin unidirectional lamina under a state of plane stress as shown in Fig. 3.9. The properties of the lamina are anisotropic so it will have modulus values of E_1 and E_2 in the fibre and transverse directions, respectively. The values of these parameters may be determined as illustrated above.

Fig. 3.9 Single thin lamina under plane stress

The direct strains in the global X and Y directions will be given by (see Appendix C for details)

$$\varepsilon_x = \frac{\sigma_x}{E_1} - \nu_{21}\frac{\sigma_y}{E_2} \tag{3.17}$$

$$\varepsilon_y = \frac{\sigma_y}{E_2} - \nu_{12}\frac{\sigma_x}{E_1} \tag{3.18}$$

and the shear strain γ_{xy} is related to the shear stress, τ_{xy}, and the shear modulus, G_{12}, by

$$\gamma_{xy} = \frac{\tau_{xy}}{G_{12}} \tag{3.19}$$

The convention normally used is that direct stresses and strains have one suffix to indicate the direction of the stress or strain. Shear stresses and strains have two suffices. The first suffix indicates the direction of the normal to the plane on which the stress acts and the second suffix indicates the direction of the stress (or strain). Poisson's Ratio has two suffices. Thus, v_{12} is the negative ratio of the strain in the 2-direction to the strain in the 1-direction for a stress applied in the 1-direction ($v_{12} = -\varepsilon_2/\varepsilon_1$ for an applied σ_1). v_{12} is sometimes referred to as the major Poisson's Ratio and v_{21} is the minor Poisson's Ratio. In an isotropic material where $v_{21} = v_{21}$, then the suffices are not needed and normally are not used.

In the analysis of composites it is convenient to use matrix notation because this simplifies the computations very considerably. Thus we may write the above equations as (see Appendix E)

$$\left\{ \begin{array}{c} \varepsilon_x \\ \varepsilon_y \\ \gamma_{xy} \end{array} \right\} = \left[\begin{array}{ccc} \dfrac{1}{E_1} & -\dfrac{v_{21}}{E_2} & 0 \\[2mm] -\dfrac{v_{12}}{E_1} & \dfrac{1}{E_2} & 0 \\[2mm] 0 & 0 & \dfrac{1}{G_{12}} \end{array} \right] \left\{ \begin{array}{c} \sigma_x \\ \sigma_y \\ \tau_{xy} \end{array} \right\}$$

or in abbreviated form

$$\{\varepsilon\} = [S]\{\sigma\} \tag{3.20}$$

where $[S]$ is called the *compliance* matrix.

In order to describe completely the state of triaxial (as opposed to biaxial) stress in an anisotropic material, the compliance matrix will have 36 terms. The reader is referred to the more advanced composites texts listed in the Bibliography if these more complex states of stress are of interest. It is conventional to be consistent and use the terminology of the more general analysis even when one is considering the simpler plane stress situation. Hence, the compliance matrix $[S]$ has the terms

$$[S] = \left[\begin{array}{ccc} S_{11} & S_{12} & S_{16} \\ S_{21} & S_{22} & S_{26} \\ S_{61} & S_{62} & S_{66} \end{array} \right]$$

where

$$S_{11} = \frac{1}{E_1}, \quad S_{22} = \frac{1}{E_2}, \quad S_{66} = \frac{1}{G_{12}}$$

$$S_{12} = S_{21} = -\frac{\nu_{12}}{E_1} = -\frac{\nu_{21}}{E_2}$$

$$S_{16} = S_{61} = S_{26} = S_{62} = 0$$

Using matrix notation, equation (3.20) may be transposed to give the stresses as a function of the strains

$$\{\sigma\} = [S]^{-1}\{\varepsilon\}$$

This may also be written as

$$\{\sigma\} = [Q]\{\varepsilon\} \tag{3.21}$$

where $[Q]$ is the *stiffness* matrix and its terms will be

$$Q_{11} = \frac{E_1}{1 - \nu_{12}\nu_{21}}, \quad Q_{22} = \frac{E_2}{1 - \nu_{21}\nu_{12}}, \quad Q_{66} = G_{12}$$

$$Q_{12} = Q_{21} = \frac{\nu_{12}E_2}{(1 - \nu_{12}\nu_{21})} = \frac{\nu_{21}E_1}{(1 - \nu_{21}\nu_{12})}$$

$$Q_{16} = Q_{61} = Q_{26} = Q_{62} = 0$$

(ii) Loading Off the Fibre Axis

The previous analysis is a preparation for the more interesting and practical situation where the applied loading axis does not coincide with the fibre axis. This is illustrated in Fig. 3.10.

The first step in the analysis of this situation is the transformation of the applied stresses on to the fibre axis. Referring to Fig. 3.10 it may be seen that σ_x and σ_y may be resolved into the x, y axes as follows (the reader may wish to refer to any standard Strength of Materials text such as Benham, Crawford and Armstrong for more details of this stress transformation):

$$\sigma_1 = \sigma_x \cos^2 \theta + \sigma_y \sin^2 \theta + 2\tau_{xy} \sin \theta \cos \theta$$

$$\sigma_2 = \sigma_x \sin^2 \theta + \sigma_y \cos^2 \theta - 2\tau_{xy} \sin \theta \cos \theta$$

$$\tau_{12} = -\sigma_x \sin \theta \cos \theta + \sigma_y \sin \theta \cos \theta + \tau_{xy}(\cos^2 \theta - \sin^2 \theta)$$

Using matrix notation

$$\begin{Bmatrix} \sigma_1 \\ \sigma_2 \\ \tau_{12} \end{Bmatrix} = \begin{bmatrix} c^2 & s^2 & 2sc \\ s^2 & c^2 & -2sc \\ -sc & sc & (c^2 - s^2) \end{bmatrix} \begin{Bmatrix} \sigma_x \\ \sigma_y \\ \tau_{xy} \end{Bmatrix} \tag{3.22}$$

where $c = \cos \theta$ and $s = \sin \theta$.

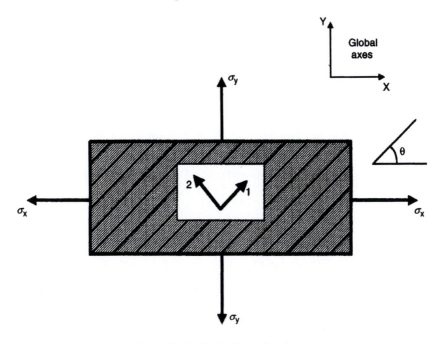

Fig. 3.10   Single thin lamina under plane stress

In shorthand form, this matrix equation may be written as

$$\{\sigma\}_{12} = [T_\sigma]\{\sigma\}_{xy}$$

where $[T_\sigma]$ is called the **Stress Transformation** matrix.

Similar transformations may be made for the strains so that

$$\left\{\begin{array}{c} \varepsilon_1 \\ \varepsilon_2 \\ \frac{1}{2}\gamma_{12} \end{array}\right\} = [T_\sigma]\left\{\begin{array}{c} \varepsilon_x \\ \varepsilon_y \\ \frac{1}{2}\gamma_{xy} \end{array}\right\} \tag{3.23}$$

In each case the stresses in the global directions x, y may also be expressed in terms of the local stresses or strains. For example,

$$\left\{\begin{array}{c} \sigma_x \\ \sigma_y \\ \tau_{xy} \end{array}\right\} = [T_\sigma]^{-1}\left\{\begin{array}{c} \sigma_1 \\ \sigma_2 \\ \tau_{12} \end{array}\right\}$$

where

$$[T_\sigma]^{-1} = \begin{bmatrix} c^2 & s^2 & -2sc \\ s^2 & c^2 & 2sc \\ sc & -sc & (c^2 - s^2) \end{bmatrix}$$

Now, in order to determine the stiffness in the global (x, y) directions for a lamina in which the fibres are aligned at an angle θ to the global x direction, it is necessary to go through the following three steps:

1. Determine the strains in the local (1,2) directions by transforming the applied strains through $\theta°$ from the global (x,y) directions.
2. Calculate the stresses in the local directions using the on-axis stiffness matrix $[Q]$ derived earlier.
3. Transform the stresses back to the global directions through an angle of $-\theta°$.

Using matrix notation, these three steps are carried out as follows:

Step 1

$$
\begin{Bmatrix} \varepsilon_1 \\ \varepsilon_2 \\ \gamma_{12} \end{Bmatrix} = \begin{bmatrix} c^2 & s^2 & sc \\ s^2 & c^2 & -sc \\ -2sc & 2sc & (c^2 - s^2) \end{bmatrix} \begin{Bmatrix} \varepsilon_x \\ \varepsilon_y \\ \gamma_{xy} \end{Bmatrix}
$$

Note the modification to $[T_\sigma]$ to give the **Strain Transformation** matrix T_ε so that we can use γ_{12} instead of $\frac{1}{2}\gamma_{12}$.

Step 2

$$
\begin{Bmatrix} \sigma_1 \\ \sigma_2 \\ \tau_{12} \end{Bmatrix} = \begin{bmatrix} Q_{11} & Q_{12} & 0 \\ Q_{21} & Q_{22} & 0 \\ 0 & 0 & Q_{66} \end{bmatrix} \begin{Bmatrix} \varepsilon_1 \\ \varepsilon_2 \\ \gamma_{12} \end{Bmatrix}
$$

Step 3

$$
\begin{Bmatrix} \sigma_x \\ \sigma_y \\ \tau_{xy} \end{Bmatrix} = \begin{bmatrix} c^2 & s^2 & -2sc \\ s^2 & c^2 & 2sc \\ sc & -sc & (c^2 - s^2) \end{bmatrix} \begin{Bmatrix} \sigma_1 \\ \sigma_2 \\ \tau_{12} \end{Bmatrix}
$$

Linking these steps together to obtain the global stresses in terms of the global strains

$$
\begin{Bmatrix} \sigma_x \\ \sigma_y \\ \tau_{xy} \end{Bmatrix} = \begin{bmatrix} c^2 & s^2 & -2sc \\ s^2 & c^2 & 2sc \\ sc & -sc & (c^2 - s^2) \end{bmatrix} \begin{bmatrix} Q_{11} & Q_{12} & 0 \\ Q_{21} & Q_{22} & 0 \\ 0 & 0 & Q_{66} \end{bmatrix}
$$
$$
\begin{bmatrix} c^2 & s^2 & sc \\ s^2 & c^2 & -sc \\ -2sc & 2sc & (c^2 - s^2) \end{bmatrix} \begin{Bmatrix} \varepsilon_x \\ \varepsilon_y \\ \gamma_{xy} \end{Bmatrix}
$$

which provides an overall stiffness matrix $[\overline{Q}]$

$$
\begin{Bmatrix} \sigma_x \\ \sigma_y \\ \tau_{xy} \end{Bmatrix} = [\overline{Q}] \begin{Bmatrix} \varepsilon_x \\ \varepsilon_y \\ \tau_{xy} \end{Bmatrix} \tag{3.24}
$$

where

$$\overline{Q}_{11} = \frac{1}{\lambda}[E_1\cos^4\theta + E_2\sin^4\theta + (2\nu_{12}E_2 + 4\lambda G_{12})\cos^2\theta\sin^2\theta]$$

$$\overline{Q}_{21} = \overline{Q}_{12} = \frac{1}{\lambda}[\nu_{12}E_2(\cos^4\theta + \sin^4\theta)$$
$$+ (E_1 + E_2 - 4\lambda G_{12})\cos^2\theta\sin^2\theta]$$

$$\overline{Q}_{61} = \overline{Q}_{16} = \frac{1}{\lambda}[\cos^3\theta\sin\theta(E_1 - \nu_{12}E_2 - 2\lambda G_{12})$$
$$- \cos\theta\sin^3\theta(E_2 - \nu_{12}E_2 - 2\lambda G_{12})]$$

$$\overline{Q}_{22} = \frac{1}{\lambda}[E_2\cos^4\theta + E_1\sin^4\theta + \sin^2\theta\cos^2\theta(2\nu_{12}E_2 + 4\lambda G_{12})]$$

$$\overline{Q}_{62} = \overline{Q}_{26} = \frac{1}{\lambda}[\cos\theta\sin^3\theta(E_1 - \nu_{12}E_2 - 2\lambda G_{12})]$$
$$- \cos^3\theta\sin\theta(E_2 - \nu_{12}E_2 - 2\lambda G_{12})]$$

$$\overline{Q}_{66} = \frac{1}{\lambda}[\sin^2\theta\cos^2\theta(E_1 + E_2 - 2\nu_{12}E_2 - 2\lambda G_{12})]$$
$$+ \lambda G_{12}(\cos^4\theta + \sin^4\theta)]$$

in which $\lambda = (1 - \nu_{12}\nu_{21})$ and $\nu_{12}E_2 = \nu_{21}E_1$.

By a similar analysis it may be shown that, for applied stresses (rather than applied strains) in the global directions, the overall compliance matrix $[\overline{S}]$ has the form

$$\begin{Bmatrix} \varepsilon_x \\ \varepsilon_y \\ \tau_{xy} \end{Bmatrix} = \begin{bmatrix} \overline{S}_{11} & \overline{S}_{12} & \overline{S}_{16} \\ \overline{S}_{21} & \overline{S}_{22} & \overline{S}_{26} \\ \overline{S}_{61} & \overline{S}_{62} & \overline{S}_{66} \end{bmatrix} \begin{Bmatrix} \sigma_x \\ \sigma_y \\ \tau_{xy} \end{Bmatrix} \tag{3.25}$$

where $[\overline{S}] = [\overline{Q}]^{-1}$ and the individual terms are

$$\overline{S}_{11} = S_{11}\cos^4\theta + S_{22}\sin^4\theta + (2S_{12} + S_{66})\cos^2\theta\sin^2\theta$$
$$\overline{S}_{21} = \overline{S}_{12} = (S_{11} + S_{22} - S_{66})\cos^2\theta\sin^2\theta + S_{12}(\cos^4\theta + \sin^4\theta)$$
$$\overline{S}_{61} = \overline{S}_{16} = (2S_{11} - 2S_{12} - S_{66})\cos^3\theta\sin\theta - (2S_{22} - 2S_{12} - S_{66})\sin^3\theta\cos\theta$$
$$\overline{S}_{22} = S_{11}\sin^4\theta + S_{22}\cos^4\theta + (2S_{12} + S_{66})\cos^2\theta\sin^2\theta$$
$$\overline{S}_{62} = \overline{S}_{26} = (2S_{11} - 2S_{12} - S_{66})\cos\theta\sin^3\theta - (2S_{22} - 2S_{12} - S_{66})\sin\theta\cos^3\theta$$
$$\overline{S}_{66} = 4(S_{11} - 2S_{12} + S_{66})\cos^2\theta\sin^2\theta + S_{66}(\cos^2\theta - \sin^2\theta)^2$$

from which, using the earlier terms from the on-axis lamina stiffness matrix $[S]$,

$$\frac{1}{E_x} = \overline{S}_{11} = \frac{c^2}{E_1}(c^2 - s^2 v_{12}) + \frac{s^2}{E_2}(s^2 - c^2 v_{21}) + \frac{c^2 s^2}{G_{12}}$$

$$\frac{1}{E_y} = \overline{S}_{22} = \frac{s^2}{E_1}(s^2 - c^2 v_{12}) + \frac{c^2}{E_2}(c^2 - s^2 v_{21}) + \frac{c^2 s^2}{G_{12}}$$

$$\frac{1}{G_{xy}} = \overline{S}_{66} = \frac{4c^2 s^2}{E_1}(1 + v_{12}) + \frac{4c^2 s^2}{E_2}(1 + v_{21}) + \frac{(c^2 - s^2)^2}{G_{12}}$$

$$\frac{v_{xy}}{E_x} = \frac{v_{yx}}{E_y} = -\overline{S}_{21} = \frac{c^2}{E_1}(c^2 v_{12} - s^2) + \frac{s^2}{E_2}(s^2 v_{21} - c^2) + \frac{c^2 s^2}{G_{12}}$$

Thus one may use the above expressions to calculate the stiffness of a uni-directional lamina when it is loaded at any angle θ to the fibre direction. If computer facilities are available for the matrix manipulation then it is not necessary to work out the individual terms as above – the required information can be determined directly from the matrices. For example, as indicated above

$$E_x = \frac{1}{\overline{S}_{11}}, \quad E_y = \frac{1}{\overline{S}_{22}}, \quad G_{xy} = \frac{1}{\overline{S}_{66}} \tag{3.26}$$

3.7 Summary of Approach to Analysis of Unidirectional Composites

1. The strains and stresses in the local (1–2) axis are related by

$$[\varepsilon]_{12} = [S][\sigma]_{12}$$

or $\qquad [\sigma]_{12} = [Q][\varepsilon]_{12} \qquad$ where $[Q] = [S]^{-1}$

2. The global $(x-y)$ stresses and strains may be related to the local (1–2) stresses and strains by

$$[\sigma]_{12} = [T_\sigma][\sigma]_{xy}$$

$$[\varepsilon]_{12} = [T_\varepsilon][\varepsilon]_{xy}$$

3. The global stresses and strains are related by

$$[\varepsilon]_{xy} = [\overline{S}][\sigma]_{xy}$$

$$[\sigma]_{xy} = [\overline{Q}][\varepsilon]_{xy}$$

where

$$[\overline{Q}] = [T_\sigma]^{-1}[Q][T_\varepsilon] \text{ and } [\overline{S}] = [\overline{Q}]^{-1}$$

In each case $[\sigma]_{12}$ and $[\varepsilon]_{12}$ are written as shorthand for

$$\begin{bmatrix} \sigma_1 \\ \sigma_2 \\ \tau_{12} \end{bmatrix} \text{ and } \begin{bmatrix} \varepsilon_1 \\ \varepsilon_2 \\ \gamma_{12} \end{bmatrix}$$

The following Examples illustrate the use of these equations

Example 3.6 A unidirectional composite consisting of carbon fibres in a PEEK matrix has the fibres aligned at 25° to the loading axis. If the fibres and matrix have the properties indicated below, calculate E_x, E_y, G_{xy}, v_{xy}, and v_{yx}.

$$E_f = 240 \text{ GN/m}^2 \qquad E_m = 3 \text{ GN/m}^2$$
$$v_f = 0.23 \qquad v_m = 0.35$$
$$V_f = 0.6$$
$$\rho_f = 1700 \text{ kg/m}^3 \qquad \rho_m = 1300 \text{ kg/m}^3$$
$$G_f = 80 \text{ GN/m}^2 \qquad G_m = 1.1 \text{ GN/m}^2$$

Solution The lamina properties in the local (1–2) directions are calculated as follows:

$$E_1 = E_f V_f + E_m.(1 - V_f), \qquad E_1 = 145.2 \text{ GN/m}^2$$

$$E_2 = \frac{E_m \cdot (1 + 2 \cdot \beta \cdot V_f)}{(1 - \beta \cdot V_f)}, \qquad E_2 = 15.33 \text{ GN/m}^2$$

Halpin–Tsai Equation

$$G_{12} = G_m \cdot \frac{(G_f + G_m) + V_f(G_f - G_m)}{(G_f + G_m) - V_f(G_f - G_m)},$$

$$G_{12} = 4.185 \text{ GN/m}^2 \qquad \text{Halpin–Tsai Equation}$$

$$v_{12} = v_f \cdot V_f + v_m \cdot (1 - V_f) \qquad v_{12} = 0.278$$

$$v_{21} = v_{12} \cdot \left(\frac{E_2}{E_1}\right) \qquad v_{21} = 0.029$$

The behaviour of the lamina when subjected to loading at θ degrees off the fibre axis is determined using matrix manipulation as follows:

Loading On-Axis Compliance Matrix   *Stiffness Matrix*

$$S = \begin{bmatrix} \dfrac{1}{E_1} & -\dfrac{v_{21}}{E_2} & 0 \\ -\dfrac{v_{12}}{E_1} & \dfrac{1}{E_2} & 0 \\ 0 & 0 & \dfrac{1}{G_{12}} \end{bmatrix} \qquad Q = S^{-1}$$

Stress Transformation Matrix   *Overall Compliance Matrix*

$$T_\sigma = \begin{bmatrix} c^2 & s^2 & 2.s.c \\ s^2 & c^2 & -2.s.c \\ -s.c & s.c & (c^2 - s^2) \end{bmatrix} \qquad \bar{S} = \bar{Q}^{-1}$$

The lamina properties are then obtained as:

$$\overline{S}_{11} = 0.041, \qquad E_x = \frac{1}{\overline{S}_{11}} \qquad E_x = 24.26 \text{ GN/m}^2$$

$$\overline{S}_{22} = 0.079, \qquad E_y = \frac{1}{\overline{S}_{22}} \qquad E_y = 12.7 \text{ GN/m}^2$$

$$\overline{S}_{66} = 0.143, \qquad G_{xy} = \frac{1}{\overline{S}_{66}} \qquad G_{xy} = 6.9 \text{ GN/m}^2$$

$$\overline{S}_{21} = 0.026, \qquad v_{xy} = -E_x \cdot \overline{S}_{21}, \qquad v_{yx} = -E_y \cdot \overline{S}_{12},$$

$$v_{xy} = 0.627, \qquad v_{yx} = -0.328$$

Alternatively the lamina properties can be calculated from the equations given above.

$$\overline{S}_{11} = \frac{c^2}{E_1} \cdot (c^2 - s^2 \cdot v_{12}) + \frac{s^2}{E_2} \cdot (s^2 - c^2 \cdot v_{21}) + \frac{c^2 \cdot s^2}{G_{12}} = 0.041,$$

$$E_x = \frac{1}{\overline{S}_{11}} = 24.26$$

$$\overline{S}_{22} = \frac{s^2}{E_1} \cdot (s^2 - c^2 \cdot v_{12}) + \frac{c^2}{E_2} \cdot (c^2 - s^2 \cdot v_{21}) + \frac{c^2 \cdot s^2}{G_{21}} = 0.079,$$

$$E_y = \frac{1}{\overline{S}_{22}} = 12.702$$

$$\overline{S}_{66} = \frac{4 \cdot c^2 \cdot s^2}{E_1} \cdot (1 + v_{12}) + \frac{4 \cdot c^2 \cdot s^2}{E_2} \cdot (1 + v_{21}) + \frac{(c^2 - s^2)^2}{G_{12}} = 0.143,$$

$$G_{xy} = \frac{1}{\overline{S}_{66}} = 6.9$$

$$\overline{S}_{12} = \frac{c^2}{E_1} \cdot (c^2 \cdot v_1 - s^2) + \frac{s^2}{E_2} \cdot (s^2 \cdot v_2 - c^2) + \frac{c^2 \cdot s^2}{G_{12}}$$

$$v_{xy} = -\overline{S}_{21} \cdot E_x, \quad v_{xy} = 0.627$$

$$v_{yx} = -\overline{S}_{12} \cdot E_y, \quad v_{yx} = 0.328$$

In this Example the angle, θ, was set at 25°. Fig. 3.11 shows the variation of the moduli and Poisson's Ratio values for all angles between 0° and 90°. This gives an interesting overview of the performance of unidirectional composites when subjected to off-axis loading.

Example 3.7 A thin unidirectional Kevlar fibre/epoxy composite has the properties listed below. If the fibres are aligned at θ to the x-axis, show how

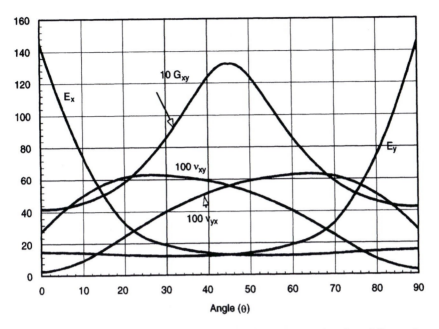

Fig. 3.11 Variation of properties in unidirectional composite as a function of fibre angle

the strains in the global $X - Y$ directions vary with θ when a stress σ is applied as shown in Fig. 3.12.

$$E_1 = 80000 \text{ MN/m}^2, \quad E_2 = 4000 \text{ MN/m}^2, \quad G_{12} = 1500 \text{ MN/m}^2$$

$$v_{12} = 0.43$$

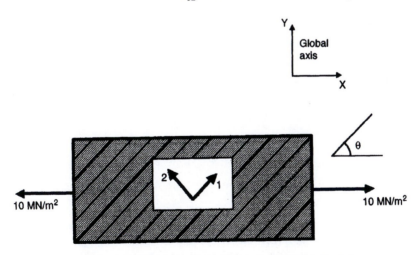

Fig. 3.12 Unidirectional Kevlar/epoxy composite (single ply)

Solution Let $\theta = 25°$ and $\sigma_x = 10$ MN/m^2 to illustrate the method of solution.

The Stress Transformation Matrix is

$$T_\sigma = \begin{bmatrix} c^2 & s^2 & 2sc \\ s^2 & c^2 & -2sc \\ -sc & sc & (c^2 - s^2) \end{bmatrix} \quad \text{or} \quad T_\sigma = \begin{pmatrix} 0.821 & 0.179 & 0.766 \\ 0.179 & 0.821 & -0.766 \\ -0.383 & 0.383 & 0.643 \end{pmatrix}$$

In order to get the strains in the global directions it is necessary to determine the overall compliance matrix $[\bar{S}]$. This is obtained as indicated above, ie

$$[\bar{S}] = [\bar{Q}]^{-1} \quad \text{where} \quad [\bar{Q}] = [T_\sigma]^{-1}[Q] \cdot [T_\epsilon]$$

The local compliance matrix is

$$S = \begin{bmatrix} \dfrac{1}{E_1} & \dfrac{-v_{12}}{E_1} & 0 \\ \dfrac{-v_{12}}{E_1} & \dfrac{1}{E_2} & 0 \\ 0 & 0 & \dfrac{1}{G_{12}} \end{bmatrix} \quad \text{and} \quad Q = S^{-1}$$

The Strain Transformation matrix is

$$T_\epsilon = \begin{bmatrix} c^2 & s^2 & sc \\ s^2 & c^2 & -sc \\ -2sc & 2sc & (c^2 - s^2) \end{bmatrix}$$

So,

$$\bar{Q} = T_\sigma^{-1} \cdot Q \cdot T_\epsilon \quad \text{and} \quad \bar{S} = \bar{Q}^{-1}$$

Then

$$\begin{bmatrix} \varepsilon_x \\ \varepsilon_y \\ \gamma_{xy} \end{bmatrix} = \bar{S} \begin{bmatrix} \sigma_x \\ \sigma_y \\ \tau_{xy} \end{bmatrix} \quad \bar{S} = \begin{bmatrix} 1.12 \cdot 10^{-4} & -6.3 \cdot 10^{-5} & -1.87 \cdot 10^{-4} \\ -6.3 \cdot 10^{-5} & 2.65 \cdot 10^{-4} & 5.89 \cdot 10^{-6} \\ -1.87 \cdot 10^{-4} & 5.89 \cdot 10^{-6} & 4.35 \cdot 10^{-4} \end{bmatrix}$$

Directly by matrix manipulation

$$\varepsilon_x = 1.126 \cdot 10^{-3} \qquad \varepsilon_y = -6.309 \cdot 10^{-4} \qquad \gamma_{xy} = -1.878 \cdot 10^{-3}$$

or by multiplying out the terms

$$\varepsilon_x = [(\bar{S}_{11}) \cdot (\sigma_x) + (\bar{S}_{12}) \cdot (\sigma_y)] + (\bar{S}_{16}) \cdot (\tau_{xy}) \qquad \varepsilon_x = 1.126 \cdot 10^{-3}$$

and similarly for the other two strains. Figure 3.13 illustrates how the strains vary for different values of θ.

Example 3.8 A thin unidirectional carbon fibre composite is loaded as shown in Fig. 3.14 and has the properties listed below. If the fibres are aligned at 35° to the x-axis, calculate the stresses parallel and perpendicular to the fibres,

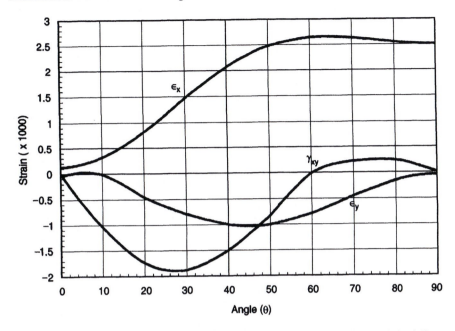

Fig. 3.13   Variation in direct and shear strains for unidirectional composite loaded axially

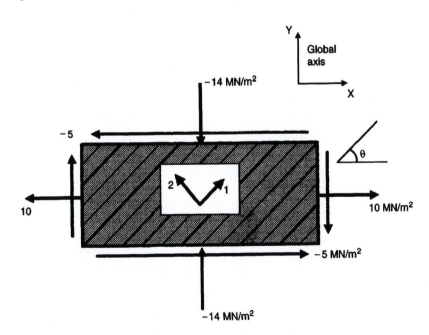

Fig. 3.14   Unidirectional carbon fibre single ply composite

when the stresses σ_x, σ_y and τ_{xy} are applied. Calculate also the strains in the global $X - Y$ directions.

$$E_1 = 125000^1 \text{ MN/m}^2, \quad E_2 = 7800 \text{ MN/m}^2, \quad G_{12} = 4400 \text{ MN/m}^2,$$

$$v_{12} = 0.34 \quad \sigma_x = 10 \text{ MN/m}^2, \quad \sigma_y = -14 \text{ MN/m}^2, \quad \tau_{xy} = -5 \text{ MN/m}^2$$

Solution The Stress Transformation Matrix is

$$T_\sigma = \begin{bmatrix} c^2 & s^2 & 2sc \\ s^2 & c^2 & -2sc \\ -sc & sc & (c^2 - s^2) \end{bmatrix} \quad \text{or} \quad T_\sigma = \begin{pmatrix} 0.671 & 0.329 & 0.94 \\ 0.329 & 0.671 & -0.94 \\ -0.47 & 0.47 & 0.342 \end{pmatrix}$$

The stresses parallel and perpendicular to the fibres are then given by

$$\begin{bmatrix} \sigma_1 \\ \sigma_2 \\ \tau_{12} \end{bmatrix} = T_\sigma \cdot \begin{bmatrix} \sigma_x \\ \sigma_y \\ \tau_{xy} \end{bmatrix}$$

So,

$$\sigma_1 = -2.59 \text{ MN/m}^2 \qquad \sigma_2 = -1.4 \text{ MN/m}^2 \qquad \tau_{12} = -12.98 \text{ MN/m}^2$$

In order to get the strains in the global directions it is necessary to determine the overall compliance matrix $[\bar{S}]$. This is obtained as indicated above, ie

$$[\bar{S}] = [\bar{Q}]^{-1} \quad \text{where} \quad [\bar{Q}] = [T_\sigma]^{-1} \cdot [Q] \cdot [T_\epsilon]$$

The local compliance matrix is

$$S = \begin{bmatrix} \dfrac{1}{E_1} & \dfrac{-v_{12}}{E_1} & 0 \\ \dfrac{-v_{12}}{E_1} & \dfrac{1}{E_2} & 0 \\ 0 & 0 & \dfrac{1}{G_{12}} \end{bmatrix} \quad \text{and} \quad Q = S^{-1}$$

The Strain Transformation matrix is

$$T_\epsilon = \begin{bmatrix} c^2 & s^2 & sc \\ s^2 & c^2 & -sc \\ -2sc & 2sc & (c^2 - s^2) \end{bmatrix}$$

So,

$$\bar{Q} = T_\sigma^{-1} \cdot Q \cdot T_\epsilon \quad \text{and} \quad \bar{S} = \bar{Q}^{-1}$$

Then

$$\begin{bmatrix} \varepsilon_x \\ \varepsilon_y \\ \gamma_{xy} \end{bmatrix} = \bar{S} \cdot \begin{bmatrix} \sigma_x \\ \sigma_y \\ \tau_{xy} \end{bmatrix} \quad \bar{S} = \begin{bmatrix} 6.64 \times 10^{-5} & -2.16 \times 10^{-5} & -7.02 \times 10^{-5} \\ -2.16 \times 10^{-5} & 1.07 \times 10^{-4} & -4.27 \times 10^{-5} \\ -7.02 \times 10^{-5} & -4.27 \times 10^{-5} & 1.51 \times 10^{-4} \end{bmatrix}$$

Directly by matrix manipulation

$$\varepsilon_x = 1.318 \times 10^{-3} \qquad \varepsilon_y = -1.509 \times 10^{-3} \qquad \gamma_{xy} = 8.626 \times 10^{-4}$$

or by multiplying out the terms

$$\varepsilon_x = [(\overline{S}_{11}) \cdot (\sigma_x) + (\overline{S}_{12}) \cdot (\sigma_y)] + (\overline{S}_{16}) \cdot (\tau_{xy}) \qquad \varepsilon_x = 1.318 \times 10^{-3}$$

and similarly for the other two strains.

3.8 General Deformation Behaviour of a Single Ply

The previous section has considered the in-plane deformations of a single ply. In practice, real engineering components are likely to be subjected to this type of loading plus (or as an alternative) bending deformations. It is convenient at this stage to consider the flexural loading of a single ply because this will develop the method of solution for multi-ply laminates.

(i) Loading on Fibre Axis

Consider a unidirectional sheet of material with the fibres aligned in the x-direction and subjected to a stress, σ_x. If the sheet has thickness, h, as shown in Fig. 3.15 and we consider unit width, then the normal axial force N_x is given by

$$N_x = \int_{-\frac{h}{2}}^{\frac{h}{2}} \sigma_x \, dz$$

Fig. 3.15   General forces and moments acting on a single ply

Or more generally for all the force components N_x, N_y and N_{xy}

$$[N] = \int_{-\frac{h}{2}}^{\frac{h}{2}} [\sigma] \, dz \tag{3.27}$$

Similarly the bending moment, M_1, per unit width is given by

$$M_x = \int_{-\frac{h}{2}}^{\frac{h}{2}} \sigma_x \cdot z \cdot dz$$

Once again, all the moments M_x, M_y and M_{xy}, can be expressed as

$$[M] = \int_{-\frac{h}{2}}^{\frac{h}{2}} [\sigma] z \, dz \tag{3.28}$$

Now in order to determine $[\sigma]$ as a function of z, consider the strain $\varepsilon(z)$ at any depth across the section. It will be made up of an in-plane component (ε) plus a bending component (z/R) which is normally written as $z.\kappa$, where κ is the curvature of bending.

Hence,

$$\varepsilon(z) = \varepsilon + z.\kappa$$

The stresses will then be given by

$$\sigma(z) = [Q] \cdot [\varepsilon] + [Q] \cdot z.[\kappa]$$

where $[Q]$ is the stiffness matrix as defined earlier.

Now, from equation (3.27), the forces $[N]$ are given by

$$[N] = \int_{-\frac{h}{2}}^{\frac{h}{2}} [Q] \cdot [\varepsilon] \cdot dz + \int_{-\frac{h}{2}}^{\frac{h}{2}} [Q] \cdot [\kappa] \cdot z \, dz$$

$$= [Q][\varepsilon]h + [Q]\frac{1}{2}\left(\frac{h^2}{4} - \frac{h^2}{4}\right)[\kappa]$$

$$[N] = [A][\varepsilon] + [B][\kappa] \tag{3.29}$$

where $[A]$ is the *Extensional Stiffness* matrix $(= [Q]h)$ and $[B]$ is a *Coupling Matrix*. It may be observed that in the above analysis $[B]$ is in fact zero for this simple single ply situation. However, its identity has been retained as it has relevance in laminate analysis, to be studied later.

Returning to equation (3.28), the moments may be written as

$$[M] = [Q][\varepsilon] \int\limits_{-\frac{h}{2}}^{\frac{h}{2}} z \, dz + [Q][\kappa] \int\limits_{-\frac{h}{2}}^{\frac{h}{2}} z^2 dz$$

$$= [B][\varepsilon] + [Q][\kappa] \cdot \frac{h^3}{12}$$

$$[M] = [B][\varepsilon] + [D][\kappa] \qquad (3.30)$$

where $[D]$ is the Bending Stiffness Matrix $(= [Q]h^3/12)$.

Equations (3.29) and (3.30) may be grouped into the following form

$$\begin{bmatrix} N \\ M \end{bmatrix} = \begin{bmatrix} A & B \\ B & D \end{bmatrix} \begin{bmatrix} \varepsilon \\ \kappa \end{bmatrix} \qquad (3.31)$$

and these are known as the *Plate Constitutive Equations*.

For this special case of a single ply, $[B] = 0$ and so the forces are related to the strains by

$$[N] = [A][\varepsilon] \qquad (3.32)$$

(which is effectively $[\sigma] = [Q][\varepsilon]$ as shown previously) or the strains are related to the forces by

$$[\varepsilon] = [a][N] \qquad \text{where } [a] = [A]^{-1}$$

(which is effectively $[\varepsilon] = [S][\sigma]$ as shown previously).

Also, for the single ply situation, the moments are related to the curvature by

$$[M] = [D][\kappa] \qquad (3.33)$$

or the curvature is related to the moments by

$$[\kappa] = [d][M] \qquad \text{where } [d] = [D]^{-1}$$

It should be noted that it is only possible to utilise $[a] = [A]^{-1}$ and $[d] = [D]^{-1}$ for the special case where $[B] = 0$. In other cases, the terms in the $[a]$ and $[d]$ matrices have to be determined from

$$\begin{bmatrix} a & b \\ \beta & d \end{bmatrix} = \begin{bmatrix} A & B \\ B & D \end{bmatrix}^{-1} \qquad (3.34)$$

(ii) Loading off Fibre Axis

For the situation where the loading is applied off the fibre axis, then the above approach involving the Plate Constitutive Equations can be used but it is necessary to use the *transformed* stiffness matrix terms $\overline{Q}$.

Hence, in the above analysis

$$[A] = [\overline{Q}] \cdot h$$
$$[D] = [\overline{Q}] \cdot h^3/12$$

The use of the Plate Constitutive Equations is illustrated in the following Examples.

Example 3.9 For the 2 mm thick unidirectional carbon fibre/PEEK composite described in Example 3.6, calculate the values of the moduli, Poisson's Ratio and strains in the global direction when a stress of $\sigma_x = 50$ MN/m^2 is applied. You should use

  (i) the lamina stiffness and compliance matrix approach and
  (ii) the Plate Constitutive Equation approach.

Solution

(i) As shown in Example 3.6, for the loading in Fig. 3.16

$$\overline{S} = \begin{bmatrix} 4.12 \times 10^{-5} & -2.58 \times 10^{-5} & -6.24 \times 10^{-5} \\ -2.58 \times 10^{-5} & 7.87 \times 10^{-5} & 1.77 \times 10^{-5} \\ -6.24 \times 10^{-5} & 1.77 \times 10^{-5} & 1.43 \times 10^{-4} \end{bmatrix} \text{m}^2/\text{MN}$$

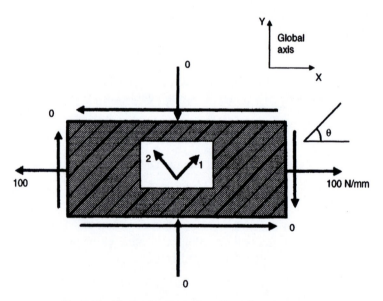

Fig. 3.16   Single ply composite subjected to plane stress

and

$$E_x = \frac{1}{\bar{S}_{11}} = 24.26 \text{ GN/m}^2, \quad E_y = \frac{1}{\bar{S}_{22}} = 12.7 \text{ GN/m}^2$$

$$G_{xy} = \frac{1}{\bar{S}_{66}} = 6.9 \text{ GN/m}^2, \quad \nu_{xy} = -E_x \bar{S}_{21} = 0.627$$

$$\nu_{yx} = -E_y \bar{S}_{12} = 0.328$$

Also,

$$\begin{bmatrix} \varepsilon_x \\ \varepsilon_y \\ \tau_{xy} \end{bmatrix} = \bar{S} \begin{bmatrix} \sigma_x \\ \sigma_y \\ \tau_{xy} \end{bmatrix}$$

So,

$$\varepsilon_x = 2.06 \times 10^{-3}, \qquad \varepsilon_y = -1.29 \times 10^{-3}, \qquad \alpha_{xy} = -3.12 \times 10^{-3}$$

Note that

$$\varepsilon_x = \frac{\sigma_x}{E_x} - \nu_{yx} \frac{\sigma_y}{E_y}$$

$$\varepsilon_y = \frac{\sigma_y}{E_y} - \nu_{xy} \frac{\sigma_x}{E_x}$$

As $\sigma_y = 0$ in this Example

$$\nu_{xy} = -\frac{\varepsilon_y}{\varepsilon_x} = 0.627 \text{ as above}$$

(ii) Alternatively using the Plate Constitutive Equation Approach.

$$N_x = \sigma_x h = 100 \text{ N/mm}, \quad N_y = 0, \quad N_{xy} = 0.$$

$$A = \bar{Q} \cdot h \quad \text{and} \quad D = \bar{Q} \cdot \frac{h^3}{12}$$

$$A = \begin{bmatrix} 2.06 \times 10^5 & 4.86 \times 10^4 & 8.37 \times 10^4 \\ 4.86 \times 10^4 & 3.76 \times 10^4 & 1.65 \times 10^4 \\ 8.37 \times 10^4 & 1.65 \times 10^4 & 4.84 \times 10^4 \end{bmatrix} \text{ N/mm}$$

$$D = \begin{bmatrix} 6.86 \times 10^4 & 1.62 \times 10^4 & 2.79 \times 10^4 \\ 1.62 \times 10^4 & 1.25 \times 10^4 & 5.51 \times 10^3 \\ 2.79 \times 10^4 & 5.51 \times 10^3 & 1.61 \times 10^4 \end{bmatrix} \text{ N mm}$$

$a = A^{-1}$ and $d = D^{-1}$ (since $B = 0$)

$$a = \begin{bmatrix} 2.06 \times 10^{-5} & -1.29 \times 10^{-5} & -3.12 \times 10^{-5} \\ -1.29 \times 10^{-5} & 3.93 \times 10^{-5} & 8.89 \times 10^{-6} \\ -3.12 \times 10^{-5} & 8.89 \times 10^{-6} & 7.16 \times 10^{-5} \end{bmatrix} \text{ mm/N}$$

$$d = \begin{bmatrix} 6.18 \times 10^{-5} & -3.87 \times 10^{-5} & 9.37 \times 10^{-5} \\ -3.87 \times 10^{-5} & 1.18 \times 10^{-4} & 2.66 \times 10^{-5} \\ -9.37 \times 10^{-5} & 2.66 \times 10^{-5} & 2.14 \times 10^{-4} \end{bmatrix} \text{ (N mm)}^{-1}$$

$$E_x = \frac{1}{a_{11} \times h}, \qquad E_y = \frac{1}{a_{22} \times h}, \qquad G_{xy} = \frac{1}{a_{66} \times h}, \qquad v_{xy} = \frac{-a_{12}}{a_{11}},$$

$$v_{yx} = \frac{-a_{12}}{a_{22}}$$

$$E_x = 24.26 \text{ GN/m}^2, \qquad E_y = 12.7 \text{ GN/m}^2$$

$$G_{xy} = 6.98 \text{ GN/m}^2, \qquad v_{xy} = 0.627, \qquad v_{yx} = 0.328$$

These values agree with those calculated above. Also, for the applied force $N_x = 50 \times 2 = 100$ N/mm.

$$\begin{bmatrix} \varepsilon_x \\ \varepsilon_y \\ \gamma_{xy} \end{bmatrix} = a \cdot \begin{bmatrix} N_x \\ N_y \\ N_{xy} \end{bmatrix} \qquad \left(= a \begin{bmatrix} \sigma_x \\ \sigma_y \\ \tau_{xy} \end{bmatrix} \cdot h \right)$$

$$\varepsilon_x = 2.061 \times 10^{-3} \qquad \varepsilon_y = -1.291 \times 10^{-3} \qquad \gamma_{xy} = -3.124 \times 10^{-3}$$

It is interesting to observe that as well as the expected axial and transverse strains arising from the applied axial stress, σ_x, we have also a shear strain. This is because in composites we can often get *coupling* between the different modes of deformation. This will also be seen later where coupling between axial and flexural deformations can occur in unsymmetric laminates. Fig. 3.17 illustrates why the shear strains arise in uniaxially stressed single ply in this Example.

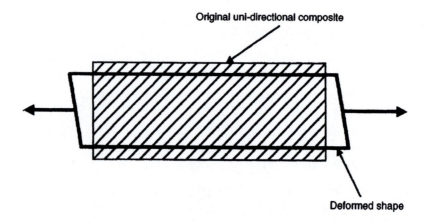

Fig. 3.17   Coupling effects between extension and shear

Example 3.10 If a moment of $M_y = 100$ Nm/m is applied to the unidirectional composite described in the previous Example, calculate the curvatures which will occur. Determine also the stress and strain distributions in the global $(x-y)$ and local $(1-2)$ directions.

Solution Using the D and d matrices from the previous Example, then for the applied moment $M_y = 100$ N:

$$\begin{bmatrix} \kappa_x \\ \kappa_y \\ \kappa_{xy} \end{bmatrix} = d \cdot \begin{bmatrix} M_x \\ M_y \\ M_{xy} \end{bmatrix} \qquad \nu_{yx} = \frac{-\kappa_x}{\kappa_y} \qquad \nu_{xy} = \frac{-\kappa_y}{\kappa_x}$$

This enables the curvatures to be determined as

$$\kappa_x = -3.87 \text{ m}^{-1}, \quad \kappa_y = 12 \text{ m}^{-1}, \quad \kappa_{xy} = 2.67 \text{ m}^{-1},$$

$$\nu_{yx} = 0.328, \quad \nu_{xy} = 3.05$$

The bending strains in the global directions are given generally by

$$\varepsilon = \kappa.Z$$

Hence

$$(\varepsilon_x)_{\max} = \pm \kappa_x \left(\frac{h}{2}\right) = \mp 3.87 \times 10^{-3}$$

$$(\varepsilon_y)_{\max} = \pm \kappa_y \left(\frac{h}{2}\right) = \pm 0.012$$

$$(\gamma_{xy})_{\max} = \pm \kappa_{xy} \left(\frac{h}{2}\right) = \pm 2.67 \times 10^{-3}$$

The stresses are then obtained from

$$\begin{bmatrix} \sigma_x \\ \sigma_y \\ \tau_{xy} \end{bmatrix} = \overline{Q} \cdot \begin{bmatrix} \varepsilon_x \\ \varepsilon_y \\ \gamma_{xy} \end{bmatrix}$$

So

$$\sigma_x = 0, \quad \sigma_y = 150 \text{ MN/m}^2, \quad \tau_{xy} = 0$$

For the local directions, the strain and stress transformation matrices can be used:

$$\begin{bmatrix} \sigma_1 \\ \sigma_2 \\ \tau_{12} \end{bmatrix} = T_\sigma \cdot \begin{bmatrix} \sigma_x \\ \sigma_y \\ \tau_{xy} \end{bmatrix}$$

So

$$\sigma_1 = 26.8 \text{ MN/m}^2, \quad \sigma_2 = 123.2 \text{ MN/m}^2, \quad \tau_{xy} = 57.5 \text{ MN/m}^2$$

$$\begin{bmatrix} \varepsilon_1 \\ \varepsilon_2 \\ \gamma_{12} \end{bmatrix} = T_\varepsilon \cdot \begin{bmatrix} \varepsilon_x \\ \varepsilon_y \\ \gamma_{xy} \end{bmatrix}$$

So

$$\varepsilon_1 = -5.14 \times 10^{-5}, \quad \varepsilon_2 = 8 \times 10^{-3}, \quad \gamma_{12} = 0.014$$

These stresses and strains are illustrated in Fig. 3.18.

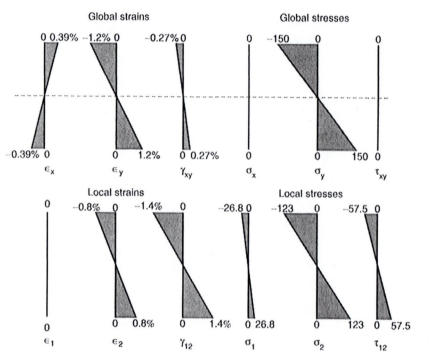

Fig. 3.18   Stresses and strains, Example 3.10

Note that if both plane stresses and moments are applied then the total stresses will be the algebraic sum of the individual stresses.

3.9 Deformation Behaviour of Laminates

(i) Laminates Made from Unidirectional Plies

The previous analysis has shown that the properties of unidirectional fibre composites are highly anisotropic. To alleviate this problem, it is common to build up laminates consisting of stacks of unidirectional lamina arranged at different orientations. Clearly many permutations are possible in terms of the numbers of layers (or plies) and the relative orientation of the fibres in each layer. At first glance it might appear that the best means of achieving a more isotropic behaviour would be to have two layers with the unidirectional fibres arranged perpendicular to each other. For example, two layers arranged at $0°$ and $90°$ to the global x-direction or at $+45°$ and $-45°$ to the x-direction might appear to offer more balanced properties in all directions. In fact the lack of symmetry about the centre plane of the laminate causes very complex behaviour in such cases.

In general it is best to aim for symmetry about the centre plane. A laminate in which the layers above the centre plane are a mirror image of those below it is described as symmetric. Thus a four stack laminate with fibres oriented at $0°, 90°, 90°$ and $0°$ is symmetric. The convention is to denote this as $[0°/90°/90°/0°]_T$ or $[0°, 90°_2, 0°]_T$ or $[0°/90°]_s$. In general terms any laminate of the type $[\theta, -\theta, -\theta, \theta]_T$ is symmetric and there may of course be any even number of layers or plies. They do not all have to be the same thickness but symmetry must be maintained. In the case of a symmetric laminate where the central ply is not repeated, this can be denoted by the use of an overbar. Thus the laminate $[45/-45/0/90/0/-45/45]_T$ can be written as $[\pm45, 0, \overline{90}]_s$.

In-plane Behaviour of a Symmetric Laminate

The in-plane stiffness behaviour of symmetric laminates may be analysed as follows. The plies in a laminate are all securely bonded together so that when the laminate is subjected to a force in the plane of the laminate, all the plies deform by the same amount. Hence, the strain is the same in every ply but because the modulus of each ply is different, the stresses are not the same. This is illustrated in Fig. 3.19.

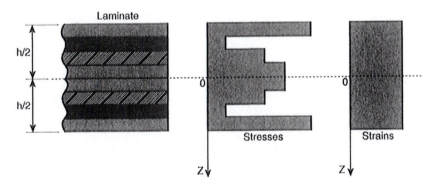

Fig. 3.19   Stresses and strains in a symmetric laminate

When external forces are applied in the global $x-y$ direction, they will equate to the summation of all the forces in the individual plies. Thus, for unit width of laminate this may be expressed in terms of the stresses or forces as

$$\bar{\sigma}_x = \frac{1}{h} \int_{-h/2}^{h/2} (\sigma_x)_f \, dZ \quad (\text{or } N_x = \int_{-h/2}^{h/2} (\sigma_x)_f \, dZ)$$

$$\bar{\sigma}_y = \frac{1}{h} \int_{-h/2}^{h/2} (\sigma_y)_f \, dZ \quad (\text{or } N_y = \int_{-h/2}^{h/2} (\sigma_y)_f \, dZ)$$

$$\bar{\tau}_{xy} = \frac{1}{h} \int_{-h/2}^{h/2} -h/2(\tau_{xy})_f \, dZ \quad (\text{or } N_{xy} = \int_{-h/2}^{h/2} (\tau_{xy})_f \, dZ)$$

where h is the thickness of the laminate. σ_x, N_x are the overall stresses or forces and $(\sigma)_f$ is the stress in the ply 'f' (see Fig. 3.20).

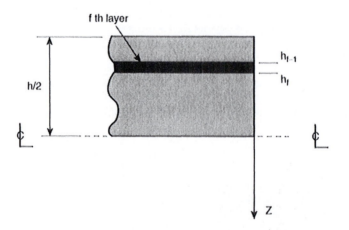

Fig. 3.20   Ply f in the laminate

In matrix form we can write

$$\left\{ \begin{array}{c} \bar{\sigma}_x \\ \bar{\sigma}_y \\ \bar{\tau}_{xy} \end{array} \right\} = \frac{1}{h} \int_{-h/2}^{h/2} \left\{ \begin{array}{c} \sigma_x \\ \sigma_y \\ \tau_{xy} \end{array} \right\}_f \, dZ$$

$$= \frac{1}{h} \int_{-h/2}^{h/2} \left[\begin{array}{ccc} \bar{Q}_{11} & \bar{Q}_{12} & \bar{Q}_{16} \\ & \bar{Q}_{22} & \bar{Q}_{26} \\ sym. & & \bar{Q}_{66} \end{array} \right] dZ \left\{ \begin{array}{c} \varepsilon_x \\ \varepsilon_y \\ \gamma_{xy} \end{array} \right\}_f \, dZ$$

As the strains are independent of Z they can be taken outside the integral:

$$\left\{\begin{array}{c}\overline{\sigma}_x \\ \overline{\sigma}_y \\ \overline{\tau}_{xy}\end{array}\right\} = \frac{1}{h}\int_{-h/2}^{h/2}\left[\begin{array}{ccc}\overline{Q}_{11} & \overline{Q}_{12} & \overline{Q}_{16} \\ & \overline{Q}_{22} & \overline{Q}_{26} \\ sym. & & \overline{Q}_{66}\end{array}\right]dZ\left\{\begin{array}{c}\varepsilon_x \\ \varepsilon_y \\ \gamma_{xy}\end{array}\right\}_f$$

$$\left\{\begin{array}{c}\overline{\sigma}_x \\ \overline{\sigma}_y \\ \overline{\tau}_{xy}\end{array}\right\} = [A]\left\{\begin{array}{c}\varepsilon_x \\ \varepsilon_y \\ \gamma_{xy}\end{array}\right\}_f\frac{1}{h} \qquad (\text{or } [N]_{xy} = [A][\varepsilon]_{xy})$$

where, for example,

$$A_{11} = \int_{-h/2}^{h/2}\overline{Q}_{11}dZ = 2.\int_{0}^{h/2}\overline{Q}_{11}dZ$$

$[A]$ is the **Extensional Stiffness Matrix** although it should be noted that it also contains shear terms.

Within a single ply, such as the fth, the $\overline{Q}$ - terms are constant so,

$$A_{11}^f = \sum \overline{Q}_{11}^f h_f$$

In overall terms

$$[A] = \sum_{f=1}^{p}\overline{Q}h_f \qquad (3.35)$$

Thus the stiffness matrix for a symmetric laminate may be obtained by adding, in proportion to the ply thickness, the corresponding terms in the stiffness matrix for each of the plies.

Having obtained all the terms for the extensional stiffness matrix $[A]$, this may then be inverted to give the compliance matrix $[a]$.

$$[a] = [A]^{-1}$$

The laminate properties may then be obtained as above from inspection of the compliance matrix.

$$E_X = \frac{1}{a_{11}h} \qquad E_y = \frac{1}{a_{22}h} \qquad G = \frac{1}{a_{66}h}$$

where h is the thickness of the laminate.

$$\nu_{xy} = \frac{-a_{12}}{a_{11}} \qquad \nu_{yx} = \frac{-a_{12}}{a_{22}}$$

3.10 Summary of Steps to Predict Stiffness of Symmetric Laminates

1. The Stiffness matrix $[\overline{Q}]$ is obtained as earlier each individual ply in the laminate.
2. The Stiffness matrix $[A]$ for the laminate is determined by adding the product of thickness and $[\overline{Q}]$ for each ply.
3. The Compliance matrix $[a]$ for the laminate is determined by inverting $[A]$ ie $[a] = [A]^{-1}$.
4. The stresses and strains in the laminate are then determined from

$$\begin{Bmatrix} \varepsilon_x \\ \varepsilon_y \\ \gamma_{xy} \end{Bmatrix} = [a] \cdot \begin{Bmatrix} \sigma_x \\ \sigma_y \\ \tau_{xy} \end{Bmatrix} h$$

Example 3.11 A series of individual plies with the properties listed below are laid in the following sequence to make a laminate

$$[0/35_2/-35_2]_s$$

Determine the moduli for the laminate in the global $X-Y$ directions and the strains in the laminate when stresses of $\sigma_x = 10$ MN/m^2, $\sigma_y = -14$ MN/m^2 and $\tau_{xy} = -5$ MN/m^2 are applied. The thickness of each is 1 mm.

$$E_1 = 125000 \text{ MN/m}^2 \quad E_2 = 7800 \text{ MN/m}^2 \quad G_{12} = 4400 \text{ MN/m}^2$$

$$\nu_{12} = 0.34$$

Solution The behaviour of each ply when subjected to loading at θ degrees off the fibre axis is determined using Matrix manipulation as follows:

Compliance Matrix   *Stiffness Matrix*

$$S = \begin{bmatrix} \dfrac{1}{E_1} & -\dfrac{\nu_{12}}{E_1} & 0 \\ \dfrac{\nu_{12}}{E_1} & \dfrac{1}{E_2} & 0 \\ 0 & 0 & \dfrac{1}{G_{12}} \end{bmatrix} \qquad\qquad Q = S^{-1}$$

Stress Transformation Matrix   *Strain Transformation Matrix*

$$[T]_\sigma = \begin{bmatrix} c^2 & s^2 & 2sc \\ s^2 & c^2 & -2sc \\ -sc & sc & (c^2-s^2) \end{bmatrix} \qquad [T]_\varepsilon = \begin{bmatrix} c^2 & s^2 & sc \\ s^2 & c^2 & -sc \\ -2sc & 2sc & (c^2-s^2) \end{bmatrix}$$

Overall Stiffness Matrix   *Overall Compliance Matrix*

$$\overline{Q}(\theta) = T_\sigma^{-1} \cdot Q \cdot T_\varepsilon \qquad\qquad \overline{S}(\theta) = \overline{Q}(\theta)^{-1}$$

$$\overline{Q}(35) = \begin{bmatrix} 6.27 \times 10^4 & 2.71 \times 10^4 & 3.66 \times 10^4 \\ 2.71 \times 10^4 & 2.21 \times 10^4 & 1.87 \times 10^4 \\ 3.66 \times 10^4 & 1.87 \times 10^4 & 2.88 \times 10^4 \end{bmatrix}$$

Hence, the Extension Stiffness matrix is given by

$$A = 2.\overline{Q}(0) + 4.\overline{Q}(35) + 4.\overline{Q}(-35)$$

$$A = \begin{bmatrix} 7.53 \times 10^5 & 2.22 \times 10^5 & 0 \\ 2.22 \times 10^5 & 1.93 \times 10^5 & 0 \\ 0 & 0 & 2.39 \times 10^5 \end{bmatrix} \text{ N mm}$$

The compliance matrix is obtained by inverting $[A]$

$$a = A^{-1}$$

$$a = \begin{bmatrix} 2.0 \times 10^{-6} & -2.31 \times 10^{-6} & 0 \\ -2.31 \times 10^{-6} & 7.84 \times 10^{-6} & 0 \\ 0 & 0 & 4.17 \times 10^{-6} \end{bmatrix} \text{ (N mm}^{-1})$$

The stiffness terms in the global directions may be obtained from:

$$E_x = \frac{1}{a_{11} \times 10}, \qquad E_x = 49.7 \text{ GN/m}^2$$

$$E_y = \frac{1}{a_{22} \times 10}, \qquad E_y = 12.8 \text{ GN/m}^2 \quad \text{and} \quad G_{xy} = \frac{1}{a_{66} \times 10} = 24 \text{ GN/m}^2$$

$$v_{xy} = \frac{a_{12}}{a_{11}} \qquad v_{xy} = 1.151$$

(note the high value of Poisson's Ratio which can be obtained in composite laminates)

$$v_{yx} = \frac{a_{12}}{a_{22}} \qquad v_{yx} = 0.295$$

and the strains may be obtained as

$$\begin{bmatrix} \varepsilon_x \\ \varepsilon_y \\ \gamma_{xy} \end{bmatrix} = a \cdot \begin{bmatrix} \sigma_x \\ \sigma_y \\ \tau_{xy} \end{bmatrix} \cdot 10$$

$$\varepsilon_x = 5.24 \times 10^{-4}, \quad \varepsilon_y = 1.33 \times 10^{-3}, \quad \gamma_{xy} = -2.08 \times 10^{-4}$$

$$\varepsilon_x = 0.052\%, \qquad \varepsilon_y = -0.133\%, \qquad \gamma_{xy} = -0.021\%$$

It may be seen from the above analysis that for cross-ply $(0/90)_s$ and symmetric angle ply $[-0/\theta]_s$ laminates, $A_{16} = A_{61} = 0$ and $A_{26} = A_{62} = 0$ (also $a_{16} = a_{61} = a_{62} = 0$). For other types of laminates this will not be the case.

3.11 General Deformation Behaviour of Laminates

The previous section has illustrated a simple convenient means of analysing in-plane loading of symmetric laminates. Many laminates are of this type and so this approach is justified. However, there are also many situations where other types of loading (including bending) are applied to laminates which may be symmetric or non-symmetric. In order to deal with these situations it is necessary to adopt a more general type of analysis.

Convention for defining thicknesses and positions of plies

In this more general analysis it is essential to be able to define the position and thickness of each ply within a laminate. The convention is that the geometrical mid-plane is taken as the datum. The top and bottom of each ply are then defined relative to this. Those above the mid-plane will have negative co-ordinates and those below will be positive. The bottom surface of the fth ply has address h_f and the top surface of this ply has address h_{f-1}. Hence the thickness of the fth ply is given by

$$h(f) = h_f - h_{f-1}$$

For the 6 ply laminate shown in Fig. 3.21, the thickness of ply 5 is given by

$$h(5) = h_5 - h_4 = 3 - 1 = 2 \text{ mm}$$

Fig. 3.21   Six ply laminate

The thickness of ply 1 is given by

$$h(1) = h_1 - h_0 = (-3) - (-6) = 3 \text{ mm}$$

Analysis of Laminates

The general deformation analysis of a laminate is very similar to the general deformation analysis for a single ply.

(i) **Force Equilibrium:** If there are F plies, then the force resultant, N_L, for the laminate is given by the sum of the forces for each ply.

$$[N]_L = \sum_{f=1}^{F} [N]_f = \sum_{f=1}^{F} \int_{h_{f-1}}^{h_f} [\sigma]_f dz$$

and using the definition for $[\sigma]$ from the analysis of a single ply,

$$[N]_L = \sum_{f=1}^{F} ([\overline{Q}] \cdot [\varepsilon] + [\overline{Q}] \cdot z \cdot [\kappa])_f dz$$

$$= \sum_{f=1}^{F} \left\{ [\overline{Q}] \cdot [\varepsilon] \int_{h_{f-1}}^{h_f} dz + [\overline{Q}] \cdot [\kappa] \int_{h_{f-1}}^{h_f} z \cdot dz \right\}_f$$

$$[N]_L = [A][\varepsilon]_L + [B][\kappa]_L \tag{3.36}$$

where

$$[A] = \sum_{f=1}^{F} [\overline{Q}]_f (h_f - h_{f-1}) \tag{3.37}$$

This is called the *Extensional Stiffness Matrix* and the similarity with that derived earlier for the single ply should be noted.

Also, the *Coupling Matrix*, $[B]$ is given by

$$[B] = \frac{1}{2} \sum_{f=1}^{F} [\overline{Q}]_f (h_f^2 - h_{f-1}^2) \tag{3.38}$$

The Coupling Matrix will be zero for a symmetrical laminate.

(ii) **Moment Equilibrium:** As in the case of the forces, the moments may be summed across F plies to give

$$[M]_L = \sum_{f=1}^{F} [M]_f = \sum_{f=1}^{F} \int_{h_{f-1}}^{h_f} [\sigma]_f dz$$

and once again using the expressions from the analysis of a single ply,

$$[M]_L = \sum_{f=1}^{F} \int_{h_{f-1}}^{h_f} ([\overline{Q}][\varepsilon]z + [\overline{Q}][\kappa]z^2)dz$$

$$= \sum_{f=1}^{F} \{[\overline{Q}][\varepsilon] \int_{h_{f-1}}^{h_f} z\,dz + [\overline{Q}][\kappa] \int_{h_{f-1}}^{h_{f-1}} z^2\,dz\}_f$$

$$[M]_L = [B][\varepsilon]_L + [D][\kappa]_L \tag{3.39}$$

where $[B]$ is as defined above and

$$[D] = \frac{1}{3}\sum_{f=1}^{F}[\overline{Q}]_f(h_f^3 - h_{f-1}^3) \tag{3.40}$$

As earlier we may group equations (3.36) and (3.39) to give the *Plate Constitutive Equation* as

$$\begin{bmatrix} N \\ M \end{bmatrix} = \begin{bmatrix} A & B \\ B & D \end{bmatrix}\begin{bmatrix} \varepsilon \\ \kappa \end{bmatrix} \tag{3.41}$$

This equation may be utilised to give elastic properties, strains, curvatures, etc. It is much more general than the approach in the previous section and can accommodate bending as well as plane stresses. Its use is illustrated in the following Examples.

Example 3.12 For the laminate $[0/35_2/-35_2]_s$ determine the elastic constants in the global directions using the Plate Constitutive Equation. When stresses of $\sigma_x = 10$ MN/m^2, $\sigma_y = -14$ MN/m^2 and $\tau_{xy} = -5$ MN/m^2 are applied, calculate the stresses and strains in each ply in the local and global directions. If a moment of $M_x = 1000$ N m/m is added, determine the new stresses, strains and curvatures in the laminate. The plies are each 1 mm thick.

$$E_1 = 125 \text{ GN/m}^2, \quad E_2 = 7.8 \text{ GN/m}^2, \quad G_{12} = 4.4 \text{ GN/m}^2, \quad \nu_{12} = 0.34$$

Solution The locations of each ply are illustrated in Fig. 3.22.
Using the definitions given above, and the $\overline{Q}$ values for each ply, we may determine the matrices A, B and D from

$$A = \sum_{f=1}^{10}\overline{Q}_f(h_f - h_{f-1}),$$

$$B = \frac{1}{2}\cdot\sum_{f=1}^{10}\overline{Q}_f\left[(h_f)^2 - (h_{f-1})^2\right], \qquad D = \frac{1}{3}\cdot\sum_{f=1}^{10}\overline{Q}_f\cdot[(h_f)^3 - (h_{f-1})^3]$$

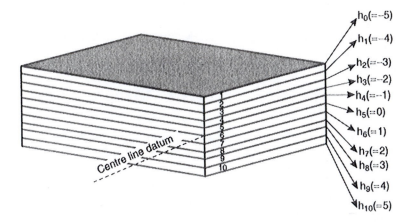

Fig. 3.22   Ten ply laminate

Then

$$\begin{bmatrix} \varepsilon_x \\ \varepsilon_y \\ \varepsilon_{xy} \end{bmatrix} = a \cdot \begin{bmatrix} N_x \\ N_y \\ N_{xy} \end{bmatrix} = a \cdot \begin{bmatrix} \sigma_x \\ \sigma_y \\ \tau_{xy} \end{bmatrix} \cdot h$$

where h = full laminate thickness = 10 mm, and $a = A^{-1}$ since $[B] = 0$.
This matrix equation gives the global strains as

$$\varepsilon_x = 5.24 \times 10^{-4}, \qquad \varepsilon_y = 1.33 \times 10^{-3}, \qquad \varepsilon_{xy} = 2.08 \times 10^{-4}$$

Also,

$$E_x = \frac{1}{a_{11} \cdot h}, \qquad E_y = \frac{1}{a_{22} \cdot h}, \qquad G_{xy} = \frac{1}{a_{66} \cdot h}$$

$$v_{xy} = \frac{-a_{12}}{a_{11}}, \qquad v_{yx} = \frac{-a_{12}}{a_{22}}$$

$$E_x = 49.7 \text{ GN/m}^2, \qquad E_y = 12.8 \text{ GN/m}^2, \qquad G_{xy} = 24 \text{ GN/m}^2$$

$$v_{xy} = 1.149 \qquad v_{yx} = 0.296$$

It may be seen that these values agree with those calculated in the previous
Example. To get the stresses in the global (xy) directions for the 'f'th ply

$$\begin{bmatrix} \sigma_x \\ \sigma_y \\ \tau_{xy} \end{bmatrix} = \overline{Q}_f \begin{bmatrix} \varepsilon_x \\ \varepsilon_y \\ \gamma_{xy} \end{bmatrix}$$

when $f = 1$, $f = 2$, $f = 9$ and $f = 10$

$$\sigma_x = 4.5 \text{ MN/m}^2, \quad \sigma_y = -11.3 \text{ MN/m}^2, \quad \tau_{xy} = -0.28 \text{ MN/m}^2$$

when $f = 3$, $f = 4$, $f = 7$ and $f = 8$

$$\sigma_x = -10.8 \text{ MN/m}^2, \quad \sigma_y = -19.2 \text{ MN/m}^2, \quad \tau_{xy} = -11.8 \text{ MN/m}^2$$

when $f = 5$ and 6

$$\sigma_x = 62.6 \text{ MN/m}^2, \quad \sigma_y = -9 \text{ MN/m}^2, \quad \tau_{xy} = -0.9 \text{ MN/m}^2.$$

Note that in the original question, the applied force per unit width in the x-direction was 100 N/mm (ie $\sigma_x \cdot (10)$). As each ply is 1 mm thick, then the above stresses are also equal to the forces per unit width for each ply. If we add the above values for all 10 plies, then it will be seen that the answer is 100 N/mm as it should be for equilibrium. Similarly, if we add N_y and N_{xy} for each ply, these come to -140 N/mm and -50 N/mm which also agree with the applied forces in these directions.

In the local (1–2) directions we can obtain the stresses and strains by using the transformation matrices. Hence, for the 'f'th ply

$$\begin{bmatrix} \sigma_1 \\ \sigma_2 \\ \tau_{12} \end{bmatrix} = T_{\sigma_f} \begin{bmatrix} \sigma_x \\ \sigma_y \\ \tau_{xy} \end{bmatrix}$$

$$\begin{bmatrix} \varepsilon_1 \\ \varepsilon_2 \\ \gamma_{12} \end{bmatrix} = T_{\varepsilon_f} \begin{bmatrix} \varepsilon_x \\ \varepsilon_y \\ \gamma_{xy} \end{bmatrix}$$

So that for $f = 1, 2, 9$ and 10

$$\sigma_1 = -0.44 \text{ MN/m}^2, \quad \sigma_2 = -6.4 \text{ MN/m}^2, \quad \tau_{xy} = 7.3 \text{ MN/m}^2$$

$$\varepsilon_1 = 1.38 \times 10^{-5}, \quad \varepsilon_2 = -8.15 \times 10^{-4}, \quad \gamma_{12} = 1.67 \times 10^{-3}$$

For $f = 3, 4, 7$ and 8

$$\sigma_1 = -0.96 \text{ MN/m}^2, \quad \sigma_2 = -5.8 \text{ MN/m}^2, \quad \tau_{12} = -7.5 \text{ MN/m}^2$$

$$\varepsilon_1 = 1.82 \times 10^{-4}, \quad \varepsilon_2 = -6.2 \times 10^{-4}, \quad \gamma_{12} = -1.8 \times 10^{-3}$$

For $f = 5$ and 6

$$\sigma_1 = 4.5 \text{ MN/m}^2, \quad \sigma_2 = -11.3 \text{ MN/m}^2, \quad \tau_{12} = -0.28 \text{ MN/m}^2$$

$$\varepsilon_1 = 5.25 \times 10^{-4}, \quad \varepsilon_2 = -1.33 \times 10^{-3}, \quad \gamma_{12} = -2.09 \times 10^{-4}$$

When the moment $M_x = 1000$ Nm/m is added, the curvatures, κ, can be calculated from

$$\begin{bmatrix} \kappa_x \\ \kappa_y \\ \kappa_{xy} \end{bmatrix} = d \cdot \begin{bmatrix} M_x \\ M_y \\ M_{xy} \end{bmatrix} \qquad \nu_{xy} = \frac{-\kappa_y}{\kappa_x} \qquad \nu_{yx} = \frac{-\kappa_x}{\kappa_y}$$

where $d = D^{-1}$ since $[B] = 0$.

$$\kappa_x = 0.435 \text{ m}^{-1}, \qquad \kappa_y = -0.457 \text{ m}^{-1}, \qquad \kappa_{xy} = -0.147 \text{ m}^{-1}$$

$$\nu_{xy} = 1.052 \qquad \nu_{yx} = 0.95$$

When the bending moment is applied the global stresses and strains in each ply may be obtained as follows:

$$\varepsilon_x = \kappa_x \cdot Z, \qquad \varepsilon_y = \kappa_y \cdot Z, \qquad \gamma_{xy} = \kappa_{xy} \cdot Z$$

At the top surface, $Z = -5$ mm

$$\varepsilon_x = -2.17 \times 10^{-3}, \qquad \varepsilon_y = 2.28 \times 10^{-3}, \qquad \gamma_{xy} = -7.34 \times 10^{-4}$$

and the stresses are given by

$$\begin{bmatrix} \sigma_x \\ \sigma_y \\ \tau_{xy} \end{bmatrix} = \overline{Q}_1 \begin{bmatrix} \varepsilon_x \\ \varepsilon_y \\ \gamma_{xy} \end{bmatrix}$$

So that

$$\sigma_x = -47 \text{ MN/m}^2, \qquad \sigma_y = 5.7 \text{ MN/m}^2, \qquad \tau_{xy} = 15.4 \text{ MN/m}^2.$$

The local stresses and strains are then obtained from the stress and strain transformations

$$\begin{bmatrix} \sigma_1 \\ \sigma_2 \\ \tau_{12} \end{bmatrix} = T_{\sigma_1} \begin{bmatrix} \sigma_x \\ \sigma_y \\ \tau_{xy} \end{bmatrix} \quad \text{and} \quad \begin{bmatrix} \varepsilon_1 \\ \varepsilon_2 \\ \gamma_{12} \end{bmatrix} = T_{\varepsilon_1} \begin{bmatrix} \varepsilon_x \\ \varepsilon_y \\ \gamma_{xy} \end{bmatrix}$$

$$\sigma_1 = -44.1 \text{ MN/m}^2, \qquad \sigma_2 = 2.8 \text{ MN/m}^2, \qquad \tau_{12} = -19.5 \text{ MN/m}^2$$

$$\varepsilon_1 = -3.6 \times 10^{-4}, \qquad \varepsilon_2 = 4.7 \times 10^{-4}, \qquad \gamma_{12} = -4.44 \times 10^{-3}$$

For the next interface, $z = -4$ mm, the new values of ε_x, ε_y and γ_{xy} can be calculated and hence the stresses in the global and local co-ordinates. $f = 1$ and $f = 2$ need to be analysed for this interface but there will be continuity across the interface because the orientation of the plies is the same in both cases. However, at $z = -3$ mm there will be a discontinuity of stresses in the global direction and discontinuity of stresses and strains in the local directions due to the difference in fibre orientation in plies 2 and 3.

The overall distribution of stresses and strains in the local and global directions is shown in Fig. 3.23. If both the normal stress and the bending are applied together then it is necessary to add the effects of each separate condition. That is, direct superposition can be used to determine the overall stresses.

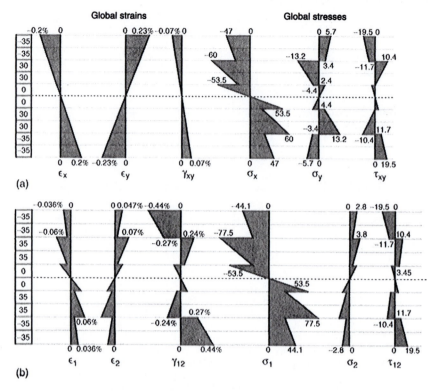

Fig. 3.23 Stresses and strains, Example 3.12: (a) global; (b) local

Note, to assist the reader the values of the terms in the matrices are

$$A = \begin{bmatrix} 7.52 \times 10^5 & 2.22 \times 10^5 & 0 \\ 2.22 \times 10^5 & 1.93 \times 10^5 & 0 \\ 0 & 0 & 2.39 \times 10^5 \end{bmatrix} \text{N/mm} \qquad B = \begin{bmatrix} 0 & 0 & 0 \\ 0 & 0 & 0 \\ 0 & 0 & 0 \end{bmatrix}$$

$$a = \begin{bmatrix} 2.01 \times 10^6 & -2.31 \times 10^{-6} & 0 \\ -2.31 \times 10^{-6} & 7.82 \times 10^{-6} & 0 \\ 0 & 0 & 4.17 \times 10^{-6} \end{bmatrix} \text{(N/mm)}^{-1}$$

$$D = \begin{bmatrix} 5.25 \times 10^6 & 2.24 \times 10^6 & -1.75 \times 10^6 \\ 2.24 \times 10^6 & 1.84 \times 10^6 & -9.042 \times 10^5 \\ -1.75 \times 10^6 & -9.04 \times 10^5 & 2.38 \times 10^6 \end{bmatrix} \text{Nmm,}$$

$$d = \begin{bmatrix} 4.34 \times 10^{-7} & -4.57 \times 10^{-7} & 1.46 \times 10^{-7} \\ -4.57 \times 10^{-7} & 1.14 \times 10^{-6} & 9.75 \times 10^{-8} \\ 1.46 \times 10^{-7} & 9.75 \times 10^{-8} & 5.63 \times 10^{-7} \end{bmatrix} \text{(Nmm)}^{-1}$$

The solution method using the Plate Constitutive Equation is therefore straightforward and very powerful. Generally a computer is needed to handle the matrix manipulation – although this is not difficult, it is quite time-consuming if it has to be done manually.

The following Example compares the behaviour of a single ply and a laminate.

Example 3.13 A single ply of carbon fibre/epoxy composite has the following properties:

$$E_1 = 175 \text{ GN/m}^2, \quad E_2 = 12 \text{ GN/m}^2, \quad G_{12} = 5 \text{ GN/m}^2, \quad \nu_{12} = 0.3$$

Plot the variations of E_x, E_y, G_{xy} and ν_{xy} for values of θ in the range 0 to 90° for (i) a single ply 0.4 mm thick and (ii) a laminate with the stacking sequence $[\pm\theta]_s$. This laminate has four plies, each 0.1 mm thick. Discuss the meaning of the results.

Solution

(i) *Single ply* The method of solution simply involves the determination of $[S], [Q], [\overline{Q}], [\overline{S}]$ and $[a]$ as illustrated previously, ie

$$|S| = \begin{vmatrix} \dfrac{1}{E_1} & -\dfrac{\nu_{12}}{E_1} & 0 \\[2mm] -\dfrac{\nu_{12}}{E_1} & \dfrac{1}{E_2} & 0 \\[2mm] 0 & 0 & \dfrac{1}{G_{12}} \end{vmatrix} \qquad \text{and} \qquad |Q| = |S|^{-1}$$

Then

$$|\overline{Q}| = T_\sigma \cdot Q \cdot T_\varepsilon \qquad \text{and} \qquad |\overline{S}| = |\overline{Q}|^{-1}$$

$$|A| = |\overline{Q}| \cdot h \qquad \text{and} \qquad |a| = |A|^{-1}$$

Then for $\theta = 25°$ (for example)

$$E_x = \frac{1}{a_{11}h} = 28.3 \text{ GN/m}^2$$

$$E_y = \frac{1}{a_{22}h} = 11.7 \text{ GN/m}^2$$

$$G_{xy} = \frac{1}{a_{66}h} = 7.3 \text{ GN/m}^2$$

$$\nu_{xy} = \frac{-a_{12}}{a_{11}} = 0.495 \qquad \text{and} \qquad \nu_{yx} = \frac{-a_{12}}{a_{22}} = 0.205$$

Fig. 3.24 shows the variation of these elastic constants for all values of θ between 0 and 90°.

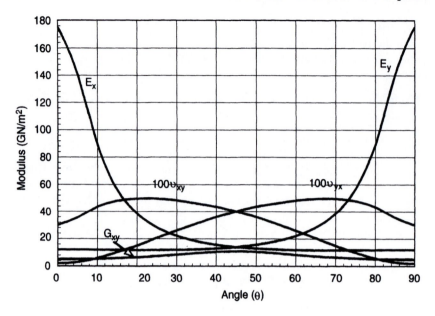

Fig. 3.24 Variation of elastic properties for a single ply of carbon/epoxy composite

(ii) Symmetric 4 Ply Laminate   The same procedure is used again except that this time the $|A|$ matrix has to be summed for all the plies, ie

$$|A| = \sum_{f=1}^{f=4} |\overline{Q}|_f \, (h_f - h_{f-1}) \text{ and } |a| = |A|^{-1}$$

where $h_0 = -0.2, h_1 = -0.1, h_2 = 0, h_3 = 0.1, h_4 = 0.2.$

Once again, for $\theta = 50°$ (for example)

$$E_x = \frac{1}{a_{11}h} = 14.6 \text{ GN/m}^2$$

$$E_y = \frac{1}{a_{22}h} = 24.8 \text{ GN/m}^2$$

$$G_{xy} = \frac{1}{a_{66}h} = 44 \text{ GN/m}^2$$

$$v_{xy} = \frac{-a_{12}}{a_{11}} = 0.615 \quad \text{and} \quad v_{yx} = \frac{-a_{12}}{a_{22}} = 1.045$$

Fig. 3.25 shows the variation of these elastic constants for values of θ in the range 0 to 90°.

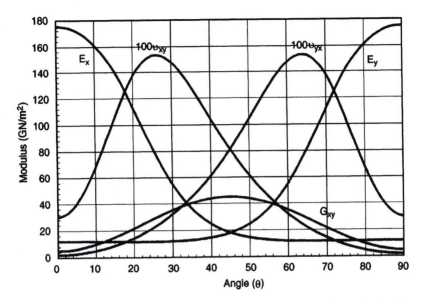

Fig. 3.25 Variation of elastic properties for a $(+/-45)$ symmetric laminate of carbon/epoxy

It is interesting to compare the behaviour of the single ply and the laminate as shown in Figs 3.24 and 3.25. Firstly it is immediately evident that the laminate offers a better balance of properties as well as improvements in absolute terms. The shear modulus in particular is much better in the laminate. Its peak value at 45° arises because shear is equivalent to a state of stress where equal tensile and compressive stresses are applied at 45° to the shear direction. Thus shear loading on a $[\pm 45]_s$ laminate is equivalent to tensile and compressive loading on a $[0/90]_s$ laminate. Thus the fibres are effectively aligned in the direction of loading and this provides the large stiffness (or modulus) which is observed.

It is also worthy of note that large values of Poisson's Ratio can occur in a laminate. In this case a peak value of over 1.5 is observed – something which would be impossible in an isotropic material. Large values of Poisson's Ratio are a characteristic of unidirectional fibre composites and arise due to the coupling effects between extension and shear which were referred to earlier.

It is also important to note that although the laminae $[\pm 45]_s$ indicates that $E_x = E_y = 18.1$ GN/m², this laminate is not isotropic or even quasi-isotropic. As shown in Chapter 2, in an isotropic material, the shear modulus is linked to the other elastic properties by the following equation

$$G_{xy} = \frac{E_x}{2(1 + v_{xy})}$$

For the $[\pm 45]_s$ laminate this would give

$$G_{xy} = \frac{18.1}{2(1 + 0.814)} = 5 \text{ GN/m}^2$$

However, Fig. 3.25 shows that $G_{xy} = 45.2 \text{ GN/m}^2$ for the $[\pm 45]_s$ laminate.

Some laminates do exhibit quasi-isotropic behaviour. The simplest one is $[0, \pm 60]_s$. For this laminate

$$E_x = E_y = 66.4 \text{ GN/m}^2 \quad \text{and} \quad v_{xy} = v_{yx} = 0.321, \quad G_{xy} = 25.1 \text{ GN/m}^2$$

(using the individual ply data in the above Example).

If we check G_{xy} from the isotropic equation we get

$$G_{xy} = \frac{E_x}{2(1 + v_{xy})} = \frac{66.4}{2(1 + 0.321)} = 25.1 \text{ GN/m}^2.$$

This agrees with the value calculated from the laminate theory.

In general any laminate with the lay-up

$$\left[0, \frac{\pi}{N}, \frac{2\pi}{N}, \cdots, \frac{N-1}{N}\pi\right]_s$$

or

$$\left[\frac{\pi}{N}, \frac{2\pi}{N}, \cdots, \pi\right]_s$$

is quasi-isotropic where N is an integer equal to 3 or greater. The angles for the plies are expressed in radians.

3.12 Analysis of Multi-layer Isotropic Materials

The Plate Constitutive equations can be used for curved plates provided the radius of curvature is large relative to the thickness (typically $r/h > 50$). They can also be used to analyse laminates made up of materials other than unidirectional fibres, eg layers which are isotropic or made from woven fabrics can be analysed by inserting the relevant properties for the local 1–2 directions. Sandwich panels can also be analysed by using a thickness and appropriate properties for the core material. These types of situation are considered in the following Examples.

Example 3.14 A blow moulded plastic bottle has its wall thickness made of three layers. The layers are:

Outside and inside skin – Material A

thickness $= 0.4$ mm, $E_1 = E_2 = 3 \text{ GN/m}^2$, $G_{12} = 1.1 \text{ GN/m}^2$, $v_{12} = 0.364$

Core – Material B

thickness $= 0.4$ mm, $E_1 = E_2 = 0.8 \text{ GN/m}^2$, $G_{12} = 0.285 \text{ GN/m}^2$, $v_{12} = 0.404$.

If the diameter of the bottle is 160 mm, calculate the hoop and axial strains in the bottle wall when an internal pressure of 200 kN/m² is applied. Calculate also the stresses in the individual layers.

Solution Although the composite wall is curved, the (r/h) value is large and so it can be analysed using the method illustrated for laminates. In this case, each ply is isotropic and so the properties do not vary with θ. It is thus necessary to get $\overline{Q}$ for each ply relative to the centre line of the wall thickness

$$h_0 = 0.6, \quad h_1 = -0.2, \quad h_2 = 0, \quad h_3 = 0.2, \quad h_4 = 0.6$$

ie the wall section is treated as if it has four plies

[material A (0.4 mm)/material B (0.2 mm)]$_s$

Also

$$\text{hoop stress, } \sigma_y = \frac{pr}{h} = \frac{0.2 \times 80}{1.2} = 13.33 \text{ MN/m}^2$$

(or $N_y = \sigma_y h = 16$ N/mm)

$$\text{axial stress, } \sigma_x = \frac{pr}{2h} = \frac{0.2 \times 80}{2 \times 1.2} = 6.67 \text{ MN/m}^2$$

(or $N_R = \sigma_R h = 8$ N/mm)

$$\overline{Q}_1 = \begin{bmatrix} 3.444 \times 10^3 & 1.244 \times 10^3 & 0 \\ 1.244 \times 10^3 & 3.444 \times 10^3 & 0 \\ 0 & 0 & 1.103 \times 10^3 \end{bmatrix} \text{N/mm}^2,$$

$$\overline{Q}_2 = \begin{bmatrix} 956.5 & 376.7 & 0 \\ 376.7 & 956.5 & 0 \\ 0 & 0 & 285.8 \end{bmatrix} \text{N/mm}^2$$

Also,

$$\overline{Q}_1 = \overline{Q}_4 \quad \text{and} \quad \overline{Q}_2 = \overline{Q}_3$$

Then the Extensional Stiffness Matrix is given by

$$A = \sum_{f=1}^{4} Q_f(h_f - h_{f-1}), \qquad a = A^{-1} \text{ (since } [B] = 0).$$

The strains are related to the forces (or stresses) by

$$\begin{bmatrix} \varepsilon_x \\ \varepsilon_y \\ \gamma_{xy} \end{bmatrix} = a \cdot \begin{bmatrix} N_x \\ N_y \\ N_{xy} \end{bmatrix}$$

Therefore the axial and hoop strains are

$$\varepsilon_x = 7.932 \times 10^{-4}, \qquad \varepsilon_y = 4.809 \times 10^{-3} (\gamma_{xy} = 0)$$

The effective elastic constants for the composite are:

$$E_x = \frac{1}{a_{11} \cdot h} \qquad E_y = \frac{1}{a_{22} \cdot h} \qquad G_{xy} = \frac{1}{a_{66} \cdot h}$$

$$v_{xy} = \frac{-a_{12}}{a_{11}} \qquad v_{yx} = \frac{-a_{12}}{a_{22}}$$

So

$$E_x = 2.26 \times 10^3 \text{ MN/m}^2 \quad E_y = 2.26 \times 10^3 \text{ MN/m}^2 \quad G_{xy} = 830.6 \text{ MN/m}^2$$

and

$$v_{xy} = 0.36, \qquad v_{yx} = 0.36$$

The stresses in the individual layers are obtained from

$$\begin{bmatrix} \sigma_{Ax} \\ \sigma_{Ay} \\ \tau_{Axy} \end{bmatrix} = \overline{Q}_1 \cdot \begin{bmatrix} \varepsilon_x \\ \varepsilon_y \\ \gamma_{xy} \end{bmatrix} \qquad \begin{bmatrix} \sigma_{Bx} \\ \sigma_{By} \\ \tau_{Bxy} \end{bmatrix} = \overline{Q}_2 \cdot \begin{bmatrix} \varepsilon_x \\ \varepsilon_y \\ \gamma_{xy} \end{bmatrix}$$

So,

$$\sigma_{Ax} = 8.71 \text{ MN/m}^2 \quad \text{and} \quad \sigma_{Ay} = 17.55 \text{ MN/m}^2$$

$$(\tau_{Axy} = 0 \text{ MN/m}^2)$$

$$\sigma_{Bx} = 2.5 \text{ MN/m}^2 \quad \text{and} \quad \sigma_{By} = 4.9 \text{ MN/m}^2$$

$$(\tau_{Bxy} = 0 \text{ MN/m}^2)$$

The values of stress may be checked by simple equilibrium analysis.

(i) Force equilibrium: $\quad N_A + N_B = N_x$

$$\sigma_{Ax} A_A + \sigma_{Bx} A_B = N_x$$

(ii) Geometry of deformation:

$$\varepsilon_A = \varepsilon_B = \varepsilon_x$$

So

$$\frac{\sigma_{Ax}}{E_A} = \frac{\sigma_{Bx}}{E_B} \quad \text{or} \quad \sigma_{Ax} = \left(\frac{E_A}{E_B}\right) \sigma_{Bx}$$

$$\left(\frac{E_A}{E_B E}\right) \sigma_B A_A + \sigma_B A_B = N_x$$

$$\sigma_{Bx} = N_x / \left(\left(\frac{E_A}{E_B}\right) A_A + A_B\right) = 2.4 \text{ N/mm}^2$$

$$\sigma_{Ax} = \left(\frac{3}{0.8}\right) 2.4 = 8.8 \text{ N/mm}^2$$

To assist the reader the values of the terms in the matrices are

$$A = \begin{bmatrix} 3.15 \times 10^3 & 1.16 \times 10^3 & 0 \\ 1.16 \times 10^3 & 3.158 \times 10^3 & 0 \\ 0 & 0 & 996.7 \end{bmatrix} \text{N/mm} \qquad B = \begin{bmatrix} 0 & 0 & 0 \\ 0 & 0 & 0 \\ 0 & 0 & 0 \end{bmatrix}$$

$$a = \begin{bmatrix} 3.67 \times 10^{-4} & -1.35 \times 10^{-4} & 0 \\ -1.35 \times 10^{-4} & 3.67 \times 10^{-4} & 0 \\ 0 & 0 & 1.003 \times 10^{-3} \end{bmatrix} \text{mm/N}$$

This type of analysis could also be used for a sandwich structure with solid skins and a foamed core. It is simply a matter of using the appropriate values of E_1, G_{12}, E_2, v_{12} for the skin and core material. This is illustrated in the following Example.

Example 3.15 A sandwich moulding is made up of solid skins with a foamed plastic core. The skins and core may be regarded as isotropic with the following the properties:

Skin – Material A, $E_1 = E_2 = 3.5$ GN/m^2, $G_{12} = 1.25$ GN/m^2, $v_{12} = 0.4$.

Foam Core – Material B, $E_1 = E_2 = 0.06$ GN/m^2, $G_{12} = 0.021$ GN/m^2,

$$v_{12} = 0.43.$$

The skins are each 1 mm thick and the core is 20 mm thick.

If moments of $M_x = 400$ Nm/m and $M_y = 300$ Nm/m are applied to the moulding, calculate the resulting curvatures and the stresses and strains in the cross-section.

Solution As the skins and core are isotropic, we only need to get the $\overline{Q}$ values for each:

Skin   *Foamed Core*

$$\overline{Q}_1 = \begin{bmatrix} 4.167 \times 10^3 & 1.667 \times 10^3 & 0 \\ 1.667 \times 10^3 & 4.167 \times 10^3 & 0 \\ 0 & 0 & 1.25 \times 10^3 \end{bmatrix} \qquad \overline{Q}_2 = \begin{bmatrix} 73.63 & 31.63 & 0 \\ 31.63 & 73.63 & 0 \\ 0 & 0 & 20.98 \end{bmatrix}$$

$$\overline{Q}_4 = \overline{Q}_1, \qquad\qquad\qquad\qquad \overline{Q}_3 = \overline{Q}_2$$

Assuming as before that there are four layers with the centre line at the mid-plane of the cross-section, then

$$h_0 = -11, \quad h_1 = -10, \quad h_2 = 0, \quad h_3 = 10, \quad h_4 = 11 \text{ mm}$$

As these are only moments applied and the section is symmetric ($B = 0$) we only need the D and d matrices. These are given by

$$D = \frac{1}{3} \sum_{f=1}^{4} \overline{Q}_f [h_f^3 - h_{f-1}^3]$$

$$D = \begin{bmatrix} 9.69 \times 10^5 & 3.89 \times 10^5 & 0 \\ 3.89 \times 10^5 & 9.69 \times 10^5 & 0 \\ 0 & 0 & 2.89 \times 10^5 \end{bmatrix} \text{Nmm,}$$

$$d = \begin{bmatrix} 1.23 \times 10^{-6} & -4.94 \times 10^{-7} & 0 \\ -4.94 \times 10^{-6} & 1.23 \times 10^{-6} & 0 \\ 0 & 0 & 3.45 \times 10^{-6} \end{bmatrix}$$

Then the curvatures, κ, are given by

$$\begin{bmatrix} \kappa_x \\ \kappa_y \\ \kappa_{xy} \end{bmatrix} = d \begin{bmatrix} M_x \\ M_y \\ M_{xy} \end{bmatrix}$$

So,

$$\kappa_x = 3.44 \times 10^{-4} \text{ mm}^{-1}, \quad \kappa_y = 1.72 \times 10^{-4} \text{ mm}^{-1}, \quad \kappa_{xy} = 0$$

The strains will be directly proportional to the distance across the thickness, ie

$$\varepsilon = \kappa \cdot Z$$

At top surface, $Z = -11$ mm, so $\varepsilon_x = (3.44 \times 10^{-4})(-11) = -3.78 \times 10^{-3}$:

$$\text{stress, } \sigma_x = E_1\varepsilon_x = 3500(3.78 \times 10^{-3}) = -13.25 \text{ MN/m}^2$$

At interface (skin side), $Z = -10$ mm, so $\varepsilon_x = (3.44 \times 10^{-4})(-10) = -3.44 \times 10^{-3}$:

$$\text{stress, } \sigma_x = E_1\varepsilon_x = 3500(3.44 \times 10^{-3}) = -12.04 \text{ MN/m}^2$$

At interface (core side), $\varepsilon_x = -3.44 \times 10^{-3}$:

$$\text{stress, } \sigma_x = 60(-3.44 \times 10^{-3}) = -0.21 \text{ MN/m}^2$$

Similarly in the y-direction,
At top surface

$$\varepsilon_y = -1.89 \times 10^{-3}, \qquad \sigma_y = -6.6 \text{ MN/m}^2$$

At interface (skin)

$$\varepsilon_y = -1.72 \times 10^{-3}, \qquad \sigma_y = -6 \text{ MN/m}^2$$

At interface (core)

$$\varepsilon_y = -1.72 \times 10^{-3}, \qquad \sigma_y = -0.1 \text{ MN/m}^2$$

In both cases the stresses and strains will be mirrored on the bottom section of the beam (ie positive instead of negative). These are illustrated in Fig. 3.26.

Fig. 3.26 Stresses and strains, Example 3.15

3.13 Analysis of Non-symmetric Laminates

Example 3.16 A unidirectional carbon fibre/PEEK laminate has the stacking sequence $[0/35_2/-35_2]_T$. If it has an in-plane stress of $\sigma_x = 100$ MN/m^2 applied, calculate the strains and curvatures in the global directions. The properties of the individual plies are

$$E_1 = 145 \text{ GN/m}^2, \quad E_2 = 15 \text{ GN/m}^2, \quad G_{12} = 4 \text{ GN/m}^2, \quad \nu_{12} = 0.278$$

Each ply has a thickness of 0.1 mm.

Solution This stacking sequence is similar to that in Example 3.11 except that in this case the laminate is not symmetrical. As shown in Fig. 3.27, the centre line of the laminate is in the middle of one of the plies and

$$h_0 = -0.25, \quad h_1 = -0.15, \quad h_2 = -0.05, \quad h_3 = 0.05,$$
$$h_4 = 0.15, \quad h_5 = 0.25$$

Fig. 3.27 Non-symmetric laminate

The approach is the same as before. $\overline{Q}$ must be determined for each layer as shown below and thus the values in the A, B and D matrices.

$$\overline{Q}_1 = \begin{bmatrix} 1.46 \times 10^5 & 4.20 \times 10^3 & 0 \\ 4.20 \times 10^3 & 1.51 \times 10^4 & 0 \\ 0 & 0 & 4 \times 10^3 \end{bmatrix}$$

$$\overline{Q}_2 = \begin{bmatrix} 7.28 \times 10^4 & 3.44 \times 10^4 & 4.17 \times 10^4 \\ 3.44 \times 10^4 & 2.80 \times 10^4 & 1.98 \times 10^4 \\ 4.17 \times 10^4 & 1.98 \times 10^4 & 3.42 \times 10^4 \end{bmatrix}$$

$$\overline{Q}_3 = \begin{bmatrix} 7.28 \times 10^4 & 3.44 \times 10^4 & 4.17 \times 10^4 \\ 3.44 \times 10^4 & 2.80 \times 10^4 & 1.98 \times 10^4 \\ 4.17 \times 10^4 & 1.98 \times 10^4 & 3.42 \times 10^4 \end{bmatrix}$$

$$\overline{Q}_4 = \begin{bmatrix} 7.28 \times 10^4 & 3.44 \times 10^4 & -4.17 \times 10^4 \\ 3.44 \times 10^4 & 2.80 \times 10^4 & -1.98 \times 10^4 \\ -4.17 \times 10^4 & -1.98 \times 10^4 & 3.42 \times 10^4 \end{bmatrix}$$

$$\overline{Q}_5 = \begin{bmatrix} 7.28 \times 10^4 & 3.44 \times 10^4 & -4.17 \times 10^4 \\ 3.44 \times 10^4 & 2.80 \times 10^4 & -1.98 \times 10^4 \\ -4.17 \times 10^4 & -1.98 \times 10^4 & 3.42 \times 10^4 \end{bmatrix}$$

$$A = \sum_{f=1}^{5} \overline{Q}_f (h_f - h_{f-1}) \qquad B = \frac{1}{2} \sum_{f=1}^{5} \overline{Q}_f \left[(h_f)^2 - (h_{f-1})^2 \right]$$

$$D = \frac{1}{3} \sum_{f=1}^{5} \overline{Q}_f \left[(h_f)^3 - (h_{f-1})^3 \right]$$

In this case $[B] \neq 0$ and so $[a]$ and $[d]$ cannot be obtained by inverting the $[A]$ and $[D]$ matrices respectively. They must be obtained from

$$\begin{bmatrix} a & b \\ \beta & d \end{bmatrix} = \begin{bmatrix} A & B \\ B & D \end{bmatrix}^{-1}$$

The in-plane strains and the moments and curvatures are all linked in this non-symmetrical lamina and are obtained from

$$\begin{bmatrix} \varepsilon_x \\ \varepsilon_y \\ \gamma_{xy} \\ \kappa_x \\ \kappa_y \\ \kappa_{xy} \end{bmatrix} = \begin{bmatrix} a_{11} & a_{12} & a_{16} & b_{11} & b_{12} & b_{16} \\ a_{21} & a_{22} & a_{26} & b_{21} & b_{22} & b_{26} \\ a_{61} & a_{62} & a_{66} & b_{61} & b_{62} & b_{66} \\ \beta_{11} & \beta_{12} & \beta_{16} & d_{11} & d_{12} & d_{16} \\ \beta_{21} & \beta_{22} & \beta_{26} & d_{21} & d_{22} & d_{26} \\ \beta_{61} & \beta_{62} & \beta_{66} & d_{61} & d_{62} & d_{66} \end{bmatrix} \begin{bmatrix} N_x \\ N_y \\ N_{xy} \\ M_x \\ M_y \\ M_{xy} \end{bmatrix}$$

where

$$N = \sigma \cdot h(N_x = 50 \text{ N/mm})$$

which gives the strains and curvatures as

$$\varepsilon_x = 3.326 \times 10^{-3}, \qquad \varepsilon_y = -2.868 \times 10^{-3}, \qquad \gamma_{xy} = -3.405 \times 10^{-5}$$

$$\kappa_x = 0.012 \text{ mm}^{-1}, \qquad \kappa_y = -0.01 \text{ mm}^{-1}, \qquad \kappa_{xy} = 0.019 \text{ mm}^{-1}$$

The important point to note from this Example is that in a non-symmetrical laminate the behaviour is very complex. It can be seen that the effect of a simple uniaxial stress, σ_x, is to produce strains and curvatures in all directions. This has relevance in a number of polymer processing situations because unbalanced cooling (for example) can result in layers which have different properties, across a moulding wall thickness. This is effectively a composite laminate structure which is likely to be non-symmetrical and complex behaviour can be expected when loading is applied.

The reader may wish to check the matrices in this Example:

$$A = \begin{bmatrix} 4.37 \times 10^4 & 1.41 \times 10^4 & 0 \\ 1.41 \times 10^4 & 1.27 \times 10^4 & 0 \\ 0 & 0 & 1.40 \times 10^4 \end{bmatrix}$$

$$a = \begin{bmatrix} 6.65 \times 10^{-5} & -5.73 \times 10^{-5} & -6.81 \times 10^{-7} \\ -5.73 \times 10^{-5} & 1.53 \times 10^{-4} & -2.05 \times 10^{-5} \\ -6.81 \times 10^{-7} & -2.05 \times 10^{-5} & 1.02 \times 10^{-4} \end{bmatrix}$$

$$B = \begin{bmatrix} -1.46 \times 10^3 & 604.5 & -1.67 \times 10^3 \\ 604.5 & 258.2 & -792 \\ -16.7 \times 10^3 & -792 & 604.4 \end{bmatrix}$$

$$b = \begin{bmatrix} 2.30 \times 10^{-4} & -2.09 \times 10^{-4} & 3.77 \times 10^{-4} \\ -2.24 \times 10^{-4} & 1.49 \times 10^{-4} & 5.34 \times 10^{-5} \\ 9.55 \times 10^{-5} & 2.01 \times 10^{-4} & -2.03 \times 10^{-4} \end{bmatrix}$$

$$\beta = \begin{bmatrix} 2.30 \times 10^{-4} & -2.24 \times 10^{-4} & 9.55 \times 10^{-5} \\ -2.09 \times 10^{-4} & 1.49 \times 10^{-4} & 2.01 \times 10^{-4} \\ 3.77 \times 10^{-4} & 5.34 \times 10^{-5} & -2.03 \times 10^{-4} \end{bmatrix}$$

$$D = \begin{bmatrix} 1.05 \times 10^{3} & 235.2 & -167.1 \\ 235.2 & 239.2 & -79.2 \\ -167.1 & -79.2 & 233.0 \end{bmatrix}$$

$$d = \begin{bmatrix} 2.15 \times 10^{-3} & -1.60 \times 10^{-3} & 1.64 \times 10^{-3} \\ -1.601 \times 10^{-3} & 6.65 \times 10^{-3} & -4.04 \times 10^{-4} \\ 1.64 \times 10^{-3} & -4.04 \times 10^{-3} & 8.74 \times 10^{-3} \end{bmatrix}$$

3.14 Analysis of Short Fibre Composites

In order to understand the effect of discontinuous fibres in a polymer matrix it is important to understand the reinforcing mechanism of fibres. Fibres exert their effect by restraining the deformation of the matrix as shown in Fig. 3.28. The external loading applied through the matrix is transferred to the fibres by shear at the fibre/matrix interface. The resultant stress distributions in the fibre and matrix are complex. In short fibres the tensile stress increases from zero at the ends to a value $(\sigma_f)_{max}$ which it would have if the fibre was continuous. This is shown in Fig. 3.29. From the previous section it may be seen that $(\sigma_f)_{max}$ may be determined from

$$\frac{(\sigma_f)_{max}}{E_f} = \frac{\sigma_c}{E_1} \tag{3.42}$$

where σ_c is the stress applied to the composite and E_1 may be determined from the rule of mixtures.

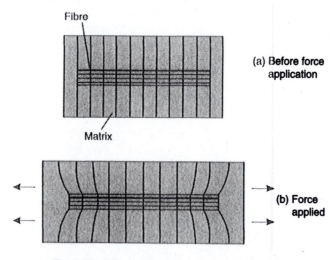

Fig. 3.28   Effect of fibre on deformation of matrix

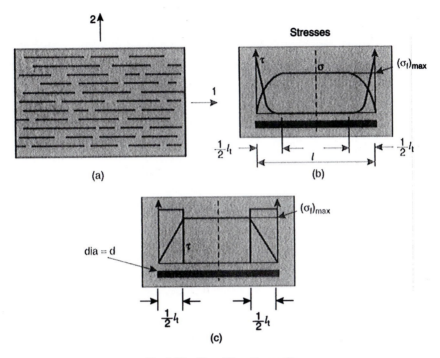

Fig. 3.29   Short Fibre Composites

The stress distribution in short fibres is often simplified to the form shown in Fig. 3.29(c)

It is evident from Fig. 3.29 that there is a minimum fibre length which will permit the fibre to achieve its full load-carrying potential. The minimum fibre length in which the maximum fibre stress, $(\sigma_f)_{max}$, can be achieved is called the **load transfer length, ℓ_t.** The value of ℓ_t may be determined from a simple force balance

force transmitted by shear at interface $= \tau_y(\ell/2)\pi d$

force exerted by fibre $\qquad\qquad\quad = \sigma_f(\pi d^2/4)$

hence,

$$\ell_t = \frac{(\sigma_f)_{max}d}{2\tau_y} \tag{3.43}$$

where, τ_y is the shear strength of the fibre/matrix interface.

The maximum value of ℓ_t will occur when $(\sigma_f)_{max}$ reaches the tensile strength of the fibre, σ_{fu}, and this is defined as the **critical fibre length, ℓ_c**

$$\ell_c = \frac{\sigma_{fu}d}{2\tau_y} \tag{3.44}$$

Example 3.17 Short carbon fibres with a diameter of 10 μm are to be used to reinforce nylon 66. If the design stress for the composite is 300 MN/m^2 and the following data is available on the fibres and nylon, calculate the load transfer length for the fibres and also the critical fibre length. The volume fraction of the fibres is to be 0.3.

	Modulus (GN/m^2)	Strength (GN/m^2)
Carbon fibres	230	2.9
Nylon 66	2.8	–

The interfacial shear strength for carbon/nylon may be taken as 4 MN/m^2

Solution

$$E_1 = E_f V_f + E_m V_m$$

$$E_1 = 230(0.3) + 2.8(0.7) = 71 \text{ GN/m}^2$$

Using (3.42)

$$(\sigma_f)_{max} = E_f \left(\frac{\sigma_c}{E_1}\right) = 230 \left(\frac{300}{71}\right) = 972 \text{ MN/m}^2$$

Using (3.43)

$$\ell_t = \frac{(\sigma_f)_{max} d}{2\tau_y} = \frac{972(10 \times 10^{-3})}{2 \times 4} = 1.2 \text{ mm}$$

Using (3.44)

$$\ell_c = \frac{\sigma_{fu} d}{2\tau_y} = \frac{2900(10 \times 10^{-3})}{2 \times 4} = 3.6 \text{ mm}$$

It may be seen from Fig. 3.29 that due to the ineffective end portions of short fibres, the average stress in the fibre will be less than in a continuous fibre. The exact value of the average stress will depend on the length of the fibres. Using the stress distributions shown in Fig. 3.29(b) the fibre stresses may be analysed as follows.

$$F_1 = \sigma_f(\pi d^2/4)$$

$$F_2 = \left[\sigma_f + \left(\frac{d\sigma_f}{dx}\right) dx\right] \frac{\pi d^2}{4}$$

$$F_3 = (\tau_y \pi d) dx$$

Now, for equilibrium of forces $F_1 = F_2 + F_3$

$$\sigma_f(\pi d^2/4) = \left(\sigma_f + \frac{d\sigma_f}{dx}dx\right)(\pi d^2/4) + (\tau\pi d)dx$$

$$(d/4)d\sigma_f = -\tau_y dx$$

Integrating this equation gives

$$(d/4)\int_0^{\sigma f} d\sigma_f = -\int_{\frac{1}{2}\ell}^{x} \tau_y dx$$

$$\sigma_f = \frac{4\tau_y\left(\frac{1}{2}\ell - x\right)}{d} \tag{3.45}$$

This is the general equation for the stress in the fibres but there are 3 cases to consider, as shown in Fig. 3.30

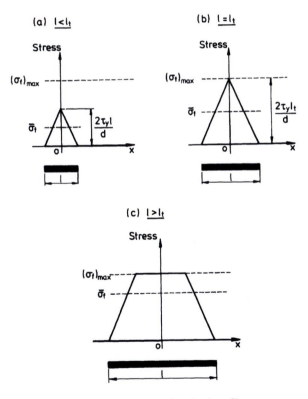

Fig. 3.30   Stress variations in short fibres

(a) Fibre lengths less than ℓ_t

In this case the peak value of stress occurs at $x = 0$, so from equation (3.45)

$$\sigma_f = \frac{2\tau_y \ell}{d}$$

The average fibre stress, $\overline{\sigma}_f$, is obtained by dividing the area under the stress/fibre length graph by the fibre length.

$$\overline{\sigma}_f = \frac{\frac{1}{2}\ell\left(\frac{2\tau_y \ell}{d}\right)}{\ell} = \frac{\tau_y \ell}{d}$$

Now from (3.6)

$$\sigma_c = \left(\frac{\tau_y \ell}{d}\right) V_f + \sigma'_m (1 - V_f) \tag{3.46}$$

(b) Fibre length equal to ℓ_t

In this case the peak stress is equal to the maximum fibre stress.
So at $x = 0$

$$\sigma_f = (\sigma_f)_{max} = \frac{2\tau_y \ell_t}{d} \tag{3.47}$$

$$\text{Average fibre stress} = \overline{\sigma}_f = \frac{\frac{1}{2}\left(\frac{2\tau_y \ell_t}{d}\right)\ell_t}{\ell_t}$$

$$\overline{\sigma}_f = \frac{\tau_y \ell_t}{d}$$

So from (3.6)

$$\sigma_c = \left(\frac{\tau_y \ell_t}{d}\right) V_f + \sigma'_m (1 - V_f) \tag{3.48}$$

(c) Fibre length greater than ℓ_t

(i) For $\frac{1}{2}\ell > x > \frac{1}{2}(\ell - \ell_t)$

$$\sigma_f = \frac{4\tau_y}{d}\left(\frac{1}{2}\ell - x\right)$$

(ii) for $\frac{1}{2}(\ell - \ell_t) > x > 0$

$$\sigma_f = \text{constant} = (\sigma_f)_{max}$$

$$\sigma_f = \frac{2\tau_y \ell_t}{d}$$

Also, as before, the average fibre stress may be obtained from

$$\bar{\sigma}_f = \frac{[(\sigma_f)_{max}](\ell - \ell_t) + [(\sigma_f)_{max}]\frac{1}{2}\ell_t}{\ell} = [(\sigma_f)_{max}]\left(1 - \frac{\ell_t}{2\ell}\right)$$

So from (3.6)

$$\sigma_c = V_f(\sigma_f)_{max}\left(1 - \frac{\ell_t}{2\ell}\right) + \sigma'_m(1 - V_f) \tag{3.49}$$

Note that in order to get the average fibre stress as close as possible to the maximum fibre stress, the fibres need to be considerably longer than the critical length. At the critical length the average fibre stress is only half of the value achieved in continuous fibres.

Experiments show that equations such as (3.49) give satisfactory agreement with the measured values of strength and modulus for polyester sheets re-inforced with chopped strands of glass fibre. Of course these strengths and modulus values are only about 20–25% of those achieved with continuous fibre reinforcement. This is because with randomly oriented short fibres only a small percentage of the fibres are aligned along the line of action of the applied stress. Also the packing efficiency is low and the generally accepted maximum value for V_f of about 0.4 is only half of that which can be achieved with continuous filaments.

In order to get the best out of fibre reinforcement it is not uncommon to try to control within close limits the fibre content which will provide maximum stiffness for a fixed weight of matrix and fibres. In flexure it has been found that optimum stiffness is achieved when the volume fraction is 0.2 for chopped strand mat (CSM) and 0.37 for continuous fibre reinforcement.

Example 3.18 Calculate the maximum and average fibre stresses for glass fibres which have a diameter of 15 μm and a length of 2.5 mm. The interfacial shear strength is 4 MN/m^2 and $\ell_t/\ell = 0.3$.

Solution Since $\ell > \ell_t$ then

$$(\sigma_f)_{max} = \frac{2\tau_y \ell_t}{d} = \frac{2\tau_y \ell}{d}\left(\frac{\ell_t}{\ell}\right) = \frac{2 \times 4 \times 2.5 \times 10^{-3} \times 0.3}{15 \times 10^{-6}}$$

$$(\sigma_f)_{max} = 400 \text{ MN/m}^2$$

Also
$$\bar{\sigma}_f = (\sigma_f)_{max}\left(1 - \frac{\ell_t}{2\ell}\right) = 400\left(1 - \frac{0.3}{2}\right)$$

$$\bar{\sigma}_f = 340 \text{ MN/m}^2$$

In practice it should be remembered that short fibres are more likely to be randomly oriented rather than aligned as illustrated in Fig. 2.35. The problem of analysing and predicting the performance of randomly oriented short fibres is complex. However, the stiffness of such systems may be predicted quite accurately using the following simple empirical relationship.

$$E_{\text{random}} = 3E_1/8 + 5E_2/8 \qquad (3.50)$$

Hull also proposed that the shear modulus and Poisson's Ratio for a random short fibre composite could be approximated by

$$G_{\text{random}} = \tfrac{1}{8}E_1 + \tfrac{1}{4}E_2 \qquad (3.51)$$

$$\nu_{\text{random}} = \frac{E_r}{2G_r} - 1 \qquad (3.52)$$

E_1 and E_2 refer to the longitudinal and transverse moduli for aligned fibre composites of the type shown in (Fig. 3.29). These values can be determined experimentally or using specifically formulated empirical equations. However, if the fibres are relatively long then equation (3.5) and (3.13) may be used. These give results which are sufficiently accurate for most practical purposes.

3.15 Creep Behaviour of Fibre Reinforced Plastics

The viscoelastic nature of the matrix in many fibre reinforced plastics causes their properties to be time and temperature dependent. Under a constant stress they exhibit creep which will be more pronounced as the temperature increases. However, since fibres exhibit negligible creep, the time dependence of the properties of fibre reinforced plastics is very much less than that for the unreinforced matrix.

3.16 Strength of Fibre Composites

Up to this stage we have considered the deformation behaviour of fibre composites. An equally important topic for the designer is avoidance of failure. If the definition of 'failure' is the attainment of a specified deformation then the earlier analysis may be used. However, if the occurrence of yield or fracture is to be predicted as an extra safeguard then it is necessary to use another approach.

In an isotropic material subjected to a uniaxial stress, failure of the latter type is straightforward to predict. The tensile strength of the material $\hat{\sigma}_T$ will be known from materials data sheets and it is simply a question of ensuring that the applied uniaxial stress does not exceed this.

If an isotropic material is subjected to multi-axial stresses then the situation is slightly more complex but there are well established procedures for predicting failure. If σ_x and σ_y are applied it is not simply a question of ensuring that neither of these exceed $\hat{\sigma}_T$. At values of σ_x and σ_y below $\hat{\sigma}_T$ there can be a plane within the material where the stress reaches $\hat{\sigma}_T$ and this will initiate failure.

A variety of methods have been suggested to deal with the prediction of failure under multi-axial stresses and some of these have been applied to composites. The main methods are

(i) *Maximum Stress Criterion*: This criterion suggests that failure of the composite will occur if any one of five events happens

$$\sigma_1 \geq \hat{\sigma}_{IT} \quad \text{or} \quad \sigma_1 \leq \hat{\sigma}_{IC} \quad \text{or} \quad \sigma_2 \geq \hat{\sigma}_{2T} \quad \text{or} \quad \sigma_2 \leq \hat{\sigma}_{2C} \quad \text{or} \quad \tau_{12} \geq \hat{\tau}_{12}$$

That is, if the local tensile, compressive or shear stresses exceed the materials tensile, compressive or shear strength then failure will occur. Some typical values for the strengths of uni-directional composites are given in Table 3.5.

Table 3.5

Typical strength properties of unidirectional fibre reinforced plastics

Material	Fibre volume fraction, V_f	$\hat{\sigma}_{IT}$ (GN/m²)	$\hat{\sigma}_{2T}$ (GN/m²)	$\hat{\tau}_{12}$ (GN/m²)	$\hat{\sigma}_{IC}$ (GN/m²)	$\hat{\sigma}_{2C}$ (GN/m²)
GFRP (E glass/epoxy)	0.6	1.4	0.05	0.04	0.22	0.1
GFRP (E glass/polyester)	0.42	0.52	0.034	–	–	–
KFRP (Kevlar 49/epoxy)	0.6	1.5	0.027	0.047	0.24	0.09
CFRP (Carbon/epoxy)	0.6	1.8	0.08	0.1	1.57	0.17
CFRP (Carbon HM/epoxy)	0.62	1.24	0.02	0.04	0.29	0.03

GFRP – Glass fibre reinforced plastic
KFRP – Kevlar fibre reinforced plastic
CRFP – Carbon fibre reinforced plastic

(ii) *Maximum Strain Criterion:* This criterion is similar to the above only it uses strain as the limiting condition rather than stress. Hence, failure is predicted to occur if

$$\varepsilon_1 \geq \hat{\varepsilon}_{IT} \quad \text{or} \quad \varepsilon_1 \leq \hat{\varepsilon}_{IC} \quad \text{or} \quad \varepsilon_2 \geq \hat{\varepsilon}_{2T} \quad \text{or} \quad \varepsilon_2 \leq \hat{\varepsilon}_{2C} \quad \text{or} \quad \gamma_{12} \geq \hat{\gamma}_{12}$$

(iii) *Tsai−Hill Criterion:* This empirical criterion defines failure as occurring if

$$\left(\frac{\sigma_1}{\hat{\sigma}_1}\right)^2 - \left(\frac{\sigma_1 \sigma_2}{\hat{\sigma}_1^2}\right) + \left(\frac{\sigma_2}{\hat{\sigma}_2}\right)^2 + \left(\frac{\tau_{12}}{\hat{\tau}_{12}}\right)^2 \geq 1 \tag{3.53}$$

The values in this equation are chosen so as to correspond with the nature of the loading. For example, if σ_1 is compressive, then $\hat{\sigma}_{IC}$ is used and so on.

In practice the second term in the above equation is found to be small relative to the others and so it is often ignored and the reduced form of the Tsai–Hill Criterion becomes

$$\left(\frac{\sigma_1}{\hat{\sigma}_1}\right)^2 + \left(\frac{\sigma_2}{\hat{\sigma}_2}\right)^2 + \left(\frac{\tau_{12}}{\hat{\tau}_{12}}\right)^2 \geq 1 \qquad\qquad (3.54)$$

3.16.1 Strength of Single Plies

These failure criteria can be applied to single ply composites as illustrated in the following Examples.

Example 3.19 A single ply Kevlar 49/epoxy composite has the following properties.

$$E_1 = 79 \text{ GN/m}^2, \quad E_2 = 4.1 \text{ GN/m}^2, \quad G_{12} = 1.5 \text{ GN/m}^2, \quad \nu_{12} = 0.43$$

$$\hat{\sigma}_{1T} = 1.5 \text{ GN/m}^2, \quad \hat{\sigma}_{2T} = 0.027 \text{ GN/m}^2, \quad \hat{\tau}_{12} = 0.047 \text{ GN/m}^2$$

$$\hat{\sigma}_{1C} = 0.24 \text{ GN/m}^2, \quad \hat{\sigma}_{2C} = 0.09 \text{ GN/m}^2.$$

If the fibres are aligned at 15° to the x-direction, calculate what tensile value of σ_x will cause failure according to (i) the Maximum Stress Criterion (ii) the Maximum Strain Criterion and (iii) the Tsai–Hill Criterion. The thickness of the composite is 1 mm.

Solution

(i) *Maximum Stress Criterion*

Consider the situation where $\sigma_x = 1 \text{ MN/m}^2$.

The stresses on the local (1–2) axes are given by

$$\begin{bmatrix} \sigma_1 \\ \sigma_2 \\ \tau_{12} \end{bmatrix} = T_\sigma \begin{bmatrix} \sigma_x \\ \sigma_y \\ \tau_{xy} \end{bmatrix}$$

Hence, $\sigma_1 = 0.93 \text{ MN/m}^2$, $\quad \sigma_2 = 0.067 \text{ MN/m}^2$, $\quad \tau_{12} = -0.25 \text{ MN/m}^2$
So

$$\frac{\hat{\sigma}_{1T}}{\sigma_1} = 1608, \quad \frac{\hat{\sigma}_{2T}}{\sigma_2} = 402, \quad \frac{\hat{\tau}_{12}}{\tau_{12}} = 188$$

Hence, a stress of $\sigma_x = 1608 \text{ MN/m}^2$ would cause failure in the local 1-direction. A stress of $\sigma_x = 402 \text{ MN/m}^2$ would cause failure in the local 2-direction and a stress of $\sigma_x = 188 \text{ MN/m}^2$ would cause shear failure in the local 1–2 directions. Clearly the latter is the limiting condition since it will occur first.

(ii) *Maximum Strain Criterion*

Once again, let $\sigma_x = 1$ MN/m². The limiting strains are given by

$$\hat{\varepsilon}_{IT} = \frac{\sigma_{IT}}{E_1} = 0.019, \qquad \hat{\varepsilon}_{IC} = \frac{\hat{\sigma}_{IC}}{E_1} = 3.04 \times 10^{-3}$$

$$\hat{\varepsilon}_{2T} = \frac{\hat{\sigma}_{2T}}{E_2} = 6.59 \times 10^{-3}, \qquad \hat{\varepsilon}_x = \frac{\hat{\sigma}_{2c}}{E_2} = 0.022$$

$$\hat{\gamma}_{12} = \frac{\hat{\tau}_{12}}{G} = 0.031$$

The strains in the local directions are obtained from

$$\begin{bmatrix} \varepsilon_1 \\ \varepsilon_2 \\ \gamma_{12} \end{bmatrix} = S. \begin{bmatrix} \sigma_1 \\ \sigma_2 \\ \tau_{12} \end{bmatrix}$$

$$\varepsilon_1 = 1.144 \times 10^{-5}, \quad \varepsilon_2 = 1.128 \times 10^{-5}, \quad \gamma_{12} = -1.688 \times 10^{-4}$$

$$\frac{\hat{\varepsilon}_{IT}}{\varepsilon_1} = 1659, \quad \frac{\hat{\varepsilon}_{2T}}{\varepsilon_2} = 583, \quad \frac{\hat{\gamma}_{12}}{\gamma_{12}} = 188$$

Thus once again, an applied stress of 188 MN/m² would cause shear failure in the local 1–2 direction.

(iii) *Tsai–Hill Criterion*

For an applied stress of 1 MN/m² and letting X be the multiplier on this stress, we can determine the value of X to make the Tsai–Hill equation become equal to 1.

$$\left(\frac{X \cdot \sigma_1}{\hat{\sigma}_{IT}} \right)^2 - \left(\frac{X^2 \cdot \sigma_1 \sigma_2}{\hat{\sigma}_{IT}} \right) + \left(\frac{X \cdot \sigma_2}{\hat{\sigma}_{2T}} \right)^2 + \left(\frac{X \cdot \tau_{12}}{\hat{\tau}_{12}} \right)^2 = 1$$

Solving this gives $X = 169$. Hence a stress of $\sigma_x = 169$ MN/m² would cause failure. It is more difficult with the Tsai–Hill criterion to identify the nature of the failure ie tensile, compression or shear. Also, it is generally found that for fibre angles in the regions 5°–15° and 40°–90°, the Tsai–Hill criterion predictions are very close to the other predictions. For angles between 15° and 40° the Tsai–Hill tends to predict more conservative (lower) stresses to cause failure.

Example 3.20 The single ply in the previous Example is subjected to the stress system

$$\sigma_x = 80 \text{ MN/m}^2, \quad \sigma_y = -40 \text{ MN/m}^2, \quad \tau_{xy} = -20 \text{ MN/m}^2$$

Determine whether failure would be expected to occur according to (a) the Maximum Stress (b) the Maximum Strain and (c) the Tsai–Hill criteria.

Solution

(a) *Maximum Stress Criterion*
   The stresses in the 1–2 directions are

$$\begin{bmatrix} \sigma_1 \\ \sigma_2 \\ \tau_{12} \end{bmatrix} = T_\sigma \begin{bmatrix} \sigma_x \\ \sigma_y \\ \tau_{xy} \end{bmatrix}$$

$$\sigma_1 = 61.9 \text{ MN/m}^2, \quad \sigma_2 = -21.9 \text{ MN/m}^2, \quad \tau_{12} = -47.3 \text{ MN/m}^2$$

$$\frac{\hat{\sigma}_{IT}}{\sigma_1} = 24.2, \quad \frac{\hat{\sigma}_x}{\sigma_2} = 4.1, \quad \frac{\hat{\tau}_{12}}{\tau_{12}} = 0.99$$

There are thus no problems in the tensile or compressive directions but the shear ratio has dropped below 1 and so failure is possible.

(b) *Maximum Strain Criterion*
   The limiting strains are as calculated in the previous Example.
   The local strains are obtained from

$$\begin{bmatrix} \varepsilon_1 \\ \varepsilon_2 \\ \gamma_{12} \end{bmatrix} = S \cdot \begin{bmatrix} \sigma_1 \\ \sigma_2 \\ \tau_{12} \end{bmatrix}$$

$$\varepsilon_1 = 9.04 \times 10^{-4}, \quad \varepsilon_2 = -5.69 \times 10^{-3}, \quad \gamma_{12} = -0.032$$

$$\frac{\hat{\varepsilon}_{IT}}{\varepsilon_1} = 21, \quad \frac{\hat{\varepsilon}_{2C}}{\varepsilon_2} = 3.86, \quad \frac{\hat{\gamma}_{12}}{\gamma_{12}} = 0.99$$

Once again failure is just possible in the shear direction.

(c) *Tsai–Hill Criterion*
   The Tsai–Hill equation gives

$$\left(\frac{\sigma_1}{\hat{\sigma}_{IT}} \right)^2 - \left(\frac{\sigma_1 \sigma_2}{\hat{\sigma}_{IT}^2} \right)^2 + \left(\frac{\sigma_2}{\hat{\sigma}_{2C}} \right)^2 + \left(\frac{\tau_{12}}{\hat{\tau}_{12}} \right)^2 = 1.08$$

As these terms equate to >1, failure is likely to occur.

3.16.2 Strength of Laminates

When a composite is made up of many plies, it is unlikely that all plies will fail simultaneously. Therefore we should expect that failure will occur in one ply before it occurs in the others. To determine which ply will fail first it is simply a question of applying the above method to each ply in turn. Thus it is necessary to determine the stresses or strains in the local (1–2) directions for each ply and then check for the possibility of failure using any or all of the above criteria. This is illustrated in the following Example.

Example 3.21 A carbon-epoxy composite has the properties listed below. If the stacking sequence is [0/-30/30]$_s$ and stresses of $\sigma_x = 400$ MN/m^2, $\sigma_y = $

160 MN/m^2 and $\tau_{xy} = -100$ MN/m^2 are applied, determine whether or not failure would be expected to occur according to (a) the Maximum Stress (b) the Maximum Strain and (c) the Tsai–Hill criteria. The thickness of each ply is 0.2 mm.

$$E_1 = 125 \text{ GN/m}^2, \quad E_2 = 9 \text{ GN/m}^2, \quad G_{12} = 4.4 \text{ GN/m}^2, \quad \nu_{12} = 0.34$$

$$\hat{\sigma}_{1T} = 1.8 \text{ GN/m}^2, \quad \hat{\sigma}_{2T} = 0.08 \text{ GN/m}^2, \quad \hat{\sigma}_{1C} = 1.57 \text{ GN/m}^2$$

$$\hat{\sigma}_{2C} = 0.17 \text{ GN/m}^2, \quad \hat{\tau}_{12} = 0.1 \text{ GN/m}^2$$

Solution It is necessary to work out the global strains for the laminate (these will be the same for each ply) and then get the local strains and stresses. Thus, for the 30° ply

$$\begin{bmatrix} \varepsilon_x \\ \varepsilon_y \\ \gamma_{xy} \end{bmatrix} = a \begin{bmatrix} N_x \\ N_y \\ N_{xy} \end{bmatrix}$$

Using $h_0 = -0.6$, $h_1 = -0.4$, $h_2 = -0.2$, $h_3 = 0$, $h_4 = 0.2$, $h_5 = 0.4$ and $h_6 = 0.6$ ($h = 1.2$ mm) gives

$$\begin{bmatrix} \varepsilon_x \\ \varepsilon_y \\ \gamma_{xy} \end{bmatrix} = \begin{bmatrix} 2.94 \\ 7.51 \\ -5.46 \end{bmatrix} \times 10^{-3}$$

Thus

$$\begin{bmatrix} \varepsilon_1 \\ \varepsilon_2 \\ \gamma_{12} \end{bmatrix} = T_\varepsilon \begin{bmatrix} \varepsilon_x \\ \varepsilon_y \\ \gamma_{xy} \end{bmatrix} = \begin{bmatrix} 1.72 \\ 8.73 \\ 1.23 \end{bmatrix} \times 10^{-3}$$

$$\begin{bmatrix} \sigma_1 \\ \sigma_y \\ \tau_{xy} \end{bmatrix} = \overline{Q}_f \begin{bmatrix} \varepsilon_x \\ \varepsilon_y \\ \tau_{xy} \end{bmatrix} \quad \text{and} \quad \begin{bmatrix} \sigma_1 \\ \sigma_2 \\ \tau_{12} \end{bmatrix} = T_\sigma \begin{bmatrix} \sigma_x \\ \sigma_y \\ \tau_{xy} \end{bmatrix}$$

So

$$\begin{bmatrix} \sigma_1 \\ \sigma_2 \\ \tau_{12} \end{bmatrix} = \begin{bmatrix} 243 \\ 84 \\ 5.4 \end{bmatrix} \text{ MN/m}^2$$

If this is repeated for each ply, then the data in Fig. 3.31 is obtained. It may be seen that failure can be expected to occur in the +30° plies in the 2-direction because the stress exceeds the boundary shown by the dotted line.

The Tsai–Hill criteria gives the following values

(i) 0° plies, 1.028
(ii) −30 plies, 0.776
(iii) 30 plies, 1.13

The failure in the 30° plies is thus confirmed. The Tsai–Hill criteria also predicts failure in the 0° plies and it may be seen in Fig. 3.31 that this is

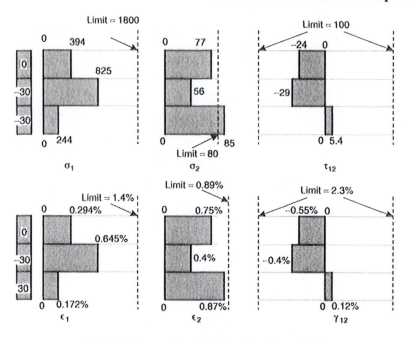

Fig. 3.31   Stress and strain in the plies, Example 3.21

probably because the stress in the 2-direction is getting very close to the limiting value.

3.17 Fatigue Behaviour of Reinforced Plastics

In common with metals and unreinforced plastics there is considerable evidence to show that reinforced plastics are susceptible to fatigue. If the matrix is thermoplastic then there is a possibility of thermal softening failures at high stresses or high cyclic frequencies as described in Section 2.21.1. However, in general, the presence of fibres reduces the hysteritic heating effect and there is a reduced tendency towards thermal softening failures. When conditions are chosen to avoid thermal softening, the normal fatigue process takes place in the form of a progressive weakening of the material due to crack initiation and propagation.

Plastics reinforced with carbon or boron are stiffer than glass reinforced plastics (grp) and they are found to be less vulnerable to fatigue. In short-fibre grp, cracks tend to develop relatively easily in the matrix and particularly at the interface close to the ends of the fibres. It is not uncommon for cracks to propagate through a thermosetting type matrix and destroy its integrity long before fracture of the moulded article occurs. With short-fibre composites it has been found that fatigue life is prolonged if the aspect ratio of the fibres is large.

The general fatigue behaviour which is observed in glass fibre reinforced plastics is illustrated in Fig. 3.32. In most grp materials, debonding occurs

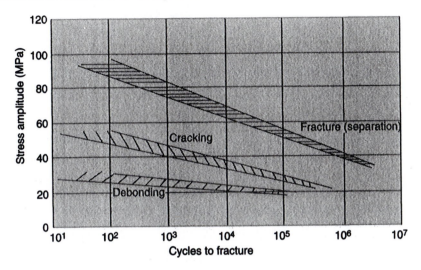

Fig. 3.32   Typical fatigue behaviour of glass reinforced polyester

after a small number of cycles, even at modest stress levels. If the material is translucent then the build-up of fatigue damage may be observed. The first signs are that the material becomes opaque each time the load is applied. Subsequently, the opacity becomes permanent and more pronounced. Eventually resin cracks become visible but the article is still capable of bearing the applied load until localised intense damage causes separation of the component. However, the appearance of the initial resin cracks may cause sufficient concern, for safety or aesthetic reasons, to limit the useful life of the component. Unlike most other materials, therefore, glass reinforced plastics give a visual warning of fatigue failure.

Since grp does not exhibit a fatigue limit it is necessary to design for a specific endurance and factors of safety in the region of 3–4 are commonly employed. Most fatigue data is for tensile loading with zero mean stress and so to allow for other values of mean stress it has been found that the empirical relationship described in Section 2.21.4 can be used. In other modes of loading (e.g. flexural or torsion) the fatigue behaviour of grp is worse than in tension. This is generally thought to be caused by the setting up of shear stresses in sections of the matrix which are unprotected by properly aligned fibres.

There is no general rule as to whether or not glass reinforcement enhances the fatigue behaviour of the base material. In some cases the matrix exhibits longer fatigue endurances than the reinforced material whereas in other cases the converse is true. In most cases the fatigue endurance of grp is reduced by the presence of moisture.

Fracture mechanics techniques, of the type described in Section 2.21.6 have been used very successfully for fibre reinforced plastics. Typical values of K for reinforced plastics are in the range $5-50$ MN m$^{-3/2}$, with carbon fibre reinforcement producing the higher values.

3.18 Impact Behaviour of Reinforced Plastics

Reinforcing fibres are brittle and if they are used in conjunction with a brittle matrix (e.g. epoxy or polyester resins) then it might be expected that the composite would have a low fracture energy. In fact this is not the case and the impact strength of most glass reinforced plastics is many times greater than the impact strengths of the fibres or the matrix. A typical impact strength for polyester resin is 2 kJ/m^2 whereas a CSM/polyester composite has impact strengths in the range $50-80$ kJ/m^2. Woven roving laminates have impact strengths in the range $100-150$ kJ/m^2. The much higher impact strengths of the composite in comparison to its component parts have been explained in terms of the energy required to cause debonding and work done against friction in pulling the fibres out of the matrix. Impact strengths are higher if the bond between the fibre and the matrix is relatively weak because if it is so strong that it cannot be broken, then cracks propagate across the matrix and fibres, with very little energy being absorbed. There is also evidence to suggest that in short-fibre reinforced plastics, maximum impact strength is achieved when the fibre length is at the critical value. There is a conflict therefore between the requirements for maximum tensile strength (long fibres and strong interfacial bond) and maximum impact strength. For this reason it is imperative that full details of the service conditions for a particular component are given in the specifications so that the sagacious designer can tailor the structure of the material accordingly.

Bibliography

Powell, P.C. *Engineering with Fibre-Polymer Laminates*, Chapman and Hall, London (1994).
Daniel, I.M. and Ishai, O. *Engineering Mechanics of Composite Materials*, Oxford University Press (1994).
Hancox, N.L. and Mayer, R.M. *Design Data for Reinforced Plastics*, Chapman and Hall, London (1993).
Mayer, R.M. *Design with Reinforced Plastics*, HMSO, London (1993).
Tsai, S.W. and Hahn, H.T. *Introduction to Composite Materials*, Technomic Westport, CT (1980).
Folkes, M.J. *Short Fibre Reinforced Thermoplastics*, Research Studies Press, Somerset (1982).
Mathews, F.L. and Rawlings, R.D. *Composite Materials: Engineering and Science*, Chapman and Hall, London (1993).
Phillips, L.N. (ed.) *Design with Advanced Composite Materials*, Design Council, London (1989).
Strong, B.A. *High Performance Engineering Thermoplastic Composites*, Technomic Lancaster, PA (1993).
Ashbee, K. *Fibre Reinforced Composites*, Technomic Lancaster, PA (1993).
Kelly, A (ed.) *Concise Encyclopedia of Compiste Materials*, Pergamon, Oxford (1994).
Stellbrink, K.K.U. *Micromechanics of Composites*, Hanser, Munich (1996).
Hull, D. *An Introduction to Composite Materials*, Cambridge University Press, (1981).
Piggott, M.R. *Load Bearing Fibre Composites*, Pergamon, Oxford (1980).
Richardson, M.O.W. *Polymer Engineering Composites*, Applied Science London (1977).
Agarwal, B. and Broutman, L.J. *Analysis and Performance of Fibre Composites*, Wiley Interscience, New York (1980).

Questions

3.1 Compare the energy absorption capabilities of composites produced using carbon fibres, aramid fibres and glass fibres. Comment on the meaning of the answer. The data in Fig. 3.2 may be used.

3.2 A hybrid composite material is made up of 20% HS carbon fibres by weight and 30% E-glass fibres by weight in an epoxy matrix. If the density of the epoxy is 1300 kg/m^3 and the data in Fig. 3.2 may be used for the fibres, calculate the density of the composite.

3.3 What weight of carbon fibres (density = 1800 kg/m^3) must be added to 1 kg of epoxy (density = 1250 kg/m^3) to produce a composite with a density of 1600 kg/m^3.

3.4 A unidirectional glass fibre/epoxy composite has a fibre volume fraction of 60%. Given the data below, calculate the density, modulus and thermal conductivity of the composite in the fibre direction.

	(kg/m^3)	E (GN/m^2)	K (W/m K)
Epoxy	1250	6.1	0.25
Glass fibre	2540	80.0	1.05

3.5 In a unidirectional Kevlar/epoxy composite the modular ratio is 20 and the epoxy occupies 60% of the volume. Calculate the modulus of the composite and the stresses in the fibres and the matrix when a stress of 50 MN/m^2 is applied to the composite. The modulus of the epoxy is 6 GN/m^2.

3.6 In a unidirectional carbon fibre/epoxy composite, the modular ratio is 40 and the fibres take up 50% of the cross-section. What percentage of the applied force is taken by the fibres?

3.7 A reinforced plastic sheet is to be made from a matrix with a tensile strength of 60 MN/m^2 and continuous glass fibres with a modulus of 76 GN/m^2. If the resin ratio by volume is 70% and the modular ratio of the composite is 25, estimate the tensile strength and modulus of the composite.

3.8 A single ply unidirectional carbon fibre reinforced PEEK material has a volume fraction of fibres of 0.58. Use the data given below to calculate the Poisson's Ratio for the composite in the fibre and transverse directions.

Material	Modulus (GN/m^2)	Poisson's Ratio
Carbon fibres (HS)	230	0.23
PEEK	3.8	0.35

3.9 A single ply unidirectional glass fibre/epoxy composite has the fibres aligned at 40° to the global x-direction. If the ply is 1.5 mm thick and it is subjected to stresses of $\sigma_x = 30$ MN/m^2 and $\sigma_y = 15$ MN/m^2, calculate the effective moduli for the ply in the x–y directions and the values of ε_x and ε_y. The properties of the ply in the fibre and transverse directions are

$$E_1 = 35 \text{ GN/m}^2, \quad E_2 = 8 \text{ GN/m}^2, \quad G_{12} = 4 \text{ GN/m}^2 \quad \text{and} \quad \nu_{12} = 0.26$$

3.10 A single ply uni-directional carbon fibre/epoxy composite has the fibres aligned at 30° to the x-direction. If the ply is 2 mm thick and it is subjected to a moment of $M_x = 180$ Nm/m and to an axial stress of $\sigma_x = 80$ MN/m^2, calculate the moduli, strains and curvatures in the x–y directions. If an additional moment of $M_y = 250$ Nm/m is added, calculate the new curvatures.

3.11 State whether the following laminates are symmetric or non-symmetric.

(i) $[0/90/45/45/90/0]_T$

(ii) $[0/90_2/45/-45/45/90_2/0]_T$

(iii) $[0/90/45/-45_6/45/90/0]_T$

3.12 A plastic composite is made up of three layers of isotropic materials as follows:

Skin layers: Material A, $E = 3.5$ GN/m^2, $G = 1.25$ GN/m^2, $v = 0.4$
Core layer: Material B, $E = 0.6$ GN/m^2, $G = 0.222$ GN/m^2, $v = 0.35$.

The skins are each 0.5 mm thick and the core is 0.4 mm thick. If an axial stress of 20 MN/m^2, a transverse stress of 30 MN/m^2 and a shear stress of 15 MN/m^2 are applied to the composite, calculate the axial and transverse stresses and strains in each layer.

3.13 A laminate is made up of plies having the following elastic constants

$$E_1 = 133 \text{ GN/m}^2, \quad E_2 = 9 \text{ GN/m}^2, \quad G_{12} = 5 \text{ GN/m}^2, \quad v_{12} = 0.31$$

If the laminate is based on $(-60, -30, 0, 30, 60, 90)_s$ and the plies are each 0.2 mm thick, calculate E_x, E_y, G_{xy} and v_{xy}. If a stress of 100 MN/m^2 is applied in the x direction, what will be the axial and lateral strains in the laminate?

3.14 A unidirectional carbon fibre/epoxy composite has the following lay-up

$$[40, -40, 40, -40]_s$$

The laminate is 8 mm thick and is subjected to stresses of $\sigma_x = 80$ MN/m^2 and $\sigma_y = 40$ MN/m^2, determine the strains in the x and y directions. The properties of a single ply are

$$E_1 = 140 \text{ GN/m}^2, \quad E_2 = 9 \text{ GN/m}^2, \quad G_{12} = 7 \text{ GN/m}^2 \quad \text{and} \quad v_{12} = 0.3$$

3.15 A filament wound composite cylindrical pressure vessel has a diameter of 1200 mm and a wall thickness of 3 mm. It is made up of 10 plies of continuous glass fibres in a polyester resin. The arrangement of the plies is $[0_3/60/-60]_s$. Calculate the axial and hoop strain in the cylinder when an internal pressure of 3 MN/m^2 is applied. The properties of the individual plies are

$$E_1 = 32 \text{ GN/m}^2, \quad E_2 = 8 \text{ GN/m}^2, \quad G_{12} = 3 \text{ GN/m}^2, \quad v_{12} = 0.3$$

3.16 Compare the response of the following laminates to an axial force N_x of 40 N/mm. The plies are each 0.1 mm thick.

(a) $[30/-30/-30/30]_T$

(b) $[30/-30/30/-30]_T$

The properties of the individual plies are

$$E_1 = 145 \text{ GN/m}^2, \quad E_2 = 15 \text{ GN/m}^2, \quad G_{12} = 4 \text{ GN/m}^2, \quad v_{12} = 0.278$$

3.17 A unidirectional carbon fibre/PEEK laminate has the stacking sequence $[0/35/-35]_T$. If the properties of the individual plies are

$$E_1 = 145 \text{ GN/m}^2, \quad E_2 = 15 \text{ GN/m}^2, \quad G_{12} = 4 \text{ GN/m}^2, \quad v_{12} = 0.278$$

If the plies are each 0.1 mm thick, calculate the strains and curvatures if an in-plane stress of 100 MN/m^2 is applied.

3.18 A single ply of carbon/epoxy composite has the properties listed below and the fibres are aligned at 25° to the x-direction. If stresses of

$$\sigma_x = 80 \text{ MN/m}^2, \quad \sigma_y = 20 \text{ MN/m}^2 \quad \text{and} \quad \tau_{xy} = -10 \text{ MN/m}^2$$

are applied, determine whether or not failure would be expected to occur according to the (a) Maximum Stress, (b) Maximum Strain and (c) Tsai–Hill criteria.

$$E_1 = 497 \text{ GN/m}^2, \quad E_2 = 5.3 \text{ GN/m}^2, \quad G_{12} = 5.6 \text{ GN/m}^2, \quad \nu_{12} = 0.31$$

$$\hat{\sigma}_{1T} = 1.24 \text{ GN/m}^2, \quad \hat{\sigma}_{2T} = 0.02 \text{ GN/m}^2, \quad \hat{\sigma}_{1C} = 0.029 \text{ GN/m}^2$$

$$\hat{\sigma}_{2C} = 0.03 \text{ GN/m}^2, \quad \hat{\tau}_{12} = 0.04 \text{ GN/m}^2$$

3.19 A single ply glass/epoxy composite has the properties listed below. If the fibres are aligned at $30°$ to the x-direction, determine the value of in-plane stresses, σ_x, which would cause failure according to (a) the Maximum Stress criterion (b) the Maximum Strain criterion and (c) the Tsai–Hill criterion.

$$E_1 = 40,000 \text{ MN/m}^2, \quad E_2 = 9000 \text{ MN/m}^2, \quad G_{12} = 4000 \text{ MN/m}^2,$$

$$\nu_{12} = 0.31, \quad \hat{\sigma}_{1T} = 1400 \text{ MN/m}^2, \quad \hat{\sigma}_{2T} = 50 \text{ MN/m}^2$$

$$\hat{\sigma}_{1C} = 220 \text{ MN/m}^2, \quad \hat{\sigma}_{2C} = 100 \text{ MN/m}^2, \quad \hat{\tau}_{12} = 40 \text{ MN/m}^2$$

3.20 A carbon/epoxy composite with the stacking arrangement $[0/-30/30]_s$ has the properties listed below. Determine the value of in-plane stress σ_x which would cause failure according to the (a) Maximum Strain (b) Maximum Stress and (c) Tsai–Hill criteria.

$$E_1 = 130 \text{ GN/m}^2, \quad E_2 = 8.8 \text{ GN/m}^2, \quad G_{12} = 6.9 \text{ GN/m}^2, \quad \nu_{12} = 0.3$$

$$\hat{\sigma}_{1T} = 1.9 \text{ GN/m}^2, \quad \hat{\sigma}_{2T} = 0.038 \text{ GN/m}^2, \quad \hat{\sigma}_{1C} = 1.7 \text{ GN/m}^2$$

$$\hat{\sigma}_x = 0.2 \text{ GN/m}^2, \quad \hat{\tau}_{12} = 0.09 \text{ GN/m}^2$$

The thickness of each ply is 0.2 mm.

3.21 A long sheet of carbon fibre reinforced epoxy is 30 mm wide, 5 mm thick and has a sharp crack 10 mm long placed centrally relative to its width and length. If the critical stress intensity factor for this material is 43 MN m$^{-3/2}$, calculate the axial tension which would cause the sheet to fracture.

3.22 A very wide sheet of grp which is known to contain intrinsic defects 1 mm long, is subjected to a fluctuating stress which varies from 0 to 80 MN/m^2. How many cycles would the sheet be expected to withstand if it is made from (a) chopped strand mat (CSM) and (b) woven roving (WR) reinforcement. The crack growth parameters C and n, and the critical stress intensity factors, K_c, for these materials are

	C	n	$K_c(\text{MN m}^{1/2})$
CSM	3.3×10^{-18}	12.7	13.5
WR	2.7×10^{-14}	6.4	26.5

3.23 If the matrix in 3.7 was reinforced with the same volume fraction of glass but in the form of randomly oriented glass fibres rather than continuous filaments, what would be the tensile strength of the composite. The fibres are 15 mm long, have an aspect ratio of 1000 and produce a reinforcement efficiency of 0.25. The fibre strength is 2 GN/m^2 and the shear strength of the interface is 4 MN/m^2.

3.24 A polyester matrix is reinforced with continuous glass fibres. A 15 mm wide beam made from this material is to be simply supported over a 300 mm length and have a point load at mid-span. For a fixed beam weight of 90 g/m investigate how the stiffness of the beam changes with the volume fraction of glass and state the optimum volume fraction. ($\rho_f = 2560 \text{ kg/m}^3$, $\rho_m = 1210 \text{ kg/m}^3$, $E_f = 76 \text{ GN/m}^2$, $E_m = 3 \text{ GN/m}^2$).

3.25 A sheet of chopped strand mat-reinforced polyester is 5 mm thick and 10 mm wide. If its modulus is 8 GN/m^2 calculate its flexural stiffness when subjected to a point load of 200 N midway along a simply supported span of 300 mm. Compare this with the stiffness of a composite beam made up of two 2.5 mm thick layers of this reinforced material separated by a 10 mm thick core of foamed plastic with a modulus of 40 MN/m^2.

3.26 A composite of gfrp skin and foamed core is to have a fixed weight of 200 g/m. If its width is 15 mm investigate how the stiffness of the composite varies with skin thickness. The density of the skin material is 1450 kg/m^3 and the density of the core material is 450 kg/m^3. State the value of skin thickness which would be best and for this thickness calculate the ratio of the weight of the skin to the total composite weight.

3.27 In a short carbon fibre reinforced nylon moulding the volume fraction of the fibres is 0.2. Assuming the fibre length is much greater that the critical fibre length, calculate the modulus of the moulding. The modulus values for the fibres and nylon are 230 GN/m^2 and 2.8 GN/m^2 respectively.

CHAPTER 4 – Processing of Plastics

4.1 Introduction

One of the most outstanding features of plastics is the ease with which they can be processed. In some cases semi-finished articles such as sheets or rods are produced and subsequently fabricated into shape using conventional methods such as welding or machining. In the majority of cases, however, the finished article, which may be quite complex in shape, is produced in a single operation. The processing stages of heating, shaping and cooling may be continuous (eg production of pipe by extrusion) or a repeated cycle of events (eg production of a telephone housing by injection moulding) but in most cases the processes may be automated and so are particularly suitable for mass production. There is a wide range of processing methods which may be used for plastics. In most cases the choice of method is based on the shape of the component and whether it is thermoplastic or thermosetting. It is important therefore that throughout the design process, the designer must have a basic understanding of the range of processing methods for plastics since an ill-conceived shape or design detail may limit the choice of moulding methods.

In this chapter each of the principal processing methods for plastics is described and where appropriate a Newtonian analysis of the process is developed. Although most polymer melt flows are in fact Non-Newtonian, the simplified analysis is useful at this stage because it illustrates the approach to the problem without concealing it by mathematical complexity. In practice the simplified analysis may provide sufficient accuracy for the engineer to make initial design decisions and at least it provides a quantitative aspect which assists in the understanding of the process. For those requiring more accurate models of plastics moulding, these are developed in Chapter 5 where the Non-Newtonian aspects of polymer melt flow are considered.

245

4.2 Extrusion

4.2.1 General Features of Single Screw Extrusion

One of the most common methods of processing plastics is **Extrusion** using a screw inside a barrel as illustrated in Fig. 4.1. The plastic, usually in the form of granules or powder, is fed from a hopper on to the screw. It is then conveyed along the barrel where it is heated by conduction from the barrel heaters and shear due to its movement along the screw flights. The depth of the screw channel is reduced along the length of the screw so as to compact the material. At the end of the extruder the melt passes through a die to produce an extrudate of the desired shape. As will be seen later, the use of different dies means that the extruder screw/barrel can be used as the basic unit of several processing techniques.

Fig. 4.1   Schematic view of single screw extruder

Basically an extruder screw has three different zones.

(a) Feed Zone The function of this zone is to preheat the plastic and convey it to the subsequent zones. The design of this section is important since the constant screw depth must supply sufficient material to the metering zone so as not to starve it, but on the other hand not supply so much material that the metering zone is overrun. The optimum design is related to the nature and shape of the feedstock, the geometry of the screw and the frictional properties of the screw and barrel in relation to the plastic. The frictional behaviour of the feed-stock material has a considerable influence on the rate of melting which can be achieved.

(b) Compression Zone In this zone the screw depth gradually decreases so as to compact the plastic. This compaction has the dual role of squeezing any trapped air pockets back into the feed zone and improving the heat transfer through the reduced thickness of material.

(c) **Metering Zone** In this section the screw depth is again constant but much less than the feed zone. In the metering zone the melt is homogenised so as to supply at a constant rate, material of uniform temperature and pressure to the die. This zone is the most straight-forward to analyse since it involves a viscous melt flowing along a uniform channel.

The pressure build-up which occurs along a screw is illustrated in Fig. 4.2. The lengths of the zones on a particular screw depend on the material to be extruded. With nylon, for example, melting takes place quickly so that the compression of the melt can be performed in one pitch of the screw. PVC on the other hand is very heat sensitive and so a compression zone which covers the whole length of the screw is preferred.

Fig. 4.2   Typical zones on a extruder screw

As plastics can have quite different viscosities, they will tend to behave differently during extrusion. Fig. 4.3 shows some typical outputs possible with different plastics in extruders with a variety of barrel diameters. This diagram is to provide a general idea of the ranking of materials – actual outputs may vary ±25% from those shown, depending on temperatures, screw speeds, etc.

Fig. 4.3   Typical extruder outputs for different plastics

In commercial extruders, additional zones may be included to improve the quality of the output. For example there may be a mixing zone consisting of screw flights of reduced or reversed pitch. The purpose of this zone is to ensure uniformity of the melt and it is sited in the metering section. Fig. 4.4 shows some designs of mixing sections in extruder screws.

Fig. 4.4   Typical designs of mixing zones

Some extruders also have a venting zone. This is principally because a number of plastics are hygroscopic – they absorb moisture from the atmosphere. If these materials are extruded wet in conventional equipment the quality of the output is not good due to trapped water vapour in the melt. One possibility is to pre-dry the feedstock to the extruder but this is expensive and can lead to contamination. Vented barrels were developed to overcome these problems. As shown in Fig. 4.5, in the first part of the screw the granules are taken in and melted, compressed and homogenised in the usual way. The melt pressure is then reduced to atmospheric pressure in the decompression zone. This allows the volatiles to escape from the melt through a special port in the barrel. The melt is then conveyed along the barrel to a second compression zone which prevents air pockets from being trapped.

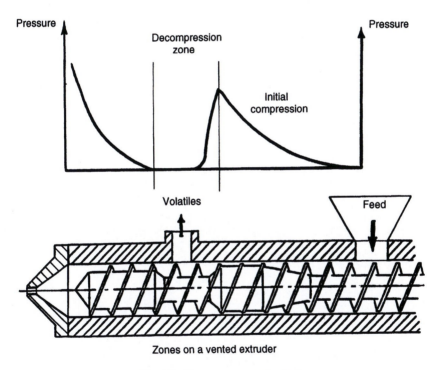

Zones on a vented extruder

Fig. 4.5   Zones on a vented extruder

The venting works because at a typical extrusion temperature of 250°C the water in the plastic exists as a vapour at a pressure of about 4 MN/m^2. At this pressure it will easily pass out of the melt and through the exit orifice. Note that since atmospheric pressure is about 0.1 MN/m^2 the application of a vacuum to the exit orifice will have little effect on the removal of volatiles.

Another feature of an extruder is the presence of a gauze filter after the screw and before the die. This effectively filters out any inhomogeneous material which might otherwise clog the die. These *screen packs* as they are called, will normally filter the melt to $120-150$ μm. However, there is conclusive evidence to show that even smaller particles than this can initiate cracks in plastics extrudates e.g. polyethylene pressure pipes. In such cases it has been found that fine melt filtration ($\simeq 45$ μm) can significantly improve the performance of the extrudate.

Since the filters by their nature tend to be flimsy they are usually supported by a breaker plate. As shown in Fig. 4.6 this consists of a large number of countersunk holes to allow passage of the melt whilst preventing dead spots where particles of melt could gather. The breaker plate also conveniently straightens out the spiralling melt flow which emerges from the screw. Since the fine mesh on the filter will gradually become blocked it is periodically removed and replaced. In many modern extruders, and particularly with the fine filter systems referred to above, the filter is changed automatically so as not to interrupt continuous extrusion.

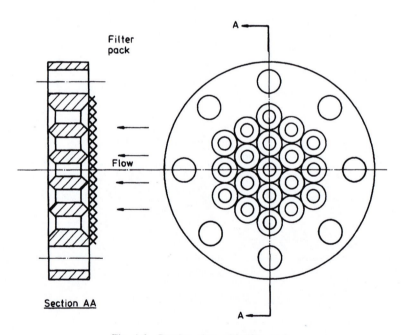

Fig. 4.6   Breaker plate with filter pack

It should also be noted that although it is not their primary function, the breaker plate and filter also assist the build-up of back pressure which improves mixing along the screw. Since the pressure at the die is important, extruders also have a valve after the breaker plate to provide the necessary control.

4.2.2 Mechanism of Flow

As the plastic moves along the screw, it melts by the following mechanism. Initially a thin film of molten material is formed at the barrel wall. As the screw rotates, it scrapes this film off and the molten plastic moves down the front face of the screw flight. When it reaches the core of the screw it sweeps up again, setting up a rotary movement in front of the leading edge of the screw flight. Initially the screw flight contains solid granules but these tend to be swept into the molten pool by the rotary movement. As the screw rotates, the material passes further along the barrel and more and more solid material is swept into the molten pool until eventually only melted material exists between the screw flights.

As the screw rotates inside the barrel, the movement of the plastic along the screw is dependent on whether or not it adheres to the screw and barrel. In theory there are two extremes. In one case the material sticks to the screw only and therefore the screw and material rotate as a solid cylinder inside the barrel. This would result in zero output and is clearly undesirable. In the second case the material slips on the screw and has a high resistance to rotation inside the barrel. This results in a purely axial movement of the melt and is the ideal situation. In practice the behaviour is somewhere between these limits as the material adheres to both the screw and the barrel. The useful output from the extruder is the result of a drag flow due to the interaction of the rotating screw and stationary barrel. This is equivalent to the flow of a viscous liquid between two parallel plates when one plate is stationary and the other is moving. Superimposed on this is a flow due to the pressure gradient which is built up along the screw. Since the high pressure is at the end of the extruder the pressure flow will reduce the output. In addition, the clearance between the screw flights and the barrel allows material to leak back along the screw and effectively reduces the output. This leakage will be worse when the screw becomes worn.

The external heating and cooling on the extruder also plays an important part in the melting process. In high output extruders the material passes along the barrel so quickly that sufficient heat for melting is generated by the shearing action and the barrel heaters are not required. In these circumstances it is the barrel cooling which is critical if excess heat is generated in the melt. In some cases the screw may also be cooled. This is not intended to influence the melt temperature but rather to reduce the frictional effect between the plastic and the screw. In all extruders, barrel cooling is essential at the feed pocket to ensure an unrestricted supply of feedstock.

The thermal state of the melt in the extruder is frequently compared with two ideal thermodynamic states. One is where the process may be regarded as

adiabatic. This means that the system is fully insulated to prevent heat gain or loss from or to the surroundings. If this ideal state was to be reached in the extruder it would be necessary for the work done on the melt to produce just the right amount of heat without the need for heating or cooling. The second ideal case is referred to as **isothermal.** In the extruder this would mean that the temperature at all points is the same and would require immediate heating or cooling from the barrel to compensate for any loss or gain of heat in the melt. In practice the thermal processes in the extruder fall somewhere between these ideals. Extruders may be run without external heating or cooling but they are not truly adiabatic since heat losses will occur. Isothermal operation along the whole length of the extruder cannot be envisaged if it is to be supplied with relatively cold granules. However, particular sections may be near isothermal and the metering zone is often considered as such for analysis.

4.2.3 Analysis of Flow in Extruder

As discussed in the previous section, it is convenient to consider the output from the extruder as consisting of three components – drag flow, pressure flow and leakage. The derivation of the equation for output assumes that in the metering zone the melt has a constant viscosity and its flow is isothermal in a wide shallow channel. These conditions are most likely to be approached in the metering zone.

   (a) Drag Flow Consider the flow of the melt between parallel plates as shown in Fig. 4.7(a).

   For the small element of fluid ABCD the volume flow rate dQ is given by

$$dQ = V \cdot dy \cdot dx \qquad (4.1)$$

Assuming the velocity gradient is linear, then

$$V = V_d \left[\frac{y}{H} \right]$$

Substituting in (4.1) and integrating over the channel depth, H, then the total drag flow, Q_d, is given by

$$Q_d = \int_0^H \int_0^T \frac{V_d y}{H} \cdot dy \cdot dx$$

$$Q_d = \tfrac{1}{2} T H V_d \qquad (4.2)$$

This may be compared to the situation in the extruder where the fluid is being dragged along by the relative movement of the screw and barrel. Fig. 4.8 shows the position of the element of fluid and (4.2) may be modified to include terms relevant to the extruder dimensions.

For example $V_d = \pi D N \cos \phi$

(a) Drag Flow

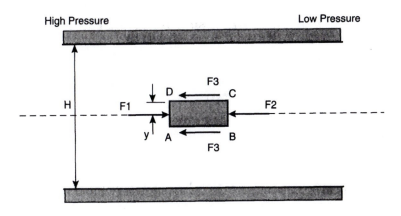

(b) Pressure Flow

In both cases, $AB = dz$, element width = dx and channel width = T

Fig. 4.7   Melt Flow between parallel plates

where N is the screw speed (in revolutions per unit time).

$$T = (\pi D \tan \phi - e) \cos \phi$$

So $$Q_d = \tfrac{1}{2}(\pi D \tan \phi - e)(\pi D N \cos^2 \phi)H$$

In most cases the term, e, is small in comparison with $(\pi D \tan \phi)$ so this expression is reduced to

$$Q_d = \tfrac{1}{2}\pi^2 D^2 N H \sin \phi \cos \phi \tag{4.3}$$

Note that the shear rate in the metering zone will be given by V_d/H.

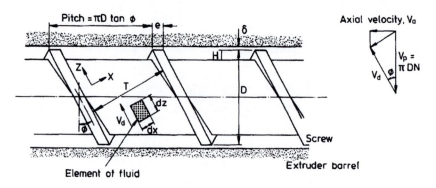

Fig. 4.8   Details of extruder screw

(b) **Pressure flow:** Consider the element of fluid shown in Fig. 4.7(b). The forces are

$$F_1 = \left(P + \frac{\partial P}{\partial z} \cdot dz\right) dy\, dx$$

$$F_2 = P \cdot dy\, dx$$

$$F_3 = \tau_y\, dz\, dx$$

where P is pressure and $d\tau$ is the shear stress acting on the element. For steady flow these forces are in equilibrium so they may be equated as follows:

$$F_1 = F_2 + 2F_3$$

which reduces to

$$y\frac{dP}{dz} = \tau_y \tag{4.4}$$

Now for a Newtonian fluid, the shear stress, τ_y, is related to the viscosity, η, and the shear rate, $\dot\gamma$, by the equation

$$\tau_y = \eta\dot\gamma = \eta\frac{dV}{dy}$$

Using this in equation (4.4)

$$y\frac{dP}{dz} = \eta\frac{dV}{dy}$$

Integrating

$$\int_0^v dV = \frac{1}{\eta}\frac{dP}{dz}\int_{H/2}^y y\, dy$$

So

$$V = \frac{1}{\eta} \frac{dp}{dz} \left(\frac{y^2}{2} - \frac{H^2}{8} \right) \tag{4.5}$$

Also, for the element of fluid of depth, dy, at distance, y, from the centre line (and whose velocity is V) the elemental flow rate, dQ, is given by

$$dQ = VT\,dy$$

This may be integrated to give the pressure flow, Q_p

$$Q_p = 2 \int_0^{H/2} \frac{1}{\eta} \frac{dP}{dz} \cdot T \left(\frac{y^2}{2} - \frac{H^2}{8} \right) dy$$

$$Q_p = -\frac{1}{12\eta} \frac{dP}{dz} \cdot TH^3 \tag{4.6}$$

Referring to the element of fluid between the screw flights as shown in Fig. 4.8, this equation may be rearranged using the following substitutions. Assuming e is small, $T = \pi D \tan \phi \cdot \cos \phi$

Also, $\sin \phi = \dfrac{dL}{dz}$ so $\dfrac{dP}{dz} = \dfrac{dP}{dL} \sin \phi$

Thus the expression for Q_p becomes

$$Q_p = -\frac{\pi D H^3 \sin^2 \phi}{12\eta} \cdot \frac{dP}{dL} \tag{4.7}$$

(c) **Leakage** The leakage flow may be considered as flow through a wide slit which has a depth, δ, a length ($e \cos \phi$) and a width of ($\pi D / \cos \phi$). Since this is a pressure flow, the derivation is similar to that described in (b). For convenience therefore the following substitutions may be made in (4.6).

$$h = \delta$$

$$T = \pi D / \cos \phi$$

$$\text{Pressure gradient} = \frac{\Delta P}{e \cos \phi} \text{ (see Fig. 4.9)}$$

So the leakage flow, Q_L, is given by

$$Q_L = \frac{\pi^2 D^2 \delta^3}{12\eta e} \tan \phi \frac{dP}{dL} \tag{4.8}$$

A factor is often required in this equation to allow for eccentricity of the screw in the barrel. Typically this increases the leakage flow by about 20%.

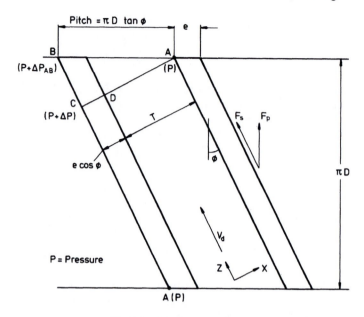

Fig. 4.9   Development of screw

The total output is the combination of drag flow, back pressure flow and leakage. So from (4.3), (4.7) and (4.8)

$$Q = \tfrac{1}{2}\pi^2 D^2 \, \text{NH} \, \sin\phi \cos\phi - \frac{\pi D H^3 \sin^2\phi}{12\eta}\frac{dP}{dL} - \frac{\pi^2 D^2 \delta^3}{12\eta e}\tan\phi\frac{dP}{dL} \qquad (4.9)$$

For many practical purposes sufficient accuracy is obtained by neglecting the leakage flow term. In addition the pressure gradient is often considered as linear so

$$\frac{dP}{dL} = \frac{P}{L}$$

where 'L' is the length of the extruder. In practice the length of an extruder screw can vary between 17 and 30 times the diameter of the barrel. The shorter the screw the cooler the melt and the faster the moulding cycle. In the above analysis, it is the melt flow which is being considered and so the relevant pressure gradient will be that in the metering zone. However, as shown in Fig. 4.2 this is often approximated by P/L. If all other physical dimensions and conditions are constant then the variation of output with screw flight angle, ϕ, can be studied. As shown in Fig. 4.10 the maximum output would be obtained if the screw flight angle was about 35°C. In practice a screw flight angle of 17.7° is frequently used because

Fig. 4.10 Variation of drag flow and pressure flow

(i) this is the angle which occurs if the pitch of the screw is equal to the diameter and so it is convenient to manufacture,

(ii) for a considerable portion of the extruder length, the screw is acting as a solids conveying device and it is known that the optimum angle in such cases is 17° to 20°.

It should also be noted that in some cases correction factors, F_d, and F_p are applied to the drag and pressure flow terms. They are to allow for edge effects and are solely dependent on the channel width, T, and channel depth, h, in the metering zone. Typical values are illustrated in Fig. 4.11.

4.2.4 Extruder/Die Characteristics

From equation (4.9) it may be seen that there are two interesting situations to consider. One is the case of free discharge where there is no pressure build up at the end of the extruder so

$$Q = Q_{max} = \tfrac{1}{2}\pi^2 D^2 N H \sin\phi\cos\phi \tag{4.10}$$

The other case is where the pressure at the end of the extruder is large enough to stop the output. From (4.9) with $Q = 0$ and ignoring the leakage flow

$$P = P_{max} = \frac{6\pi D L N \eta}{H^2 \tan\phi} \tag{4.11}$$

In Fig. 4.12 these points are shown as the limits of the screw characteristic. It is interesting to note that when a die is coupled to the extruder their requirements

Fig. 4.11   Flow correction factors as a function of screw geometry

are conflicting. The extruder has a high output if the pressure at its outlet is low. However, the outlet from the extruder is the inlet to the die and the output of the latter increases with inlet pressure. As will be seen later the output, Q, of a Newtonian fluid from a die is given by a relation of the form

$$Q = KP \qquad (4.12)$$

where $K = \dfrac{\pi R^4}{8\eta L_d}$ for a capillary die of radius R and length L_d.

Equation (4.12) enables the die characteristics to be plotted on Fig. 4.12 and the intersection of the two characteristics is the operating point of the extruder. This plot is useful in that it shows the effect which changes in various parameters will have on output. For example, increasing screw speed, N, will move the extruder characteristic upward. Similarly an increase in the die radius, R, would increase the slope of the die characteristic and in both cases the extruder output would increase.

The operating point for an extruder/die combination may also be determined from equations (4.9) and (4.12) – ignoring leakage flow

$$Q = \tfrac{1}{2}\pi^2 D^2 N H \sin\phi\cos\phi - \frac{\pi D H^3 \sin^2\phi}{12\eta}\frac{P}{L} = \frac{\pi R^4}{8\eta L_d}\cdot P$$

So for a capillary die, the pressure at the operating point is given by

$$P_{\mathrm{OP}} = \left\{ \frac{2\pi\eta D^2 N H \sin\phi\cos\phi}{(R^4/2L_d) + (DH^3 \sin^2\phi)/3L} \right\} \qquad (4.13)$$

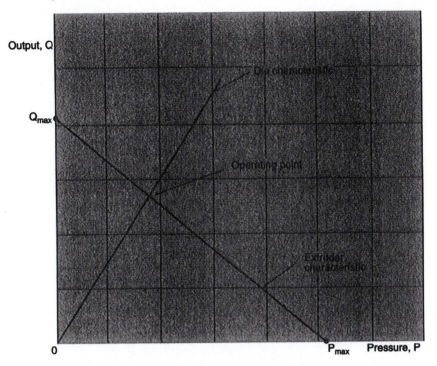

Fig. 4.12   Extruder and die characteristics

4.2.5 Other Die Geometries

For other die geometries it is necessary to use the appropriate form of equation
(4.12). The equations for a capillary and a slit die are derived in Chapter 5.
For other geometries it is possible to use the empirical equation which was
developed by Boussinesq. This has the form

$$Q = \frac{Fbd^3}{12\eta L_d} \cdot P \qquad (4.14)$$

where b is the greatest dimension of the cross-section

d is the least dimension of the cross-section

F is a non-dimensional factor as given in Fig. 4.13.

Using equation (4.14) it is possible to modify the expression for the operating
pressure to the more general form

$$P_{OP} = \left\{ \frac{2\pi\eta D^2 NH \sin\phi \cos\phi}{\dfrac{Fbd^3}{3\pi L_d} + (DH^3 \sin^2\phi/3L)} \right\} \qquad (4.15)$$

Fig. 4.13   Flow coefficient as a function of channel geometry

For a capillary die, one may obtain a value of F from Fig. 4.13 as 0.295 and substituting $b = d = 2R$, this equation reduces to the same form as equation (4.13).

Example 4.1 A single screw extruder is to be designed with the following characteristics.

L/D ratio $= 24$, screw flight angle $= 17.7°$
Max. screw speed $= 100$ rev/min, screw diameter $= 40$ mm
flight depth (metering zone) $= 3$ mm.

If the extruder is to be used to process polymer melts with a maximum melt viscosity of 500 Ns/m^2, calculate a suitable wall thickness for the extruder barrel based on the von Mises yield criterion. The tensile yield stress for the barrel metal is 925 MN/m^2 and a factor of safety of 2.5 should be used.

Solution The maximum pressure which occurs in the extruder barrel is when there is no output. Therefore the design needs to consider this worst case blockage situation. As given by equation (4.11)

$$P_{max} = \frac{6\pi DL\eta}{H^2 \tan \phi}$$

$$= \frac{6\pi \times 40 \times (24 \times 10) \times (100/60) \times 500}{(3)^2 \tan 17.7^0} = 210 \text{ MN/m}^2$$

The von Mises criterion relates the tensile yield stress of a material to a state of multi-axial stress in a component made from the material. In a cylinder (the barrel of the extruder in this case), the principal stresses which exist as a result of an internal pressure are

$$\text{hoop stress, } \sigma_1 = \frac{P_{max}D}{2h}$$

$$\text{axial stress, } \sigma_2 = \frac{P_{max}D}{4h}$$

where h = wall thickness of the barrel.

The von Mises criterion simply states that yielding (failure) will occur if

$$\left(\frac{\sigma_Y}{FS}\right)^2 \leq \sigma_1^2 + \sigma_2^2 - \sigma_1\sigma_2$$

where σ_Y = tensile yield stress of material

FS = factor of safety.

In this case, therefore

$$\left(\frac{925}{2.5}\right)^2 = \left(\frac{(210)40}{2h}\right)^2 + \left(\frac{(210)40}{4h}\right)^2 - \frac{(210)^2(40)^2}{8h^2}$$

$$h = 9.8 \text{ mm}$$

Hence a barrel wall thickness of 10 mm would be appropriate.

Example 4.2 A single screw extruder is to be used to manufacture a nylon rod 5 mm in diameter at a production rate of 1.5 m/min. Using the following information, calculate the required screw speed.

Nylon	Extruder	Die
Viscosity = 42.0 Ns/m^2	Diameter = 30 mm	Length = 4 mm
Density (solid) = 1140 kg/m^3	Length = 750 mm	Diameter = 5 mm
Density (melt) = 790 kg/m^3	Screw flight angle = 17.7°	
	Metering channel	
	depth = 2.5 mm	

Die swelling effects may be ignored and the melt viscosity can be assumed to be constant.

Solution The output rate of solid rod = speed × cross-sectional area
$$= 1.5 \times \pi(2.5 \times 10^{-3})^2/60$$
$$= 49.1 \times 10^{-6} \text{ m}^3/\text{s}$$

As the solid material is more dense than the melt, the melt flow rate must be greater in the ratio of the solid/melt densities. Therefore

$$\text{Melt flow rate through die} = 49.1 \times 10^{-6} \left(\frac{1140}{790}\right) = 70.8 \times 10^{-6} \text{ m}^3/\text{s}$$

The pressure necessary to achieve this flow rate through the die is obtained from

$$Q = \frac{\pi P R^4}{8\eta L_d}$$

$$P = \frac{8 \times 420 \times 4 \times 10^{-3} \times 70.8 \times 10^{-6}}{\pi (2.5 \times 10^{-3})^4} = 7.8 \text{ MN/m}^2$$

At the operating point, the die output and the extruder output will be the same. Hence

$$Q = 70.8 \times 10^{-6} = \tfrac{1}{2}\pi^2 (30 \times 10^{-3})^2 N (2.5 \times 10^{-3}) \sin 17.7 \cos 17.7$$

$$- \frac{\pi (30 \times 10^{-3})(2.6 \times 10^{-3})^3 \sin 17}{12 \times 420} \left(\frac{7.8 \times 10^6}{0.75} \right)$$

$$N = 22 \text{ rev/min}$$

4.2.6 General Features of Twin Screw Extruders

In recent years there has been a steady increase in the use of extruders which have two screws rotating in a heated barrel. These machines permit a wider range of possibilities in terms of output rates, mixing efficiency, heat generation, etc compared with a single screw extruder. The output of a twin screw extruder can be typically three times that of a single screw extruder of the same diameter and speed. Although the term 'twin-screw' is used almost universally for extruders having two screws, the screws need not be identical. There are in fact a large variety of machine types. Fig. 4.14 illustrates some of the possibilities with counter-rotating and co-rotating screws. In addition the screws may be conjugated or non-conjugated. A non-conjugated screw configuration is one in which the screw flights are a loose fit into one another so that there is ample space for material between the screw flights (see Fig. 4.15).

(a) Counter-rotating   (b) Co-rotating   (c) Counter-rotating   (d) Co-rotating
    (intermeshing)   (intermeshing)   (non-intermeshing)   (non-intermeshing)

Fig. 4.14   Different types of twin screw extruder

(a) Non-conjugated screws showing some
passages around each screw

(b) Conjugated screws showing closed
passages around each screw

Fig. 4.15   Two types of twin screw extruder

In a counter-rotating twin screw extruder the material is sheared and pressurised in a mechanism similar to calendering (see Section 4.5), i.e. the material is effectively squeezed between counter-rotating rolls. In a co-rotating system the material is transferred from one screw to the other in a figure-of-eight pattern as shown in Fig. 4.16. This type of arrangement is particularly suitable for heat sensitive materials because the material is conveyed through the extruder quickly with little possibility of entrapment. The movement around the screws is slower if the screws are conjugated but the propulsive action is greater.

Table 4.1

Comparison of single-screw, co-rotating and counter-rotating twin-screw extruders

| Type | Single screw | Co-rotating screw | | Counter-rotating twin screw |
		Low speed type	High speed type	
Principle	Friction between cylinder and materials and the same between material and screw	Mainly depend on the frictional action as in the case of single screw extruder		Forced mechanical conveyance based on gear pump principle
Conveying efficiency	Low	Medium		High
Mixing efficiency	Low	Medium/High		High
Shearing action	High	Medium	High	Low
Self-cleaning effect	Slight	Medium/High	High	Low
Energy efficiency	Low	Medium/High		High
Heat generation	High	Medium	High	Low
Temp distribution	Wide	Medium	Narrow	Narrow
Max. revolving speed (rpm)	100–300	25–35	250–300	35–45
Max. effective length of screw L/D	30–32	7–18	30–40	10–21

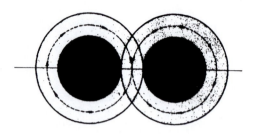

Fig. 4.16   Material flow path with co-rotating scres

The following table compares the single screw extruder with the main types of twin screw extruders.

4.2.7 Processing Methods Based on the Extruder

Extrusion is an extremely versatile process in that it can be adapted, by the use of appropriate dies, to produce a wide range of products. Some of the more common of these production techniques will now be described.

(a) Granule Production/Compounding

In the simplest case an extruder may be used to convert polymer formulations and additives into a form (usually granules) which is more convenient for use in other processing methods, such as injection moulding. In the extruder the feedstock is melted, homogenised and forced through a capillary shaped die. It emerges as a continuous lace which is cooled in a long water bath so that it may be chopped into short granules and packed into sacks. The haul-off apparatus shown in Fig. 4.17 is used to draw down the extrudate to the required dimensions. The granules are typically 3 mm diameter and about 4 mm long. In most cases a multi-hole die is used to increase the production rate.

(b) Profile Production

Extrusion, by its nature, is ideally suited to the production of continuous lengths of plastic mouldings with a uniform cross-section. Therefore as well as producing the laces as described in the previous section, the simple operation of a die change can provide a wide range of profiled shapes such as pipes, sheets, rods, curtain track, edging strips, window frames, etc (see Fig. 4.18).

The successful manufacture of profiled sections depends to a very large extent on good die design. Generally this is not straightforward, even for a simple cross-section such as a square, due to the interacting effects of post-extrusion swelling and the flow characteristics of complex viscoelastic fluids. Most dies are designed from experience to give approximately the correct shape and then *sizing* units are used to control precisely the desired shape. The extrudate is then cooled as quickly as possible. This is usually done in a water bath the length of which depends on the section and the material being cooled. For example,

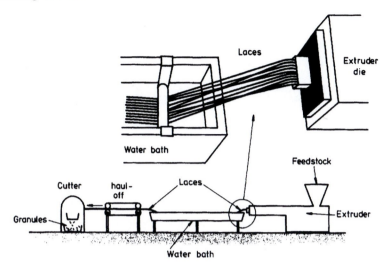

Fig. 4.17   Use of extruder to produce granules

Fig. 4.18   (a) Extruded panel sections (b) Extruded window profile

longer baths are needed for crystalline plastics since the recrystallisation is exothermic.

The storage facilities at the end of the profile production line depend on the type of product (see Fig. 4.19). If it is rigid then the cooled extrudate may be cut to size on a guillotine for stacking. If the extrudate is flexible then it can be stored on drums.

(c) Film Blowing

Although plastic sheet and film may be produced using a slit die, by far the most common method nowadays is the film blowing process illustrated in Fig. 4.20. The molten plastic from the extruder passes through an annular die and emerges as a thin tube. A supply of air to the inside of the tube prevents it from collapsing and indeed may be used to inflate it to a larger diameter.

Fig. 4.19(a)   Sheet extrusion (1) thick sheet (2) thin sheet

Fig. 4.19(b)   Pipe extrusion (1) rigid pipe (2) flexible pipe

Fig. 4.20   Film blowing process

Initially the bubble consists of molten plastic but a jet of air around the outside of the tube promotes cooling and at a certain distance from the die exit, a freeze line can be identified. Eventually the cooled film passes through collapsing guides and nip rolls before being taken off to storage drums or, for example, gussetted and cut to length for plastic bags. Most commercial systems are provided with twin storage facilities so that a full drum may be removed without stopping the process.

The major advantage of film blowing is the ease with which biaxial orientation can be introduced into the film. The pressure of the air in the bubble determines the *blow-up* and this controls the circumferential orientation. In addition, axial orientation may be introduced by increasing the nip roll speed relative to the linear velocity of the bubble. This is referred to as *draw-down*.

It is possible to make a simple estimate of the orientation in blown film by considering only the effects due to the inflation of the bubble. Since the volume flow rate is the same for the plastic in the die and in the bubble, then for unit time

$$\pi D_d h_d L_d = \pi D_b h_b L_b$$

where D, h and L refer to diameter, thickness and length respectively and the subscript 'd' is for the die and 'b' is for the bubble.

So the orientation in the machine direction, O_{MD}, is given by

$$O_{MD} = \frac{L_b}{L_d} = \frac{D_d h_d}{h_b D_b} = \frac{h_d}{h_b B_R}$$

where $B_R =$ blow-up ratio (D_b/D_d)

Also the orientation in the transverse direction, O_{TD}, is given by

$$O_{TD} = \frac{D_b}{D_d} = B_R$$

Therefore the ratio of the orientations may be expressed as

$$\frac{O_{MD}}{O_{TD}} = \frac{h_d}{h_b (B_R)^2} \tag{4.16}$$

Example 4.3 A plastic shrink wrapping with a thickness of 0.05 mm is to be produced using an annular die with a die gap of 0.8 mm. Assuming that the inflation of the bubble dominates the orientation in the film, determine the blow-up ratio required to give uniform biaxial orientation.

Solution Since $O_{MD} = O_{TD}$

then the blow-up ratio, $$B_R = \sqrt{\frac{h_d}{h_b}}$$

$$= \sqrt{\frac{0.8}{0.05}} = 4$$

Common blow-up ratios are in the range 1.5 to 4.5.

This example illustrates the simplified approach to film blowing. Unfortunately in practice the situation is more complex in that the film thickness is influenced by draw-down, relaxation of induced stresses/strains and melt flow phenomena such as die swell. In fact the situation is similar to that described for blow moulding (see below) and the type of analysis outlined in that section could be used to allow for the effects of die swell. However, since the most practical problems in film blowing require iterative type solutions involving melt flow characteristics, volume flow rates, swell ratios, etc the study of these is delayed until Chapter 5 where a more rigorous approach to polymer flow has been adopted.

(d) Blow Moulding

This process evolved originally from glass blowing technology. It was developed as a method for producing hollow plastic articles (such as bottles and barrels) and although this is still the largest application area for the process, nowadays a wide range of technical mouldings can also be made by this method e.g. rear spoilers on cars and videotape cassettes. There is also a number of variations on the original process but we will start by considering the conventional extrusion blow moulding process.

Extrusion Blow Moulding

Initially a molten tube of plastic called the *Parison* is extruded through an annular die. A mould then closes round the parison and a jet of gas inflates it to take up the shape of the mould. This is illustrated in Fig. 4.21(a). Although this process is principally used for the production of bottles (for washing-up liquid, disinfectant, soft drinks, etc.) it is not restricted to small hollow articles. Domestic cold water storage tanks, large storage drums and 200

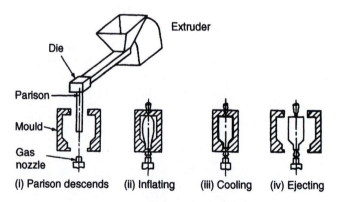

Fig. 4.21   Stages in blow moulding

gallon containers have been blow-moulded. The main materials used are PVC, polyethylene, polypropylene and PET.

The conventional extrusion blow moulding process may be continuous or intermittent. In the former method the extruder continuously supplies molten polymer through the annular die. In most cases the mould assembly moves relative to the die. When the mould has closed around the parison, a hot knife separates the latter from the extruder and the mould moves away for inflation, cooling and ejection of the moulding. Meanwhile the next parison will have been produced and this mould may move back to collect it or, in multi-mould systems, this would have been picked up by another mould. Alternatively in some machines the mould assembly is fixed and the required length of parison is cut off and transported to the mould by a robot arm.

In the intermittent processes, single or multiple parisons are extruded using a reciprocating screw or ram accumulator. In the former system the screw moves forward to extrude the parisons and then screws back to prepare the charge of molten plastic for the next shot. In the other system the screw extruder supplies a constant output to an accumulator. A ram then pushes melt from the accumulator to produce a parison as required.

Although it may appear straightforward, in fact the geometry of the parison is complex. In the first place its dimensions will be greater than those of the die due to the phenomenon of post extrusion swelling (see Chapter 5). Secondly there may be deformities (e.g. curtaining) due to flow defects. Thirdly, since most machines extrude the parison vertically downwards, during the delay between extrusion and inflation, the weight of the parison causes sagging or draw-down. This sagging limits the length of articles which can be produced from a free hanging parison. The complex combination of swelling and thinning makes it difficult to produce articles with a uniform wall thickness. This is particularly true when the cylindrical parison is inflated into an irregularly shaped mould because the uneven drawing causes additional thinning. In most cases therefore to blow mould successfully it is necessary to program the output rate or die gap to produce a controlled non-uniform distribution of thickness in the parison which will give a uniform thickness in the inflated article.

During moulding, the inflation rate and pressure must be carefully selected so that the parison does not burst. Inflation of the parison is generally fast but the overall cycle time is dictated by the cooling of the melt when it touches the mould. Various methods have been tried in order to improve the cooling rate e.g. injection of liquid carbon dioxide, cold air or high pressure moist air. These usually provide a significant reduction in cycle times but since the cooling rate affects the mechanical properties and dimensional stability of the moulding it is necessary to try to optimise the cooling in terms of production rate and quality.

Extrusion blow moulding is continually developing to be capable of producing even more complex shapes. These include unsymmetrical geometries and double wall mouldings. In recent years there have also been considerable developments in the use of in-the-mould transfers. This technology enables lables to be attached to bottles and containers as they are being moulded. Fig. 4.22 illustrates three stages in the blow moulding of a complex container.

Parison pinched as it emerges from die and then stretched between two halves of open mould

Mould closes to deform parison into shape

Side cores rotate inwards to complete forming

Fig. 4.22   Stages in blow moulding of complex hollow container

Analysis of Blow Moulding

As mentioned previously, when the molten plastic emerges from the die it swells due to the recovery of elastic deformations in the melt. It will be shown later that the following relationship applies:

$$B_{SH} = B_{ST}^2 \text{ (from Chapter 5)}$$

where B_{SH} = swelling of the thickness $(= h_1/h_d)$

B_{ST} = swelling of the diameter $(= D_1/D_d)$

therefore $$\frac{h_1}{h_d} = \left(\frac{D_1}{D_d}\right)^2$$

$$h_1 = h_d(B_{ST})^2 \tag{4.17}$$

Now consider the situation where the parison is inflated to fill a cylindrical die of diameter, D_m. Assuming constancy of volume and neglecting draw-down effects, then from Fig. 4.23

$$\pi D_1 h_1 = \pi D_m h$$

$$h = \frac{D_1}{D_m} h_1$$

$$= \frac{D_1}{D_m}(h_d \cdot B_{ST}^2)$$

$$= \frac{B_{ST} \cdot D_d}{D_m}(h_d \cdot B_{ST}^2)$$

Fig. 4.23   Analysis of blow moulding

$$h = B_{ST}^3 h_d \left(\frac{D_d}{D_m}\right) \tag{4.18}$$

This expression therefore enables the thickness of the moulded article to be calculated from a knowledge of the die dimensions, the swelling ratio and the mould diameter. The following example illustrates the use of this analysis. A further example on blow moulding may be found towards the end of Chapter 5 where there is also an example to illustrate how the amount of sagging of the parison may be estimated.

Example 4.4 A blow moulding die has an outside diameter of 30 mm and an inside diameter of 27 mm. The parison is inflated with a pressure of 0.4 MN/m² to produce a plastic bottle of diameter 50 mm. If the extrusion rate used causes a thickness swelling ratio of 2, estimate the wall thickness of the bottle. Comment on the suitability of the production conditions if melt fracture occurs at a stress of 6 MN/m².

Solution
From equation (4.18)

$$\text{wall thickness, } h = B_{ST}^3 h_d \left(\frac{D_d}{D_m}\right)$$

Now
$$h_d = \tfrac{1}{2}(30 - 27) = 1.5 \text{ mm}$$

$$B_{ST} = \sqrt{B_{SH}} = \sqrt{2} = 1.414$$

$$D_d = \tfrac{1}{2}(30 + 27) = 28.5 \text{ mm}$$

So
$$h = (1.414)^3(1.5)\left(\frac{28.5}{50}\right) = 2.42 \text{ mm}$$

The maximum stress in the inflated parison will be the hoop stress, σ_θ, which is given by

$$\sigma_\theta = \frac{PD_m}{2h} = \frac{0.4 \times 50}{2 \times 2.42}$$

$$= 4.13 \text{ MN/m}^2$$

Since this is less than the melt fracture stress (6 MN/m^2) these production conditions would be suitable. These are more worked examples on extrusion blow moulding towards the end of Chapter 5.

Extrusion Stretch Blow Moulding

Molecular orientation has a very large effect on the properties of a moulded article. During conventional blow moulding the inflation of the parison causes molecular orientation in the hoop direction. However, bi-axial stretching of the plastic before it starts to cool in the mould has been found to provide even more significant improvements in the quality of blow-moulded bottles. Advantages claimed include improved mechanical properties, greater clarity and superior permeation characteristics. Cost savings can also be achieved through the use of lower material grades or thinner wall sections.

Biaxial orientation may be achieved in blow moulding by

(a) stretching the extruded parison longitudinally before it is clamped by the mould and inflated. This is based on the Neck Ring process developed as early as the 1950s. In this case, molten plastic is extruded into a ring mould which forms the neck of the bottle and the parison is then stretched. After the mould closes around the parison, inflation of the bottle occurs in the normal way. The principle is illustrated in Fig. 4.24.

Extrusion / Injection of
plastic into pull rod

Stretching of
parison

Inflation and
ejection

Fig. 4.24   Neck ring stretch blow moulding

(b) producing a preform 'bottle' in one mould and then stretching this longitudinally prior to inflation in the full size bottle mould. This is illustrated in Fig. 4.25.

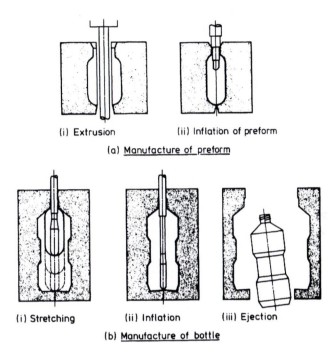

(i) Extrusion   (ii) Inflation of preform

(a) Manufacture of preform

(i) Stretching   (ii) Inflation   (iii) Ejection

(b) Manufacture of bottle

Fig. 4.25   Extrusion stretch blow moulding

Injection Stretch Blow Moulding

This is another method which is used to produce biaxially oriented blow moulded containers. However, as it involves injection moulding, the description of this process will be considered in more detail later (Section 4.3.9).

(e) Extrusion Coating Processes

There are many applications in which it is necessary to put a plastic coating on to paper or metal sheets and the extruder provides an ideal way of doing this. Normally a thin film of plastic is extruded from a slit die and is immediately brought into contact with the medium to be coated. The composite is then passed between rollers to ensure proper adhesion at the interface and to control the thickness of the coating (see Fig. 4.26).

Another major type of coating process is wire covering. The tremendous demand for insulated cables in the electrical industry means that large tonnages of plastic are used in this application. Basically a bare wire, which may be heated or have its surface primed, is drawn through a special die attached

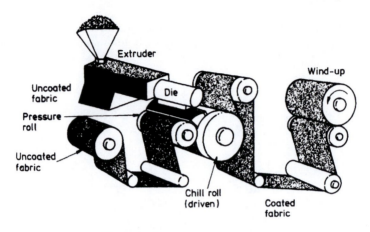

Fig. 4.26   Extrusion coating process

Fig. 4.27   Wire Covering Die

to an extruder (see Fig. 4.27). The drawing speed may be anywhere between 1 m/min and 1000 m/min depending on the diameter of the wire. When the wire emerges from the die it has a coating of plastic, the thickness of which depends on the speed of the wire and the extrusion conditions. It then passes into a cooling trough which may extend for a linear distance of several hundred metres. The coated wire is then wound on to storage drums.

Wire covering can be analysed in a very similar manner to that described for extrusion. The coating on the wire arises from two effects:

(a) *Drag Flow* due to the movement of the wire
(b) *Pressure Flow* due to the pressure difference between the extruder exit and the die exit.

From (4.2) the drag flow, Q_d, is given by

$$Q_d = \tfrac{1}{2}THV_d \qquad \text{where } T = 2\pi\left(R + \frac{h}{2}\right)$$

From (4.6) the pressure flow, Q_p, is given by

$$Q_p = \frac{1}{12\eta}\frac{dP}{dz}\cdot TH^3$$

So combining these two equations, the total output, Q, is given by

$$Q = \tfrac{1}{2}THV_d + \frac{TH^3}{12\eta}\cdot\frac{P}{L} \tag{4.19}$$

This must be equal to the volume of coating on the wire so

$$Q = \pi V_d((R+h)^2 - R^2)$$

$$Q = \pi V_d h(2R + h) \tag{4.20}$$

Combining equations (4.19) and (4.20)

$$\pi V_d h(2R + h) = \tfrac{1}{2}THV_d + \frac{TH^3}{12\eta}\cdot\frac{P}{L}$$

from which

$$P = \frac{6\eta LV_d}{H^3}(2H - H) \tag{4.21}$$

This is an expression for the pressure necessary at the extruder exit and therefore enables the appropriate extrusion conditions to be set.

(f) Recent Developments in Extrusion Technology

(i) **Co-Extrusion** As a result of the wide range of requirements which occur in practice it is not surprising that in many cases there is no individual plastic which has the correct combination of properties to satisfy a particular need. Therefore it is becoming very common in the manufacture of articles such as packaging film, yoghurt containers, refrigerator liners, gaskets and window frames that a multi-layer plastic composite will be used. This is particularly true for extruded film and thermoforming sheets (see Section 4.4). In co-extrusion two or more polymers are combined in a single process to produce a multi-layer film. These co-extruded films can either be produced by a blown film or a cast film process as illustrated in Figs 4.28(a) and (b). The cast process using a slot die and chill roll to cool the film, produces a film with good clarity and high gloss. The film blowing process, however, produces a stronger film due to the transverse orientation which can be introduced and this process offers more flexibility in terms of film thickness.

Fig. 4.28   Co-extrusion of plastic film

In most cases there is insufficient adhesion between the basic polymers and so it is necessary to have an adhesive film between each of the layers. Recent investigations of co-extrusion have been centred on methods of avoiding the need for the adhesive layer. The most successful seems to be the development of reactive bonding processes in which the co-extruded layers are chemically cross-linked together.

The main reason for producing multi-layer co-extruded films is to get materials with better barrier properties – particularly in regard to gas permeation. The following Table shows the effects which can be achieved. Data on permeability of plastics are also given in Figs 1.13 and 1.14.

Table 4.2

Transmission rates for a range of plastics

| | | | Transmission rates | |
Polymer	Layer distribution (μm)	Density (kg/m^3)	Oxygen (cm^3/m^2 24 hr atm)	Water vapour (g/m^2 24 hr)
ABS	1000	1050	30	2
uPVC	1000	1390	5	0.75
Polypropylene	1000	910	60	0.25
PET	1000	1360	1	2
LDPE	1000	920	140	0.5
HDPE	1000	960	60	0.3
PS/EVOH*/PE	825/25/150	1050	5†	1.6
PS/PVdC/PE	825/50/125	1070	1	0.4
PP/EVOH/PP	300/40/660	930	1†	0.25

*EVOH ethyl vinyl alcohol.
†Depends on humidity.

(ii) Highly Oriented Grids: Net-like polymer grids have become an extremely important development – particularly to civil engineers. The attraction in civil engineering applications is that the open grid structure permits soil particles to interlock through the apertures thus providing an extremely strong reinforcement to the soil. These geogrids under the trade name 'Tensar' are now widely used for road and runway construction, embankment supports, landslide repairs, etc.

The polymer grid achieves its very high strength due to the orientation of the polymer molecules during its manufacture. The process of manufacture is illustrated in Fig. 4.29. An extruded sheet, produced to a very fine tolerance and with a controlled structure, has a pattern of holes stamped into it. The hole shapes and pattern can be altered depending on the performance required of the finished product. The perforated sheet is then stretched in one direction to give thin sections of highly orientated polymer with the tensile strength of mild steel. This type of grid can be used in applications where uniaxial strength is required. In other cases, where biaxial strength is necessary, the sheet is subjected to a second stretching operation in the transverse direction. The advantages of highly oriented grids are that they are light and very easy to

Fig. 4.29   Tensar manufacturing process

handle. The advantage of obtaining a highly oriented molecular structure is also readily apparent when one compares the stiffness of a HDPE grid ($\simeq 10$ GN/m^2) with the stiffness of unoriented HDPE ($\simeq 1$ GN/m^2).

(iii) Reactive Extrusion: The most recent development in extrusion is the use of the extruder as a 'mini-reactor'. Reactive extrusion is the name given to the process whereby the plastic is manufactured in the extruder from base chemicals and once produced it passes through a die of the desired shape. Currently this process is being used the manufacture of low tonnage materials (<5000 tonnes p.a.) where the cost of a full size reactor run could not be justified. In the future it may be simply part of the production line.

4.3 Injection Moulding

4.3.1 Introduction

One of the most common processing methods for plastics is injection moulding. Nowadays every home, every vehicle, every office, every factory contains a multitude of different types of articles which have been injection moulded. These include such things as electric drill casings, yoghurt cartons, television housings, combs, syringes, paint brush handles, crash helmets, gearwheels, typewriters, fascia panels, reflectors, telephones, brief cases – the list is endless.

The original injection moulding machines were based on the pressure die casting technique for metals. The first machine is reported to have been patented in the United States in 1872, specifically for use with Celluloid. This was an important invention but probably before its time because in the following years very few developments in injection moulding processes were reported and it was not until the 1920s, in Germany, that a renewed interest was taken in the process. The first German machines were very simple pieces of equipment and relied totally on manual operation. Levers were used to clamp the mould and inject the melted plastic with the result that the pressures which could be attained were not very high. Subsequent improvements led to the use of pneumatic cylinders for clamping the injection which not only lifted some of the burden off the operator but also meant that higher pressures could be used.

The next major development in injection moulding, i.e. the introduction of hydraulically operated machines, did not occur until the late 1930s when a wide range of thermoplastics started to become available. However, these machines still tended to be hybrids based on die casting technology and the design of injection moulding machines for plastics was not taken really seriously until the 1950s when a new generation of equipment was developed. These machines catered more closely for the particular properties of polymer melts and modern machines are of the same basic design although of course the control systems are very much more sophisticated nowadays.

In principle, injection moulding is a simple process. A thermoplastic, in the form of granules or powder, passes from a feed hopper into the barrel where it is heated so that it becomes soft. It is then forced through a nozzle into a relatively cold mould which is clamped tightly closed. When the plastic has had sufficient time to become solid the mould opens, the article is ejected and the cycle is repeated. The major advantages of the process include its versatility in moulding a wide range of products, the ease with which automation can be introduced, the possibility of high production rates and the manufacture of articles with close tolerances. The basic injection moulding concept can also be adapted for use with thermosetting materials.

4.3.2 Details of the Process

The earliest injection moulding machines were of the plunger type as illustrated in Fig. 4.30 and there are still many of these machines in use today. A predetermined quantity of moulding material drops from the feed hopper into the barrel. The plunger then conveys the material along the barrel where it is heated by conduction from the external heaters. The material is thus plasticised under pressure so that it may be forced through the nozzle into the mould cavity. In order to split up the mass of material in the barrel and improve the heat transfer, a torpedo is fitted in the barrel as shown.

Fig. 4.30   Plunger type injection moulding machine

Unfortunately there are a number of inherent disadvantages with this type of machine which can make it difficult to produce consistent moulding. The main problems are:

(a) There is little mixing or homogenisation of the molten plastic.
(b) It is difficult to meter accurately the shot size. Since metering is on a volume basis, any variation in the density of the material will alter the shot weight.
(c) Since the plunger is compressing material which is in a variety of forms (varying from a solid granule to a viscous melt) the pressure at the nozzle can vary quite considerably from cycle to cycle.
(d) The presence of the torpedo causes a significant pressure loss.
(e) The flow properties of the melt are pressure sensitive and since the pressure is erratic, this amplifies the variability in mould filling.

Some of the disadvantages of the plunger machine may be overcome by using a pre-plasticising system. This type of machine has two barrels. Raw material is fed into the first barrel where an extruder screw or plunger plasticises the material and feeds it through a non-return valve into the other barrel. A plunger in the second barrel then forces the melt through a nozzle and into the mould. In this system there is much better homogenisation because the melt has to pass through the small opening connecting the two barrels. The shot size can - also be metered more accurately since the volume of material fed to the second barrel can be controlled by a limit switch on its plunger. Another advantage is that there is no longer a need for the torpedo on the main injection cylinder.

However, nowadays this type of machine is seldom used because it is considerably more complicated and more expensive than necessary. One area of application where it is still in use is for large mouldings because a large volume of plastic can be plasticised prior to injection using the primary cylinder plunger.

For normal injection moulding, however, the market is now dominated by the reciprocating screw type of injection moulding machine. This was a major breakthrough in machine design and yet the principle is simple. An extruder type screw in a heated barrel performs a dual role. On the one hand it rotates in the normal way to transport, melt and pressurize the material in the barrel but it is also capable, whilst not rotating, of moving forward like a plunger

Fig. 4.31   Typical cycle in reciprocating screw injection moulding machine

to inject melt into the mould. A typical injection moulding machine cycle is illustrated in Fig. 4.31. It involves the following stages:

(a) After the mould closes, the screw (not rotating) pushes forward to inject melt into the cooled mould. The air inside the mould will be pushed out through small vents at the furthest extremities of the melt flow path.

(b) When the cavity is filled, the screw continues to push forward to apply a holding pressure (see Fig. 4.31). This has the effect of squeezing extra melt into the cavity to compensate for the shrinkage of the plastic as it cools. This holding pressure is only effective as long as the gate(s) remain open.

(c) Once the gate(s) freeze, no more melt can enter the mould and so the screw-back commences. At this stage the screw starts to rotate and draw in new plastic from the hopper. This is conveyed to the front of the screw but as the mould cavity is filled with plastic, the effect is to push the screw backwards. This prepares the next shot by accumulating the desired amount of plastic in front of the screw. At a pre-set point in time, the screw stops rotating and the machine sits waiting for the solidification of the moulding and runner system to be completed.

(d) When the moulding has cooled to a temperature where it is solid enough to retain its shape, the mould opens and the moulding is ejected. The mould then closes and the cycle is repeated (see Fig. 4.32).

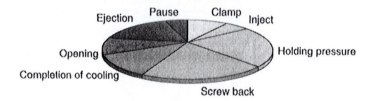

Fig. 4.32   Stages during injection moulding

There are a number of important features in reciprocating screw injection moulding machines and these will now be considered in turn.

Screws The screws used in these machines are basically the same as those described earlier for extrusion. The compression ratios are usually in the range 2.5:1 to 4:1 and the most common L/D ratios are in the range 15 to 20. Some screws are capable of injecting the plastic at pressures up to 200 MN/m^2. One important difference from an extruder screw is the presence of a back-flow check valve at the end of the screw as illustrated in Fig. 4.33. The purpose of this valve is to stop any back flow across the flights of the screw when it is acting as a plunger. When material is being conveyed forward by the rotation of the screw, the valve opens as shown. One exception is when injection moulding

(a) Valve closed   (b) Valve open

Fig. 4.33   Typical check valve

heat-sensitive materials such as PVC. In such cases there is no check valve because this would provide sites where material could get clogged and would degrade.

Barrels and Heaters These are also similar to those in extruder machines. In recent years, vented barrels have become available to facilitate the moulding of water sensitive plastics without the need for pre-drying. Water sensitivity in plastics can take several forms. If the plastic absorbs water then dimensional changes will occur, just as with wood or paper. The plastic will also be plasticised by the water so that there will be property changes such as a reduction in modulus and an increase in toughness. All these effects produced by water absorption are reversible.

Another event which may occur is *hydrolysis*. This is a chemical reaction between the plastic and water. It occurs extremely slowly at room temperature but can be significant at moulding temperatures. Hydrolysis causes degradation, reduction in properties (such as impact strength) and it is irreversible. Table 4.3 indicates the sensitivity of plastics to moisture. Note that generally extrusion requires a lower moisture content than injection moulding to produce good quality products.

Table 4.3

Water sensitivity of some common plastics

Drying not required (Materials do not hydrolyse)	Drying required	
	Absorbs only	Hydrolyses
Polyethylene	Acrylic (0.02/0.08)*	PET (0.002/0.002)
Polypropylene	ABS (0.02/0.08)	Polycarbonate (0.01/0.02)
Polystyrene	SAN (0.02/0.08)	Nylon 66 (0.08/0.15)
PVC		

*Required maximum moisture content for extrusion/injection moulding (%)

Nozzles The nozzle is screwed into the end of the barrel and provides the means by which the melt can leave the barrel and enter the mould. It is also a region where the melt can be heated both by friction and conduction from a heater band before entering the relatively cold channels in the mould. Contact with the mould causes heat transfer from the nozzle and in cases where this is excessive it is advisable to withdraw the nozzle from the mould during the screw-back part of the moulding cycle. Otherwise the plastic may freeze off in the nozzle.

There are several types of nozzle. The simplest is an open nozzle as shown in Fig. 4.34(a). This is used whenever possible because pressure drops can be minimised and there are no hold up points where the melt can stagnate and decompose. However, if the melt viscosity is low then leakage will occur from this type of nozzle particularly if the barrel/nozzle assembly retracts from the mould each cycle. The solution is to use a shut-off nozzle of which there are many types. Fig. 4.34(b) shows a nozzle which is shut off by external means. Fig. 4.34(c) shows a nozzle with a spring loaded needle valve which opens when the melt pressure exceeds a certain value or alternatively when the nozzle is pressed up against the mould. Most of the shut-off nozzles have the disadvantage that they restrict the flow of the material and provide undersirable stagnation sites. For this reason they should not be used with heat sensitive materials such as PVC.

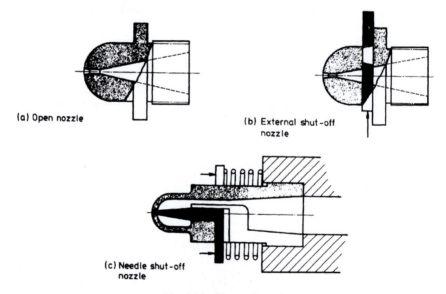

(a) Open nozzle

(b) External shut-off nozzle

(c) Needle shut-off nozzle

Fig. 4.34   Types of nozzle

Clamping Systems In order to keep the mould halves tightly closed when the melt is being injected under high pressures it is necessary to have a clamping system. This may be either (a) hydraulic or (b) mechanical (toggle) – or some combination of the two.

In the hydraulic system, oil under pressure is introduced behind a piston connected to the moving platen of the machine. This causes the mould to close and the clamp force can be adjusted so that there is no leakage of molten plastic from the mould.

The toggle is a mechanical device used to amplify force. Toggle mechanisms tend to be preferred for high speed machines and where the clamping force is relatively small. The two main advantages of the toggle system are that it is more economical to run the small hydraulic cylinder and since the toggle is self locking it is not necessary to maintain the hydraulic pressure throughout the moulding cycle. On the other hand the toggle system has the disadvantages that there is no indication of the clamping force and the additional moving parts increase maintenance costs.

4.3.3 Moulds

In the simplest case an injection mould (or 'tool') consists of two halves into which the impression of the part to be moulded is cut. The mating surfaces of the mould halves are accurately machined so that no leakage of plastic can occur at the split line. If leakage does occur the flash on the moulding is unsightly and expensive to remove. A typical injection mould is illustrated in Fig. 4.35. It may by seen that in order to facilitate mounting the mould in the machine and cooling and ejection of the moulding, several additions are made to the basic mould halves. Firstly, backing plates permit the mould to be bolted on to the machine platens. Secondly, channels are machined into the mould to allow the mould temperature to be controlled. Thirdly, ejector pins are included to that the moulded part can be freed from the mould. In most cases the ejector pins are operated by the shoulder screw hitting a stop when the mould opens. The mould cavity is joined to the machine nozzle by means of the **sprue.** The sprue anchor pin then has the function of pulling the sprue away from the nozzle and ensuring that the moulded part remains on the moving half of the mould, when the mould opens. For multi-cavity moulds the impressions are joined to the sprue by **runners** – channels cut in one or both halves of the mould through which the plastic will flow without restriction. A narrow constriction between the runner and the cavity allows the moulding to be easily separated from the runner and sprue. This constriction is called the **gate.**

A production injection mould is a piece of high precision engineering manufactured to very close tolerances by skilled craftsmen. A typical mould can be considered to consist of (i) the cavity and core and (ii) the remainder of the mould (often referred to as the bolster). Of these two, the latter is the more straightforward because although it needs to be accurately made, in general, conventional machine tools can be used. The cavity and core, however, may be quite complex in shape and so they often need special techniques. These can include casting, electro-deposition, hobbing, pressure casting, spark erosion and NC machining.

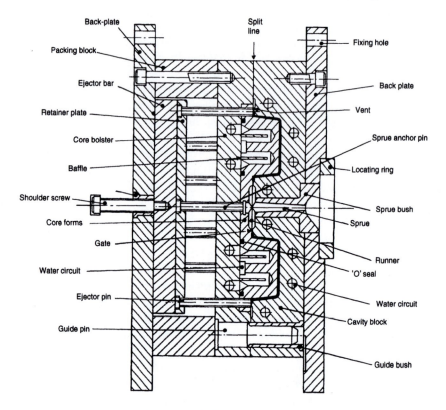

Fig. 4.35   Details of injection mould

Finishing and polishing the mould surfaces is also extremely important because the melt will tend to reproduce every detail on the surface of the mould. Finally the mould will have to be hardened to make it stand up to the treatment it receives in service. As a result of all the time and effort which goes into mould manufacture, it is sometimes found that a very complex mould costs more than the moulding machine on which it is used. Several features of the mould are worthy of special mention.

(a) Gates: As mentioned earlier the gate is the small orifice which connects the runner to the cavity. It has a number of functions. Firstly, it provides a convenient weak link by which the moulding can be broken off from the runner system. In some moulds the degating may be automatic when the mould opens. The gate also acts like a valve in that it allows molten plastic to fill the mould but being small it usually freezes off first. The cavity is thus sealed off from the runner system which prevents material being sucked out of the cavity during screw-back. As a general rule, small gates are preferable because no finishing is required if the moulding is separated cleanly from the runner. So for the initial trials on a mould the gates are made as small as possible and are only opened up if there are mould filling problems.

In a multi-cavity mould it is not always possible to arrange for the runner length to each cavity to be the same. This means that cavities close to the sprue would be filled quickly whereas cavities remote from the sprue receive the melt later and at a reduced pressure. To alleviate this problem it is common to use small gates close to the sprue and progressively increase the dimensions of the gates further along the runners. This has the effect of balancing the fill of the cavities. If a single cavity mould is multi-gated then here again it may be beneficial to balance the flow by using various gate sizes.

Examples of gates which are in common use are shown in Fig. 4.36. Sprue gates are used when the sprue bush can feed directly into the mould cavity as, for example, with single symmetrical moulding such as buckets. Pin gates are particularly successful because they cause high shear rates which reduce the viscosity of the plastic and so the mould fills more easily. The side gate is the most common type of gate and is a simple rectangular section feeding into the side of the cavity. A particular attraction of this type of gate is that mould filling can be improved by increasing the width of the gate but the freeze time is unaffected because the depth is unchanged.

(a) Sprue gate   (b) Pin gate   (c) Side gate

Fig. 4.36   Types of gate

(b) Runners: The runner is the flow path by which the molten plastic travels from the sprue (i.e. the moulding machine) to the gates (i.e. the cavity). To prevent the runner freezing off prematurely, its surface area should be small so as to minimise heat transfer to the mould. However, the cross sectional area of the runner should be large so that it presents little resistance to the flow of the plastic but not so large that the cycle time needs to be extended to allow the runner to solidify for ejection. A good indication of the efficiency of a runner is, therefore, the ratio of its cross-sectional area to its surface area. For example, a semi-circular channel cut into one half of the mould is convenient to machine but it only has an area ratio of 0.153 D where D is the diameter of the semi-circle. A full round runner, on the other hand, has a ratio of 0.25 D. A square section also has this ratio but is seldom used because it is difficult to eject. A compromise is a trapezoidal section (cut into one half of the mould) or a hexagonal section.

(c) Sprues: The sprue is the channel along which the molten plastic first enters the mould. It delivers the melt from the nozzle to the runner system. The sprue is incorporated in a hardened steel bush which has a seat designed to provide a good seal with the nozzle. Since it is important that the sprue is pulled out when the mould opens it is tapered as shown in Fig. 4.35 and there is a sprue pulling device mounted directly opposite the sprue entry. This can take many forms but typically it would be an undercut or reversed taper to provide a key for the plastic on the moving half of the mould. Since the sprue, like the runner system, is effectively waste it should not be made excessively long.

(d) Venting: Before the plastic melt is injected, the cavity in the closed mould contains air. When the melt enters the mould, if the air cannot escape it become compressed. At worst this may affect the mould filling, but in any case the sudden compression of the air causes considerable heating. This may be sufficient to burn the plastic and the mould surface at local hot spots. To alleviate this problem, vents are machined into the mating surfaces of the mould to allow the air to escape. The vent channel must be small so that molten plastic will not flow along it and cause unsightly flash on the moulded article. Typically a vent is about 0.025 mm deep and several millimeters wide. Away from the cavity the depth of the vent can be increased so that there is minimum resistance to the flow of the gases out of the mould.

(e) Mould Temperature Control: For efficient moulding, the temperature of the mould should be controlled and this is normally done by passing a fluid through a suitably arranged channel in the mould. The rate at which the moulding cools affects the total cycle time as well as the surface finish, tolerances, distortion and internal stresses of the moulded article. High mould temperatures improve surface gloss and tend to eliminate voids. However, the possibility of flashing is increased and sink marks are likely to occur. If the mould temperature is too low then the material may freeze in the cavity before it is filled. In most cases the mould temperatures used are a compromise based on experience. In Chapter 5 we will consider ways of estimating the time taken for a moulding to cool down in a mould.

Example 4.5 The runner lay-out for an eight cavity mould is illustrated in Fig. 4.37. If the mould is to be designed so that the pressure at the gate is the same in all cases, determine the radius of the runner in section A. The flow may be assumed to be isothermal.

Solution Although this runner system is symmetrical, it is not balanced. If the runner had the same diameter throughout all sections, then the mouldings close to the sprue would fill first and would be over-packed before the outermost

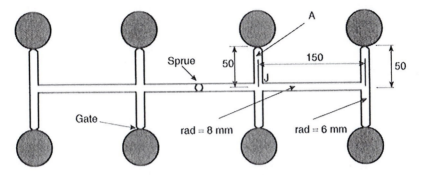

Fig. 4.37   Lay-out for eight cavity mould

cavities were filled. In a good mould design, all the cavities fill simultaneously at the same pressure. In this case it is necessary to ensure that the pressure drop in Sections 1 and 3 is the same as the pressure drop in Section 2.

It will be shown in Chapter 5 that the pressure drop, ΔP, for isothermal flow in a circular section channel is given by

$$\Delta P = \frac{8\eta LQ}{\pi R^4} \qquad (4.22)$$

where
$$\eta = \text{viscosity of the plastic}$$
$$L = \text{length of channel}$$
$$Q = \text{volume flow rate}$$
$$R = \text{radius of channel}$$

If the volume flow rate towards point J is q (ie the input at the sprue is $2q$) then at J the flow will split as follows:

$$\text{Flow along runner 1} = xq$$
$$\text{Flow along runner 2} = 1/2(1 - x)q$$
$$\text{Flow along runner 3} = 1/2xq$$

where
$$x = \frac{A_1}{A_1 + 2A_2} = \frac{R_1^2}{R_1^2 + 2R_2^2}$$

(A refers to the area of the relevant runner).
Using equation (4.22) we can write

$$\text{Pressure loss in runner 1} = \frac{8\eta L_1 xq}{\pi R_1^4}$$

$$\text{Pressure loss in runner } 2 = \frac{8\eta L_2(1-x)q}{2\pi R_2^4}$$

$$\text{Pressure loss in runner } 3 = \frac{8\eta L_3 x q}{2\pi R_3^4}$$

Thus, equating pressure losses after point J

$$\frac{8\eta L_2(1-x)q}{2\pi R_2^4} = \frac{8\eta L_1 x q}{\pi R_1^4} + \frac{8\eta L_3 x q}{2\pi R_3^4}$$

Substituting for x and rearranging to get R_2

$$R_2 = \frac{R_1 R_3^2 \sqrt{2L_2}}{\sqrt{2L_1 R_3^4 + L_3 R_1^4}}$$

For the dimensions given:

$$R_2 = 3.8 \text{ mm}$$

In practice there are a number of other factors to be taken into account. For example, the above analysis assumes that this plastic is Newtonian, ie that it has a constant viscosity, η. In reality the plastic melt is non-Newtonian so that the viscosity will change with the different shear rates in each of the three runner sections analysed. In addition, the melt flow into the mould will not be isothermal – the plastic melt immediately in contact with the mould will solidify. This will continuously reduce the effective runner cross-section for the melt coming along behind. The effects of non-Newtonian and non-isothermal behaviour are dealt with in Chapter 5.

Multi-Daylight Moulds

This type of mould, also often referred to as a three plate mould, is used when it is desired to have the runner system in a different plane from the parting line of the moulding. This would be the case in a multi-cavity mould where it was desirable to have a central feed to each cavity (see Fig. 4.38). In this type of mould there is automatic degating and the runner system and sprue are ejected separately from the moulding.

Hot Runner Moulds

The runners and sprues are necessary in a mould but they are not part of the end-product. Unfortunately, it is not economically viable to discard them so they must be re-ground for subsequent reprocessing. Regrinding is expensive and can introduce contamination into the material so that any system which avoids the accumulation of runners and sprues is attractive. A system has been developed to do this and it is really a logical extension of three plate moulding. In this system, strategically placed heaters and insulation in the mould keep the

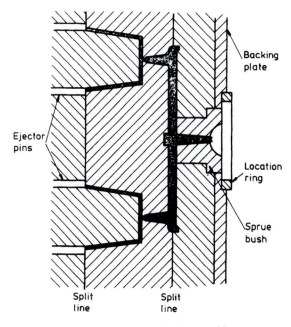

Fig. 4.38 Typical 3-plate mould

plastic in the runner at the injection temperature. During each cycle therefore the component is ejected but the melt in the runner channel is retained and injected into the cavity during the next shot. A typical mould layout is shown in Fig. 4.39.

Additional advantages of hot runner moulds are (i) elimination of trimming and (ii) possibility of faster cycle times because the runner system does not have to freeze off. However, these have to be weighed against the disadvantages of the system. Since the hot runner mould is more complex than a conventional mould it will be more expensive. Also there are many areas in the hot runner manifold where material can get trapped. This means that problems can be experienced during colour or grade changes because it is difficult to remove all of the previous material. As a practical point it should also be realised that the system only works as long as the runner remains molten. If the runner system freezes off then the hot runner manifold needs to be dismantled to remove the runners. Note also that hot runner system are not suitable for heat sensitive materials such as PVC.

Insulated Runner Moulds

This is similar in concept to the hot runner mould system. In this case, instead of having a specially heated manifold in the mould, large runners (13–25 mm diameter) are used. The relatively cold mould causes a frozen skin to form in

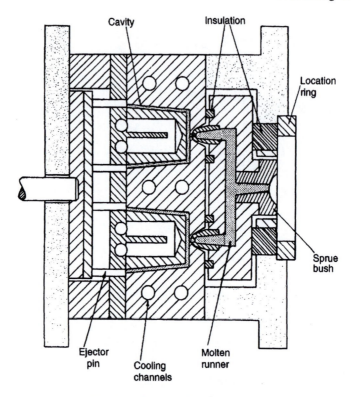

Fig. 4.39   Layout of hot runner mould

the runner which then insulates its core so that this remains molten. As in the previous case the runner remains in the mould when the moulding is ejected and the molten part of the runner is then injected into the cavity for the next shot. If an undue delay causes the whole runner to freeze off then it may be ejected and when moulding is restarted the insulation layer soon forms again. This type of system is widely used for moulding of fast cycling products such as flower pots and disposable goods. The main disadvantage of the system is that it is not suitable for polymers or pigments which have a low thermal stability or high viscosity, as some of the material may remain in a semi-molten form in the runner system for long periods of time.

A recent development of the insulated runner principle is the *distribution tube system*. This overcomes the possibility of freezing-off by insertion of heated tubes into the runners. However, this system still relies on a thick layer of polymer forming an insulation layer on the wall of the runner and so this system is not suitable for heat sensitive materials.

Note that both the insulated runner and the distribution tube systems rely on a cartridge heater in the gate area to prevent premature freezing off at the gate (see Fig. 4.40).

(a) Insulated runner system   (b) Distributed runner system

Fig. 4.40   Insulated and distributed runner systems

Mould Clamping Force

In order to prevent 'flashing', i.e. a thin film of plastic escaping out of the mould cavity at the parting line, it is necessary to keep the mould tightly closed during injection of the molten plastic. Before setting up a mould on a machine it is always worthwhile to check that there is sufficient clamping force available on the machine. To do this it is necessary to be able to estimate what clamping force will be needed. The relationship between mould area and clamp requirements has occupied the minds of moulders for many years. Practical experience suggests that the clamping pressure over the projected area of the moulding should be between 10 and 50 MN/m^2 depending on factors such as shape, thickness, and type of material. The mould clamping force may also be estimated in the following way. Consider the moulding of a disc which is centre gated as shown in Fig. 4.41(a). The force on the shaded element is given by

$$\text{Force,} \quad F = \int_0^R P_r 2\pi r \, dr \tag{4.23}$$

The cavity pressure will vary across the disc and it is necessary to make some assumption about this variation. Experimental studies have suggested that an empirical relation of the form

$$P_r = P_0 \left(1 - \left(\frac{r}{R} \right)^m \right) \tag{4.24}$$

is most satisfactory. P_0 is the pressure at the gate and m is a constant which is usually between 0.3 and 0.75. It will be shown later (Chapter 5) that 'm' is in fact equal to $(1-n)$ where 'n' is the index in the Power Law expression for polymer melt flow.

Substituting (4.24) in (4.23) then

$$F = \int_0^R P_0 \left(1 - \left(\frac{r}{R} \right)^m \right) 2\pi r \, dr$$

$$F = \pi R^2 P_0 \left(\frac{m}{m+2} \right) \tag{4.25}$$

This is a simple convenient expression for estimating the clamping force required for the disc. The same expression may also be used for more complex shapes where the projected area may be approximated as a circle. It will also give sufficiently accurate estimates for a square plate when the radius, R, in Fig. 4.41(a) is taken as half of the diagonal.

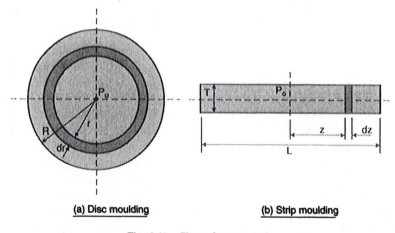

(a) Disc moulding   (b) Strip moulding

Fig. 4.41   Clamp force analysis

An alternative way of looking at this equation is that the clamping pressure, based on the projected area of the moulding, is given by

$$\text{Clamping pressure} = \left(\frac{m}{m+2} \right) \times \text{Injection pressure}$$

For any particular material the ratio $(m/(m+2))$ may be determined from the flow curves and it will be temperature and (to some extent) pressure dependent.

In practice the clamping pressure will also depend on the geometry of the cavity. In particular the flow ratio (flow length/channel lateral dimension) is important. Fig. 4.42 illustrates typical variations in the Mean Effective Pressure in the cavity for different thicknesses and flow ratios. The data used here is typical for easy flow materials such as polyethylene, polypropylene and polystyrene. To calculate the clamp force, simply multiply the appropriate Mean Effective Pressure by the projected area of the moulding. In practice it is

Fig. 4.42 Claming pressures for different cavity geometries (typical values for easy flow materials)

prudent to increase this value by 10–20% due to the uncertainties associated with specific moulds.

For plastics other than the easy flow materials referred to above, it would be normal to apply a factor to allow for the higher viscosity. Typical viscosity factors are given below.

Material	Viscosity Factor
Polyethylene, polypropylene, polystyrene	1
Nylon 66	1.2 → 1.4
ABS	1.3 → 1.4
Acrylic	1.5 → 1.7
PVC	1.6 → 1.8
Polycarbonate	1.7 → 2.0

Example 4.6 The mould shown in Fig. 4.35 produces four cup shaped ABS mouldings. The depth of the cups is 60 mm, the diameter at the is 90 mm and the wall thickness is 1.0 mm. The distance from the sprue to the cavity is 40 mm and the runner diameter is 6 mm. Calculate the clamp force necessary on the moulding machine and estimate how the clamp force would change if the mould was designed so as to feed the cups through a pin gate in the centre of the base (as illustrated in Fig. 4.38). The clamp pressure data in Fig. 4.42 should be used and the taper on the side of the cups may be ignored.

Solution
(a) Within the cavity, the maximum flow length for the plastic melt will be from the gate, along the side of the cup and across the base of the cup, ie

$$\text{Flow length} = 60 + 90 = 150 \text{ mm}$$

The thickness of the moulding is 1 mm, hence the flow ratio $= 150/1 = 150$. From Fig. 4.42 at this thickness and flow ratio, the mean effective pressure is 75 MN/m^2.

Allowing an extra 15% for uncertainties and applying the viscosity factor of 1.4 for ABS, then the appropriate mean effective pressure is $75 \times 1.15 \times 1.4 = 120$ MN/m^2. For each cavity, the projected area is $(\pi/4)(90)^2 = 6360$ mm$^2 = 6.36 \times 10^{-3}$ m^2.

Hence, clamp force per cavity $= 120 \times 6.36 \times 10^{-3} = 763$ kN.

The projected area of the runners is $4 \times 40 \times 6 = 960$ mm^2.

Assuming that the mean effective pressure also applies to the runner system, then

$$\text{clamp force for runners} = 120 \times 0.96 \times 10^{-3} = 115 \text{ kN}$$

Hence total clamp force for 4 cavities and 1 runner system is given by

$$\text{Total clamp force} = (4 \times 763) + 115 = 3167 \text{ kN}$$

The required clamp force is therefore 317 tonnes.

(b) If a pin gate in the middle of the base is used instead of an edge gate, then the flow ratio will be different. In this case

$$\text{flow length} = \tfrac{1}{2}(90) + 60 = 105$$

This is also the flow ratio, so from Fig. 4.42 the mean effective pressure is 50 MN/m^2. Applying the viscosity factor, etc as above, then

$$\text{Clamp force per cavity} = 50 \times 1.15 \times 1.4 \times \left(\frac{\pi}{4}\right)(90)^2 \times 10^{-6} = 512 \text{ kN}$$

In this case the runner system will be almost totally in the 'shadow' of the projected area of the cavities and so they can be ignored.

Hence, total clamp force $= 4 \times 512 = 2048$ kN $= 205$ tonnes.

Another common shape which is moulded is a thin rectangular strip. Consider the centre gated strip as shown in Fig. 4.41(b). In the same way as before the clamping force, F, is given by

$$F = 2 \int_{0}^{L/2} P_z T \, dz \tag{4.26}$$

$$F = 2 \int_{0}^{L/2} P_0 \left(1 - \left(\frac{z}{L/2}\right)^m\right) T \, dz$$

i.e. $$F = TP_0 L \left(\frac{m}{m+1} \right) \tag{4.27}$$

The calculation of clamp force is considered in more detail in Chapter 5.

4.3.4 Structural Foam Injection Moulding

Foamed thermoplastic articles have a cellular core with a relatively dense (solid) skin. The foam effect is achieved by the dispersion of inert gas throughout the molten resin directly before moulding. Introduction of the gas is usually carried out either by pre-blending the resin with a chemical blowing agent which releases gas when heated or by direct injection of the gas (usually nitrogen).

When the compressed gas/resin mixture is rapidly injected into the mould cavity, the gas expands explosively and forces the material into all parts of the mould.

The advantages of these types of foam moulding are

(a) for a given weight they are many times more rigid than a solid moulding
(b) they are almost completely free from orientation effects and the shrinkage is uniform
(c) very thick sections can be moulded without sink marks.

Foamed plastic articles may be produced with good results using normal screw-type injection moulding machines (see Fig. 4.43(a)). However, the limitations on shot size, injection speed and platen area imposed by conventional

(a) Standard injection moulding press

(b) Specialised foam moulding press

Fig. 4.43   Structural foam moulding equipment

injection equipment prevent the full large-part capabilities of structural foam from being realised. Specialised foam moulding machines currently in use can produce parts weighing in excess of 50 kg (see Fig. 4.43(b)).

Wall sections in foam moulding are thicker than in solid material. Longer cycle times can therefore be expected due to both the wall thickness and the low thermal conductivity of the cellular material. In contrast, however, the injection pressures in foam moulding are low when compared with conventional injection moulding. This means that less clamping force is needed per unit area of moulding and mould costs are less because lower strength mould materials may be used.

4.3.5 Sandwich Moulding

This is an injection moulding method which permits material costs to be reduced in large mouldings. In most mouldings it is the outer surface of an article which is important in terms of performance in service. If an article has to be thick in order that it will have adequate flexural stiffness then the material within the core of the article is wasted because its only function is to keep the outer surfaces apart. The philosophy of sandwich moulding is that two different materials (or two forms of the same material) should be used for the core and skin. That is, an expensive high performance material is used for the skin and a low-cost commodity or recycled plastic is used for the core. The way that this can be achieved is illustrated in Fig. 4.44.

Fig. 4.44   Stages in sandwich moulding process

Initially the skin material is injected but not sufficient to fill the mould. The core material is then injected and it flows laminarly into the interior of the core. This continues until the cavity is filled as shown in Fig. 4.44(c). Finally the nozzle valve rotates so that the skin material is injected into the sprue thereby clearing the valve of core material in preparation for the next shot. In a number of cases the core material is foamed to produce a sandwich section with a thin solid skin and a cellular core.

It is interesting that in the latest applications of sandwich moulding it is the core material which is being regarded as the critical component. This is to meet design requirements for computers, electronic equipment and some automotive parts. In these applications there is a growing demand for covers and housings with electromagnetic interference (EMI) shielding. The necessity of using a plastic with a high loading of conductive filler (usually carbon black) means that surface finish is poor and unattractive. To overcome this the sandwich moulding technique can be used in that a good quality surface can be moulded using a different plastic.

4.3.6 Gas Injection Moulding

In recent years major developments have been made in the use of an inert gas to act as the core in an injection moulded plastic product. This offers many advantages including greater stiffness/weight ratios and reduced moulded-in stresses and distortion.

The first stage of the cycle is the flow of molten polymer into the mould cavity through a standard feed system. Before this flow of polymer is complete, the injection of a predetermined quantity of gas into the melt begins through a special nozzle located within the cavity or feed system as shown in Fig. 4.45. The timing, pressure and speed of the gas injection is critical.

The pressure at the polymer gate remains high and, therefore, the gas chooses a natural path through the hotter and less viscous parts of the polymer melt towards the lower pressure areas. The flow of gas cores out a hollow centre extending from its point of entry towards the last point of fill. By controlling the amount of gas injected into the hollow core, the pressure on the cooling polymer is controlled and maintained until the moulding is packed. The final stage is the withdrawal of the gas nozzle, prior to mould opening, which allows the gas held in the hollow core to vent.

The gas injection process overcomes many of the limitations of injection mouldings such as moulded-in stress and distortion. These limitations are caused by laminar flow and variation in pressure throughout the moulding. With the gas injection process, laminar flow is considerably reduced and a uniform pressure is maintained. The difficulty of transmitting a very high pressure uniformly throughout a moulding can also cause inconsistent volumetric shrinkage of the polymer, and this leads to isolated surface sink marks. Whilst cycle times are comparable with those of conventional injection moulding, clamping forces are much lower. Also, by using gas to core out the polymer instead of mixing with it, gas-injection overcomes a number of shortcomings of the structural foam process. In particular there are no surface imperfections

Fig. 4.45 Stages in the gas injection moulding of an automotive handle (courtesy of Cinpress Ltd)

(caused by escaping gas bubbles in structural foam moulding) and cycle times are lower because thinner sections are being cooled.

4.3.7 Shear Controlled Orientation in Injection Moulding (SCORIM)

One of the major innovations in recent years is the use of pulsed pressure through the gates to introduce and control the orientation of the structure (or fillers) in injection moulded products. A special manifold is attached to the machine nozzle as illustrated in Fig. 4.46. This diagram relates to the *double live feed* of melt although up to four pistons, capable of applying oscillating pressure may be used.

Fig. 4.46 One embodiment of SCORIM where the device (B) for producing shear during solidification, by the action of pistons (D) and (E), is placed between the injection moulding machine barrel (A) and the mould (C) (Courtesy of Brunel University)

Shear controlled orientation in injection moulding (SCORIM) is based on the progressive application of macroscopic shears at the melt-solid interface during solidification in the moulding of a polymer matrix.

Macroscopic shears of specified magnitude and direction, applied at the melt-solid interface provide several advantages:

  (i) Enhanced polymer matrix or fibre alignment by design in moulded polymers or fibre reinforced polymers.
 (ii) Elimination of mechanical discontinuities that result from the initial mould filling process, including internal weld lines.
(iii) Reduction in the detrimental effects of a change in moulded section thickness.
 (iv) Elimination or reduction in defects resulting from the moulding of thick sectioned components.

4.3.8 Reaction Injection Moulding

Although there have been for many years a number of moulding methods (such as hand lay-up of glass fibres in polyester and compression moulding of thermosets or rubber) in which the plastic material is manufactured at the same time as it is being shaped into the final article, it is only recently that this concept has been applied in an injection moulding type process. In Reaction Injection Moulding (RIM), liquid reactants are brought together just prior to being injected into the mould. In-mould polymerisation then takes place which forms the plastic at the same time as the moulding is being produced. In some cases reinforcing fillers are incorporated in one of the reactants and this is referred to as Reinforced Reaction Injection Moulding (RRIM)

The basic RIM process is illustrated in Fig. 4.47. A range of plastics lend themselves to the type of fast polymerisation reaction which is required in this process – polyesters, epoxies, nylons and vinyl monomers. However, by far the most commonly used material is polyurethane. The components A and B are an isocyanate and a polyol and these are kept circulating in their separate systems until an injection shot is required. At this point the two reactants are brought together in the mixing head and injected into the mould.

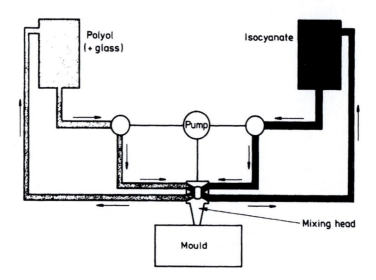

Fig. 4.47   Schematic view of reaction injection moulding

Since the reactants have a low viscosity, the injection pressures are relatively low in the RIM process. Thus, comparing a conventional injection moulding machine with a RIM machine having the same clamp force, the RIM machine could produce a moulding with a much greater projected area (typically about 10 times greater). Therefore the RIM process is particularly suitable for large area mouldings such as car bumpers and body panels. Another consequence of the low injection pressures is that mould materials other than steel may be considered. Aluminium has been used successfully and this permits weight savings in large moulds. Moulds are also less expensive than injection moulds but they must not be regarded as cheap. RIM moulds require careful design and, in particular, a good surface finish because the expansion of the material in the mould during polymerisation causes every detail on the surface of the mould to be reproduced on the moulding.

4.3.9 Injection Blow Moulding

In Section 4.2.7 we considered the process of extrusion blow moulding which is used to produce hollow articles such as bottles. At that time it was mentioned that if molecular orientation can be introduced to the moulding then the properties are significantly improved. In recent years the process of injection blow moulding has been developed to achieve this objective. It is now very widely used for the manufacture of bottles for soft drinks.

The steps in the process are illustrated in Fig. 4.48. Initially a preform is injection moulded. This is subsequently inflated in a blow mould in order to produce the bottle shape. In most cases the second stage inflation step occurs immediately after the injection moulding step but in some cases the preforms are removed from the injection moulding machine and subsequently re-heated for inflation.

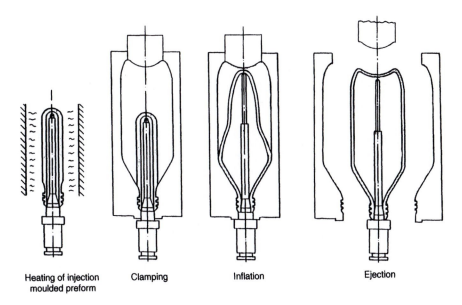

Heating of injection   Clamping   Inflation   Ejection
moulded preform

Fig. 4.48   Injection blow moulding process

The advantages of injection blow moulding are that

(i) the injection moulded parison may have a carefully controlled wall thickness profile to ensure a uniform wall thickness in the inflated bottle.
(ii) it is possible to have intricate detail in the bottle neck.
(iii) there is no trimming or flash (compare with extrusion blow moulding).

A variation of this basic concept is the *Injection Orientation Blow Moulding* technique developed in the 1960s in the USA but upgraded for commercial use in the 1980s by AOKI in Japan. The principle is very similar to that described above and is illustrated in Fig. 4.49. It may be seen that the method essentially combines injection moulding, blow moulding and thermoforming to manufacture high quality containers.

Fig. 4.49   Injection orientation strech blow moulding

4.3.10 Injection Moulding of Thermosetting Materials

In the past the thought of injection moulding thermosets was not very attractive. This was because early trials had shown that the feed-stock was not of a consistent quality which meant that continual alterations to the machine settings were necessary. Also, any undue delays could cause premature curing of the resin and consequent blockages in the system could be difficult to remove. However, in recent years the processing characteristics of thermosets have been improved considerably so that injection moulding is likely to become one of the major production methods for these materials. The injection moulding of fibre reinforced thermosets, such as DMC (Section 4.10.2), is also becoming very common.

Nowadays, the injection moulder can be supplied with uniform quality granules which consist of partially polymerised resin, fillers and additives. The formulation of the material is such that it will flow easily in the barrel with a slow rate of polymerisation. The curing is then completed rapidly in the mould.

In most respects the process is similar to the injection moulding of thermoplastics and the sequence of operations in a single cycle is as described earlier.
For thermosets a special barrel and screw are used. The screw is of approximately constant depth over its whole length and there is no check value which
might cause material blockages (see Fig. 4.50). The barrel is only kept warm
(80–110°C) rather than very hot as with thermoplastics because the material
must not cure in this section of the machine. Also, the increased viscosity
of the thermosetting materials means that higher screw torques and injection
pressures (up to 200 MN/m^2 are needed).

Fig. 4.50   Injection moulding of thermosets and rubbers

On the mould side of the machine the major difference is that the mould
is maintained very hot (150–200°C) rather than being cooled as is the case
with thermoplastics. This is to accelerate the curing of the material once it has
taken up the shape of the cavity. Another difference is that, as thermosetting
materials are abrasive and require higher injection pressures, harder steels with
extra wear resistance should be used for mould manufacture. As a result of the
abrasive nature of the thermosets, hydraulic mould clamping is preferred to a
toggle system because the inevitable dust from the moulding powder increases
the wear in the linkages of the latter.

When moulding thermosetting articles, the problem of material wastage in
sprues and runners is much more severe because these cannot be reused. It is
desirable therefore to keep the sprue and runner sections of the mould cool
so that these do not cure with the moulding. They can then be retained in the
mould during the ejection stage and then injected into the cavity to form the
next moulding. This is analogous to the hot runner system described earlier for
thermoplastics.

The advantages of injection moulding thermosets are as follows:

(a) fast cyclic times (see Table 4.4)
(b) efficient metering of material
(c) efficient pre-heating of material
(d) thinner flash – easier finishing
(e) lower mould costs (fewer impressions).

Table 4.4

For the same part, injection moulding of thermosets can offer up to 25% production increase and lower part-costs than compression.

Compression moulding	Minutes
Open mould, unload piece	0.105
Mould cleaning	0.140
Close machine, start pressure	0.100
Moulding cycle time	2.230
Total compression cycle	2.575
Injection moulding	
Unload piece, open/close machine	0.100
Moulding cycle time	1.900
Total injection cycle	2.000

4.4 Thermoforming

When a thermoplastic sheet is heated it becomes soft and pliable and the techniques for shaping this sheet are known as thermoforming. This method of manufacturing plastic articles developed in the 1950s but limitations such as poor wall thickness distribution and large peripheral waste restricted its use to simple packaging applications. In recent years, however, there have been major advances in machine design and material availability with the result that although packaging is still the major market sector for the process, a wide range of other products are made by thermoforming. These include aircraft window reveals, refrigerator liners, baths, switch panels, car bumpers, motorbike fairings etc.

The term 'thermoforming' incoroporates a wide range of possibilities for sheet forming but basically there are two sub-divisions – vacuum forming and pressure forming.

(a) Vacuum Forming

In this processing method a sheet of thermoplastic material is heated and then shaped by reducing the air pressure between it and a mould. The simplest type of vacuum forming is illustrated in Fig. 4.51(a). This is referred to as *Negative Forming* and is capable of providing a depth of draw which is 1/3–1/2 of the maximum width. The principle is very simple. A sheet of plastic, which may range in thickness from 0.025 mm to 6.5 mm, is clamped over the open mould. A heater panel is then placed above the sheet and when sufficient softening has occurred the heater is removed and the vacuum is applied. For the thicker sheets it is essential to have heating from both sides.

In some cases Negative Forming would not be suitable because, for example, the shape formed in Fig. 4.51 would have a wall thickness in the corners which is considerably less than that close to the clamp. If this was not acceptable then

Fig. 4.51   Vacuum forming process

the same basic shape could be produced by *Positive Forming*. In this case a male (positive) mould is pushed into the heated sheet before the vacuum is applied. This gives a better distribution of material and deeper shapes can be formed – depth to width ratios of 1:1 are possible. This thermoforming method is also referred to as *Drape Forming*. Another alternative would be to have a female mould as in Fig. 4.51 but after the heating stage and before the vacuum is applied, a plug comes down and guides the sheet into the cavity. When the vacuum is applied the base of the moulding is subjected to less draw and the result is a more uniform wall thickness distribution. This is called *Plug Assisted Forming*. Note that both Positive Forming and Plug Assisted Forming effectively apply a pre-stretch to the plastic sheet which improves the performance of the material quite apart from the improved wall thickness distribution.

In the packaging industry *skin* and *blister* vacuum machines are used. Skin packaging involves the encapsulation of articles between a tight, flexible transparent skin and a rigid backing which is usually cardboard. Blister packs are preformed foils which are sealed to a rigid backing card when the goods have been inserted.

The heaters used in thermoforming are usually of the infra red type with typical loadings of between 10 and 30 kW/m². Normally extra heat is concentrated at the clamped edges of the sheet to compensate for the additional heat losses in this region. The key to successful vacuum forming is achieving uniform heating over the sheet. One of the major attractions of vacuum forming is that since only atmospheric pressure is used to do the shaping, the moulds do not have to be very strong. Materials such as plaster, wood and thermosetting resins have all been used successfully. However, in long production runs mould cooling becomes essential in which case a metal mould is necessary. Experience has shown that the most satisfactory metal is undoubtedly aluminium. It is easily shaped, has good thermal conductivity, can be highly polished and has an almost unlimited life.

Materials which can be vacuum formed satisfactorily include polystyrene, ABS, PVC, acrylic, polycarbonate, polypropylene and high and low density polyethylene. Co-extruded sheets of different plastics and multi-colour laminates are also widely used nowadays. One of the most recent developments is the thermoforming of crystallisable PET for high temperature applications such as oven trays. The PET sheet is manufactured in the amorphous form and then during thermoforming it is permitted to crystallise. The resulting moulding is thus capable of remaining stiff at elevated temperatures.

(b) Pressure Forming

This is generally similar to vacuum forming except that pressure is applied above the sheet rather than vacuum below it. This advantage of this is that higher pressures can be used to form the sheet. A typical system is illustrated in Fig. 4.52 and in recent times this has become attractive as an alternative to injection moulding for moulding large area articles such as machine housings.

Fig. 4.52   Pressure forming process

(c) Matched Die Forming

A variation of thermoforming which does not involve gas pressure or vacuum is matched die forming. The concept is very simple and is illustrated in Fig. 4.53. The plastic sheet is heated as described previously and is then sandwiched between two halves of a mould. Very precise detail can be reproduced using this thermoforming method but the moulds need to be more robust than for the more conventional process involving gas pressure or vacuum.

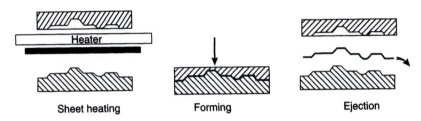

Sheet heating   Forming   Ejection

Fig. 4.53   Thermoforming between matched dies

(d) Dual-Sheet Thermoforming

This technique, also known as Twin-Sheet Forming, is a recent development. It is essentially a hybrid of blow moulding and thermoforming. Two heated sheets are placed between two mould halves and clamped as shown in Fig. 4.54. An inflation tube at the parting line then injects gas under pressure so that the sheets are forced out against the mould. Alternatively, a vacuum can be drawn between the plastic sheet and the mould in each half of the system. This technique has interesting possibilities for further development and will compete with blow moulding, injection moulding and rotational moulding in a number of market sectors. It can be noted that the two mould halves can be of different shapes and the two plastic sheets could be of different materials, provided a good weld can be obtained at the parting line.

4.4.1 Analysis of Thermoforming

If a thermoplastic sheet is softened by heat and then pressure is applied to one of the sides so as to generate a freely blown surface, it will be found that the shape so formed has a uniform thickness. If this was the case during thermoforming, then a simple volume balance between the original sheet and the final shape could provide the wall thickness of the end product.

$$A_i h_i = A_f h_f \tag{4.28}$$

where A = surface area, and h = wall thickness ('i' and 'f' refer to initial and final conditions).

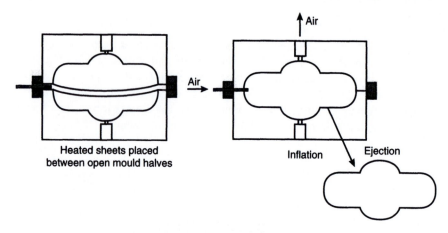

Fig. 4.54 Dual sheet forming

Example 4.7 A rectangular box 150 mm long, 100 mm wide and 60 mm deep is to be thermoformed from a flat sheet 150 mm × 100 mm × 2 mm. Estimate the average thickness of the walls of the final product if (a) conventional vacuum forming is used and (b) plug assisted moulding is used (the plug being 140 mm × 90 mm).

Solution
(a) The initial volume of the sheet is given by

$$A_i h_i = 150 \times 100 \times 2 = 3 \times 10^4 \text{ mm}^3$$

The surface area of the final product is

$$A_f = (150 \times 100) + 2(100 \times 60) + 2(150 \times 60)$$

$$= 4.5 \times 10^4 \text{ mm}^2$$

Therefore, from equation (4.28)

$$h_f = \frac{3 \times 10^4}{4.5 \times 10^4} = 0.67 \text{ mm}$$

(b) If plug assist is used then it could be assumed that over the area 140 mm × 90 mm, the wall thickness will remain at 2 mm. The volume of this part of the moulding will be

$$\text{Vol} = 140 \times 90 \times 2 = 2.52 \times 10^4 \text{ mm}^3$$

This would leave a volume of $(3 \times 10^4 - 2.52 \times 10^4)$ to form the walls. The area of the walls is

$$A_w = (2 \times 100 \times 60) + (2 \times 150 \times 60) = 3 \times 10^4 \text{ mm}^2$$

This ignores a small area in the base of the box, outside the edges of the plug. Hence, the thickness of the walls in this case would be

$$h_w = \frac{(3 \times 10^4) - (2.52 \times 10^4)}{3 \times 10^4} = 0.16 \text{ mm}$$

These calculations can give a useful first approximation of the dimensions of a thermoformed part. However, they will not be strictly accurate because in a real situation, when the plastic sheet is being stretched down into the cold mould it will freeze off at whatever thickness it has reached when it touches the mould.

Consider the thermoforming of a plastic sheet of thickness, h_0, into a conical mould as shown in Fig. 4.55(a). At this moment in time, t, the plastic is in contact with the mould for a distance, S, and the remainder of the sheet is in the form of a spherical dome of radius, R, and thickness, h. From the geometry of the mould the radius is given by

$$R = \frac{H - S \sin \alpha}{\sin \alpha \tan \alpha} \qquad (4.29)$$

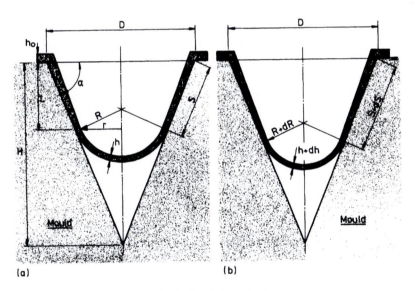

(a)   (b)

Fig. 4.55   Analysis of thermo forming

Also the surface area, A, of the spherical bubble is given by

$$A = 2\pi R^2 (1 - \cos \alpha) \qquad (4.30)$$

At a subsequent time, $(t + dt)$, the sheet will be formed to the shape shown in Fig. 4.55(b). The change in thickness of the sheet in this period of time may be estimated by assuming that the volume remains constant.

$$2\pi R^2(1 - \cos\alpha)h = 2\pi(R + dR)^2(1 - \cos\alpha)(h + dh) + 2\pi rh\,dS\sin\alpha$$

Substituting for $r(= R\sin\alpha)$ and for R from (4.29) this equation may be reduced to the form

$$\frac{dh}{h} = \left[2 - \left(\frac{\sin^2\alpha\tan\alpha}{1 - \cos\alpha}\right)\right] \cdot \frac{\sin\alpha\,dS}{(H - S\sin\alpha)} \tag{4.31}$$

This equation may be integrated with the boundary condition that $h = h_1$ at $S = 0$. As a result the thickness, h, at a distance, S, along the side of the conical mould is given by

$$h = h_1\left(\frac{H - S\sin\alpha}{H}\right)^{\sec\alpha - 1} \tag{4.32}$$

Now consider again the boundary condition referred to above. At the point when the softened sheet first enters the mould it forms part of a spherical bubble which does not touch the sides of the cone. The volume balance is therefore

$$\left(\frac{D^2}{4}\right)h_0 = \frac{2(D/2)^2(1 - \cos\alpha)h_1}{\sin^2\alpha}$$

So,
$$h_1 = \frac{\sin^2\alpha}{2(1 - \cos\alpha)} \cdot h_0$$

Making the substitution for h_1 in (4.32)

$$h = \frac{\sin^2\alpha}{2(1 - \cos\alpha)}\left[\frac{H - S \cdot \sin\alpha}{H}\right]^{\sec\alpha - 1} \cdot h_0$$

or
$$h/h_0 = \left(\frac{1 + \cos\alpha}{2}\right)\left[\frac{H - L}{H}\right]^{\sec\alpha - 1} \tag{4.33}$$

This equation may also be used to calculate the wall thickness distribution in deep truncated cone shapes but note that its derivation is only valid up to the point when the spherical bubble touches the centre of the base. Thereafter the analysis involves a volume balance with freezing-off on the base and sides of the cone.

Example 4.8 A small flower pot as shown in Fig. 4.56 is to be thermoformed using negative forming from a flat plastic sheet 2.5 mm thick. If the diameter of the top of the pot is 70 mm, the diameter of the base is 45 mm and the depth is 67 mm estimate the wall thickness of the pot at a point 40 mm from the top. Calculate also the draw ratio for this moulding.

Fig. 4.56   Thermoformed flower pot

Solution

(a)
$$\alpha = \tan^{-1}\left(\frac{67}{12.5}\right) = 79.4°$$

Using the terminology from Fig. 4.39(b)

$$H = 35 \tan \alpha = 187.6 \text{ mm}$$

From equation (4.33)

$$h/h_0 = \left(\frac{1 + \cos 79.4°}{2}\right)\left(\frac{187.6 - 40}{187.6}\right)^{(\sec 79.4)-1} = 0.203$$

$$h = 0.203 \times 2.5 = 0.51 \text{ mm}$$

(b) The *draw-ratio* for a thermoformed moulding is the ratio of the area of the product to the initial area of the sheet. In this case therefore

$$\text{Draw ratio} = \frac{\pi\sqrt{[(R - r)^2 + h^2]}(R + r) + \pi r^2}{\pi R^2}$$

$$= \frac{\pi\sqrt{[(35 - 22.5)^2 + 67^2]}(35 + 22.5) + \pi(22.5)^2}{\pi(35)^2}$$

$$= 3.6$$

4.5 Calendering

Calendering is a method of producing plastic film and sheet by squeezing the plastic through the gap (or 'nip') between two counter-rotating cylinders. The art of forming a sheet in this way can be traced to the paper, textile and metal industries. The first development of the technique for polymeric materials was in the middle 19th century when it was used for mixing additives into rubber. The subsequent application to plastics was not a complete success because the early machines did not have sufficient accuracy or control over such things as cylinder temperature and the gap between the rolls. Therefore acceptance of the technique as a viable production method was slow until the 1930s when special equipment was developed specifically for the new plastic materials. As well as being able to maintain accurately roll temperature in the region of 200°C these new machines had power assisted nip adjustment and the facility to adjust the rotational speed of each roll independently. These developments are still the main features of modern calendering equipment.

Calenders vary in respect of the number of rolls and of the arrangement of the rolls relative to one another. One typical arrangement is shown in Fig. 4.57 – the inverted L-type. Although the calendering operation as illustrated here looks very straightforward it is not quite as simple as that. In the production plant a lot of ancillary equipment is needed in order to prepare the plastic material for the calender rolls and to handle the sheet after the calendering operation. A typical sheet production unit would start with premixing of the polymer, plasticiser, pigment, etc in a ribbon mixer followed by gelation of the premix in a Banbury Mixer and/or a short screw extruder. At various stages, strainers and metal detectors are used to remove any foreign matter. These preliminary operations result in a material with a dough-like consistency which is then supplied to the calender rolls for shaping into sheets.

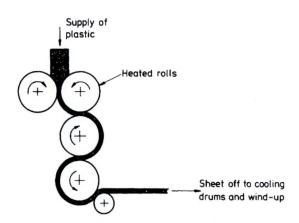

Fig. 4.57   Typical arrangement of calender rolls

However, even then the process is not complete. Since the hot plastic tends to cling to the calender rolls it is necessary to peel it off using a high speed roll of smaller diameter located as shown in Fig. 4.57. When the sheet leaves the calender it passes between embossing rolls and then on to cooling drums before being trimmed and stored on drums. For thin sheets the speed of the winding drum can be adjusted to control the drawdown. Outputs vary in the range 0.1–2 m/s depending on the sheet thickness.

Calendering can achieve surprising accuracy on the thickness of a sheet. Typically the tolerance is ± 0.005 mm but to achieve this it is essential to have very close control over roll temperatures, speeds and proximity. In addition, the dimensions of the rolls must be very precise. The production of the rolls is akin to the manufacture of an injection moulding tool in the sense that very high machining skills are required. The particular features of a calender roll are a uniform specified surface finish, minimal eccentricity and a special barrel profile ('crown') to compensate for roll deflection under the very high presurres developed between the rolls.

Since calendering is a method of producing sheet/film it must be considered to be in direct competition with extrusion based processes. In general, film blowing and die extrusion methods are preferred for materials such as polyethylene, polypropylene and polystyrene but calendering has the major advantage of causing very little thermal degradation and so it is widely used for heat sensitive materials such as PVC.

4.5.1 Analysis of Calendering

A detailed analysis of the flow of molten plastic between two rotating rolls is very complex but fortunately sufficient accuracy for many purposes can be achieved by using a simple Newtonian model. The assumptions made are that

(a) the flow is steady and laminar
(b) the flow is isothermal
(c) the fluid is incompressible
(d) there is no slip between the fluid and the rolls.

If the clearance between the rolls is small in relation to their radius then at any section x the problem may be analysed as the flow between parallel plates at a distance h apart. The velocity profile at any section is thus made up of a drag flow component and a pressure flow component.

For a fluid between two parallel plates, each moving at a velocity V_d, the drag flow velocity is equal to V_d. In the case of a calender with rolls of radius, R, rotating at a speed, N, the drag velocity will thus be given by $2\pi RN$.

The velocity component due to pressure flow between two parallel plates has already been determined in Section 4.2.3(b).

$$V_p = \frac{1}{2\eta}\frac{dP}{dx}(y^2 - (h/2)^2)$$

Therefore the total velocity at any section is given by

$$V = V_d + \frac{1}{2\eta}\frac{dP}{dx}[y^2 - (h/2)^2]$$

Considering unit width of the calender rolls the total throughput, Q, is given by

$$Q = 2 \int_0^{h/2} V \, dy$$

$$= 2 \int_0^{h/2} \left[V_d + \frac{1}{2\eta} \frac{dP}{dx} (y^2 - (h/2)^2) \right] dy$$

$$= h \left(V_d - \frac{h^2}{12\eta} \frac{dP}{dx} \right) \tag{4.34}$$

Since the output is given by $V_d H$

then
$$V_d H = h \left(V_d - \frac{h^2}{12\eta} \frac{dP}{dx} \right) \tag{4.35}$$

From this it may be seen that $\dfrac{dP}{dx} = 0$ at $h = H$.

To determine the shape of the pressure profile it is necessary to express h as a function of x. From the equation of a circle it may be seen that

$$h = H_0 + 2(R - (R^2 - x^2)^{1/2}) \tag{4.36}$$

However, in the analysis of calendering this equation is found to be difficult to work with and a useful approximation is obtained by expanding $(R^2 - x^2)^{1/2}$ using the binomial series and retaining only the first two terms. This gives

$$h = H_0 \left(1 + \frac{x^2}{H_0 R} \right) \tag{4.37}$$

Therefore as shown earlier dP/dx will be zero at

$$H = H_0 \left(1 + \frac{x^2}{H_0 R} \right)$$

$$x = \pm \sqrt{(H - H_0)R'} \tag{4.38}$$

This gives a pressure profile of the general shape shown in Fig. 4.58. The value of the maximum pressure may be obtained by rearranging (4.35) and substituting for h from (4.37)

$$\frac{dP}{dx} = \frac{12\eta V_d \left(H_0 - H + \dfrac{x^2}{R} \right)}{\left(H_0 + \dfrac{x^2}{R} \right)^3} \tag{4.39}$$

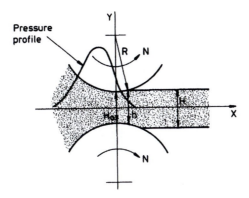

Fig. 4.58   Melt flow between calender rolls

If this equation is integrated and the value of x from (4.38) substituted then the maximum pressure may be obtained as

$$P_{max} = \frac{3\eta V_d}{H_0}\left(2\omega - \frac{(4H_0 - 3H)}{H_0}\left(\omega + \sqrt{\frac{R}{H_0}}\tan^{-1}\sqrt{\left(\frac{H - H_0}{H}\right)}\right)\right)$$
(4.40)

where
$$\omega = \frac{\sqrt{(H - H_0)R}}{H}$$
(4.41)

Example 4.9 A calender having rolls of diameter 0.4 m produces plastic sheet 2 m wide at the rate of 1300 kg/hour. If the nip between rolls is 10 mm and the exit velocity of the sheet is 0.01 m/s estimate the position and magnitude of the maximum pressure. The density of the material is 1400 kg/m³ and its viscosity is 10^4 Ns/m².

Solution Flow rate, $Q = 1300$ kg/hour $= 0.258 \times 10^{-3}$ m³/s

but
$$Q = HWV_d \text{ where } W = \text{width of sheet}$$

So
$$H = \frac{0.258 \times 10^{-3}}{2 \times 0.01} = 12.9 \text{ mm}$$

The distance upstream of the nip at which the pressure is a maximum is given by equation (4.38)

$$x = \sqrt{(12.9 - 10)200} = 24.08 \text{ mm}$$

Also from (4.37)

$$P_{max} = \frac{3 \times 10^4 \times 0.01}{10 \times 10^{-3}}\{(2 \times 1.865) - 0.13[1.865 + (4.45)(0.494)]\}$$

$$= 96 \text{ kN/m}^2$$

4.6 Rotational Moulding

Rotational moulding, like blow moulding, is used to produce hollow plastic articles. However, the principles in each method are quite different. In rotational moulding a carefully weighed charge of plastic powder is placed in one half of a metal mould. The mould halves are then clamped together and heated in an oven. During the heating stage the mould is rotated about two axes at right angles to each other. After a time the plastic will be sufficiently softened to form a homogeneous layer on the surface of the mould. The latter is then cooled while still being rotated. The final stage is to take the moulded article from the mould.

The process was originally developed in the 1940s for use with vinyl plastisols in liquid form. It was not until the 1950s that polyethylene powders were successfully moulded in this way. Nowadays a range of materials such as nylon, polycarbonate, ABS, high impact polystyrene and polypropylene can be moulded but by far the most common material is polyethylene.

The process is attractive for a number of reasons. Firstly, since it is a low pressure process the moulds are generally simple and relatively inexpensive. Also the moulded articles can have a very uniform thickness, can contain reinforcement, are virtually strain free and their surface can be textured if desired. The use of this moulding method is growing steadily because although the cycle times are slow compared with injection or blow moulding, it can produce very large, thick walled articles which could not be produced economically by any other technique. Wall thicknesses of 10 mm are not a problem for rotationally moulded articles.

There is a variety of ways in which the cycle of events described above may be carried out. For example, in some cases (particularly for very large articles) the whole process takes place in one oven. However, a more common set-up is illustrated in Fig. 4.59. The mould is on the end of an arm which first carries the cold mould containing the powder into a heated oven. During heating the mould rotates about the arm (major) axis and also about its own (minor) axis (see Fig. 4.60). After a pre-set time in the oven the arm brings the mould into a cooling chamber. The rate of cooling is very important. Clearly, fast cooling is desirable for economic reasons but this may cause problems such as warping. Normally therefore the mould is initially cooled using blown air and this is followed by a water spray. The rate of cooling has such a major effect on product quality that even the direction of the air jets on the mould during the initial gradual cooling stage can decide the success or otherwise of the process. As shown in Fig. 4.59 there are normally three arms (mould holders) in a complete system so that as one is being heated another is being cooled and so on. In many machines the arms are fixed rigidly together and so the slowest event (heating, cooling or charging/discharging) dictates when the moulds progress to the next station. In some modern machines, the arms are independent so that if cooling is completed then that arm can leave the cooling bay whilst the other arms remain in position.

It is important to realise that rotational moulding is not a centrifugal casting technique. The rotational speeds are generally below 20 rev/min with the ratio of speeds about the major and minor axes being typically 4 to 1. Also since all mould surfaces are not equidistant from the centre of rotation any centrifugal forces generated would tend to cause large variations in wall thickness. In fact in order to ensure uniformity of all thickness it is normal design practice to arrange that the point of intersection of the major and minor axis does not coincide with the centroid of the mould.

The heating of rotational moulds may be achieved using infra-red, hot liquid, open gas flame or hot-air convection. However, the latter method is the most common. The oven temperature is usually in the range 250–450°C and since the mould is cool when it enters the oven it takes a certain time to get up to a temperature which will melt the plastic. This time may be estimated as follows.

When the mould is placed in the heated oven, the heat input (or loss) per unit time must be equal to the change in internal energy of the material (in this case the mould).

$$hA(T_0 - T) = \rho C_p V \left(\frac{dT}{dt} \right) \tag{4.42}$$

where h is the convective heat transfer coefficient
   A is the surface area of mould
   T_0 is the temperature of the oven
   T_t is the temperature of the mould at time t
   ρ is the density of the mould material
   C_p is the specific heat of the mould material
   V is the volume of the walls of the mould
and t is time

Rearranging this equation and integrating then

$$hA \int_0^t dt = \rho C_p V \int_{T_i}^{T_t} \frac{dT}{(T_0 - T)}$$

$$hAt = -\rho C_p V \log_e \left(\frac{T_0 - T_t}{T_0 - T_i} \right)$$

$$\left(\frac{T_0 - T_t}{T_0 - T_i} \right) = e^{-h\beta t/\rho C_p} \tag{4.43}$$

where T_i is the initial temperature of the mould and β is the surface area to volume ratio (A/V).

This equation suggests that there is an exponential rise in mould temperature when it enters the oven, and in practice this is often found to be the case.

Fig. 4.59   Typical rotational moulding process

Fig. 4.60   Typical 'off-set arm' rotation

Fig. 4.61 illustrates typical temperature profiles during the rotational moulding of polyethylene. With typical values of oven temperatures and data for an aluminium mould

$$T_0 = 300°C, \qquad T_i = 30°C, \qquad T_t = 20°C$$

$$h = 22 \text{ W/m}^2\text{K} \qquad C_p = 917 \text{ J/kg K}, \qquad \rho = 2700 \text{ kg/m}^3$$

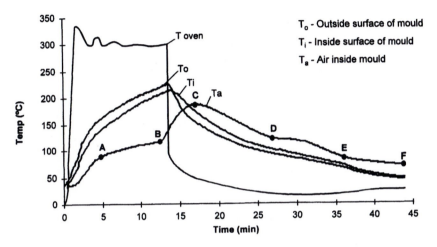

Fig. 4.61   Temperature profiles during rotational moulding

then for an aluminium cube mould 330 mm side and 6 mm thick, as was used to produce Fig. 4.61 then

$$t = \frac{-\rho C_p}{\beta h} \log_e \left\{ \frac{T_0 - T_t}{T_0 - T_i} \right\} = \frac{-2700 \times 917}{1000 \times 22} \log_e \left\{ \frac{330 - 220}{330 - 30} \right\}$$

$$t = 1.9 \text{ minutes}$$

For a steel mould of the same dimensions and thickness, a quick calculation ($h = 11$ W/m^{2}K, $C_p = 480$ J/kg K and $\rho = 7850$ kg/m^3) shows that the steel mould would take three times longer to heat up. However, in practice, steel moulds are less than a third of the thickness of aluminium. Therefore, although aluminium has a better thermal conductivity, steel moulds tend to heat up more quickly because they are thinner.

It is important to note that the above calculation is an approximation for the time taken to heat the mould to any desired temperature. Fig. 4.61 shows that in practice it takes considerably longer for the mould temperature to get to 220°C. This is because although initially the mould temperature is rising at the rate predicted in the above calculation, once the plastic starts to melt, it absorbs a significant amount of the thermal energy input.

Fig. 4.61 illustrates that the mould temperature is quite different from the set oven temperature (330°C) or indeed the actual oven temperature, throughout the moulding cycle. An even more important observation is that in order to control the rotational moulding process it is desirable to monitor the temperature of the air inside the mould. This is possible because there is normally a vent tube through the mould wall in order to ensure equal pressures inside and outside the mould. This vent tube provides an easy access for a thermocouple to measure the internal air temperature.

The internal air temperature characteristic has a unique shape which shows clearly what is happening at all stages throughout the process. Up to point A in Fig. 4.61 there is simply powder tumbling about inside the mould. At point A the mould has become sufficiently hot that plastic starts to melt and stick to the mould. The melting process absorbs energy and so over the region AB, the internal air temperature rises less quickly. It may also be seen that the temperature of the mould now starts to rise less quickly. At B all the plastic has melted and so a larger proportion of the thermal energy input goes to heating the inner air. This temperature rises more rapidly again, at a rate similar to that in the initial phase of the process.

Over the region BC the melt is effectively sintering because at B it is a powdery mass loosely bonded together whereas at C it has become a uniform melt. The value of temperature at C is very important because if the oven period is too short, then the material will not have sintered properly and there will be an excess of pin-holes. These are caused where the powder particles have fused together and trapped a pocket of air. If the oven period is too long then the pin-holes will all have disappeared but thermal/oxidative degradation will have started at the inner surface of the moulding. Extensive tests have shown that this is a source of brittleness in the mouldings and so the correct choice of temperature at C is a very important quality control parameter. For most grades of polyethylene the optimum temperature is in the region of 200°C ±5°C.

Once the mould is removed from the oven the mould starts to cool at a rate determined by the type of cooling – blown air (slow) or water spray (fast). There may be a overshoot in the internal air temperature due to the thermal momentum of the melt. This overshoot will depend on the wall thickness of the plastic product. In Fig. 4.61 it may be seen that the inner air temperature continues to rise for several minutes after the mould has been taken out of the oven (at about 13.5 minutes).

During cooling, a point D is reached where the internal air temperature decreases less quickly for a period. This represents the solidification of the plastic and because this process is exothermic, the inner air cannot cool so quickly. Once solidification is complete, the inner air cools more rapidly again. Another kink (point E) may appear in this cooling curve and, if so, it represents the point where the moulding has separated from the mould wall. In practice this is an important point to keep consistent because it affects shrinkage, warpage, etc in the final product. Once the moulding separates from the mould, it will cool more slowly and will tend to be more crystalline, have greater shrinkage and lower impact strength.

Developments in rotational moulding are continuing, with the ever increasing use of features such as

(i) mould pressurisation (to consolidate the melt, remove pin-holes, reduce cycle times and provide more consistent mould release),

(ii) internal heating/cooling (to increase cycle times and reduce warpage effects).

In overall terms the disadvantages of rotational moulding are its relative slowness and the limited choice of plastics which are commercially available in powder form with the correct additive package. However, the advantages of rotational moulding in terms of stress-free moulding, low mould costs, fast lead times and easy control over wall thickness distribution (relative to blow moulding) means that currently rotational moulding is the fastest growing sector of the plastics processing industry. Typical annual growth rates are between 10 and 12% p.a.

4.6.1 Slush Moulding

This is a method for making hollow articles using liquid plastics, particularly PVC plastisols. A shell-like mould is heated to a pre-determined temperature (typically 130°C for plastisols) and the liquid is then poured into the mould to completely fill it. A period of time is allowed to elapse until the required thickness of plastic gels. The excess liquid is then poured out and the plastic skin remaining in the mould is cured in an oven. The moulding is then taken from the mould.

It should be noted that when the plastisol liquid gels it has sufficient strength to remain in position on the inside surface of the mould. However, it has insufficient tear strength to be useful and so it has to go through the higher temperature curing stage to provide the necessary toughness and strength in the end-product. The mould is not rotated during slush moulding.

4.7 Compression Moulding

Compression moulding is one of the most common methods used to produce articles from thermosetting plastics. The process can also be used for thermoplastics but this is less common – the most familiar example is the production of LP records. The moulding operation as used for thermosets is illustrated in Fig. 4.62. A pre-weighed charge of partially polymerised thermoset is placed in the lower half of a heated mould and the upper half is then forced down. This causes the material to be squeezed out to take the shape of the mould. The application of the heat and pressure accelerates the polymerisation of the

Fig. 4.62   Principle of compression moulding

thermoset and once the crosslinking ('curing') is completed the article is solid and may be ejected while still very hot. Mould temperatures are usually in the range of 130–200°C. Cycle times may be long (possibly several minutes) so it is desirable to have multi-cavity moulds to increase production rates. As a result, moulds usually have a large projected area so the closing force needed could be in the region of 100–500 tonnes to give the 7–25 MN/m² cavity pressure needed. It should also be noted that compression moulding is also used for Dough Moulding Compounds (DMC) – these will be considered in Section 4.10.2

During compression moulding, the charge of material may be put into the mould either as a powder or a preformed 'cake'. In both cases the material is preheated to reduce the temperature difference between it and the mould. If the material is at a uniform temperature in the mould then the process may be analysed as follows.

Consider a 'cake' of moulding resin between the compression platens as shown in Fig. 4.63. When a constant force, F, is applied to the upper platen the resin flows as a result of a pressure gradient. If the flow is assumed Newtonian then the pressure flow equation derived in Section 4.2.3 may be used

$$\text{flow rate, } Q_p = \frac{1}{12\eta}\left(\frac{dP}{dz}\right)TH^3 \tag{4.6}$$

For the annular element of radius, r, in Fig. 4.63 it is more convenient to use cylindrical co-ordinates so this equation may be rewritten as

$$Q_p = \frac{1}{12\eta}\left(\frac{dP}{dr}\right) \cdot (2\pi r)H^3$$

Now if the top platen moves down by a distance, dH, the volume displaced is $(\pi r^2 dH)$ and the volume flow rate is $\pi r^2(dH/dt)$.

Therefore $$\pi r^2\left(\frac{dH}{dt}\right) = \frac{1}{12\eta}\left(\frac{dP}{dr}\right) \cdot (2\pi r)H^3$$

Fig. 4.63   Analysis of compression moulding

$$\frac{12\eta}{H^3} \cdot \frac{dH}{dt} = \frac{2}{r}\frac{dP}{dr} \tag{4.44}$$

This simple differential equation is separable and so each side may be solved in turn.

Let

$$\frac{2}{r}\frac{dP}{dr} = A \text{ where } A = f(H)$$

so

$$\int_0^P dP = \frac{A}{2} \int_R^r r\,dr$$

or

$$P = \frac{A}{4}(r^2 - R^2)$$

Now the force on the element is $2\pi r\,dr(P)$ so the total force, F, is given by integrating across the platen surface.

$$F = \int_0^R 2\pi r \left(\frac{A}{4}\right)(r^2 - R^2)dr = -\frac{\pi A R^4}{8}$$

This may be rearranged to give

$$A = -\frac{8F}{\pi R^4} = -\frac{8\pi F H^2}{V^2}$$

where $V = \pi R^2 H$

Substituting for A in (4.44)

$$-\frac{8\pi F H^2}{V^2} = \frac{12\eta}{H^3}\frac{dH}{dt}$$

So
$$-\int_0^t \frac{2\pi F}{3\eta V^2}dt = \int_{H_0}^H \frac{dH}{H^5}$$

$$\frac{2\pi Ft}{3\eta V^2} = \frac{1}{4}\left(\frac{1}{H^4} - \frac{1}{H_0^4}\right)$$

Since $H_0 \gg H$ then $(1/H_0^4)$ may be neglected. As a result the compaction force F, is given by

$$F = \frac{3\eta V^2}{8\pi t H^4} \qquad (4.45)$$

where H is the platen separation at time, t.

Example 4.10 A circular plate with a diameter of 0.3 m is to be compression moulded from phenol formaldehyde. If the preform is cylindrical with a diameter of 50 mm and a depth of 36 mm estimate the platen force needed to produce the plate in 10 seconds. The viscosity of the phenol may be taken as 10^3 Ns/m^2.

Solution

Volume,
$$V = \pi\left(\frac{50}{2}\right)^2 \times 36 = \pi\left(\frac{300}{2}\right)^2 H$$

So
$$H = 1 \text{ mm}$$

From (4.45) $F = \dfrac{3\eta V^2}{8\pi t H^4} = \dfrac{3 \times 10^3 \times (\pi \times 625 \times 36)^2}{10^6 \times 8\pi \times 10 \times (1)^4} = 59.6 \text{ kN}$

4.8 Transfer Moulding

Transfer moulding is similar to compression moulding except that instead of the moulding material being pressurized in the cavity, it is pressurized in a separate chamber and then forced through an opening and into a closed mould. Transfer moulds usually have multi-cavities as shown in Fig. 4.64. The advantages of transfer moulding are that the preheating of the material and injection through a narrow orifice improves the temperature distribution in the material and accelerates the crosslinking reaction. As a result the cycle times are reduced and there is less distortion of the mouldings. The improved flow of the material also means that more intricate shapes can be produced.

The success of transfer moulding prompted further developments in this area and clearly it was only a relatively small step to an injection moulding process for thermosets as described in Section 4.3.10.

(a) Preform in position   (b) Material forced into cavities

Fig. 4.64   Transfer moulding of thermosetting materials

4.9 Processing Reinforced Thermoplastics

Fibre reinforced thermoplastics can be processed using most of the conventional thermoplastic processing methods described earlier. Extrusion, rotational moulding, blow moulding and thermoforming of short fibre reinforced thermoplastics are all possible, but the most important commercial technique is injection moulding. In most respects this process is similar to the moulding of un-reinforced thermoplastics but there are a number of important differences. For example the melt viscosity of a reinforced plastic is generally higher than the unreinforced material. As a result the injection pressures need to be higher, by up to 80% in some cases. In addition the cycle times are generally lower because the greater stiffness of the material allows it to be ejected from the mould at a higher temperature than normal. However, the increased stiffness can also hamper ejection from the mould so it is important to have adequate taper on side walls of the cavity and a sufficient number of strategically placed ejector pins. Where possible a reciprocating screw machine is preferred to a plunger machine because of the better mixing, homogenisation, metering and temperature control of the melt. However, particular attention needs to be paid to such things as screw speed and back pressure because these will tend to break up the fibres and thus affect the mechanical properties of the mouldings.

A practical difficulty which arises during injection moulding of reinforced plastics is the increased wear of the moulding machine and mould due to the abrasive nature of the fibres. However, if hardened tool steels are used in the manufacture of screws, barrels and mould cavities then the problem may be negligible.

An inherent problem with all of the above moulding methods is that they must, by their nature, use short fibres (typically 0.2–0.4 mm long). As a result the full potential of the reinforcing fibres is not realised (see Section 2.8.5). In recent years therefore, there have been a number of developments in reinforced thermoplastics to try to overcome these problems. One approach has been to produce continuous fibre tapes or mats which can be embedded in a thermoplastic matrix. The best known materials of this type are the Aromatic Polymer Composites (APC) and the glass mat reinforced thermoplastics (GMT). One of the most interesting of these consists of unidirectional carbon fibres in a matrix of polyetheretherketone (PEEK). The material comes in the form of a wide tape which may be arranged in layers in one half of a mould to align the unidirectional fibres in the desired directions. The assembly is then pressurised between the two matched halves of the heated mould. The result is a laminated thermoplastic composite containing continuous fibres aligned to give maximum strength and stiffness in the desired directions.

Another recent development has been the arrival of special injection moulding grades of thermoplastics containing long fibres. At the granule production stage the thermoplastic lace contains continuous fibres and to achieve this it is produced by pultrusion (see Section 4.10.3) rather than the conventional compounding extruder. The result is that the granules contain fibres of the same length as the granule ($\simeq$10 mm).

These long fibres give better product performance although injection moulding machine modifications may be necessary to prevent fibre damage and reduce undesirable fibre orientation effects in the mould.

4.10 Processing Reinforced Thermosets

There is a variety of ways in which fibre reinforcement may be introduced into thermosetting materials and as a result there is a range of different methods used to process these materials. In many cases the reinforcement is introduced during the fabrication process so that its extent can be controlled by the moulder. Before looking at the possible manufacturing methods for fibre reinforced thermosetting articles it is worth considering the semantics of fibre technology. Because of their fibre form, reinforcing materials have borrowed some of their terminology from the textile industry.

Filament This is a single fibre which is continuous or at least very long compared with its diameter.

Yarn or Roving Continuous bundle of filaments generally fewer than 10,000 in number.

Tow A large bundle of fibres generally 10,000 or more, not twisted.

Fabric, Cloth or Mat Woven strands of filament. The weave pattern used depends on the flexibility and balance of strength properties required in the warp and fill directions. Fig. 4.65 shows a plain weave in which the strength is uniform in both directions. The warp direction refers to the direction parallel to

Fig. 4.65   Plain weave fibre fabric

the length of the fabric. Fabrics are usually designated in terms of the number of yarns of filament per unit length of warp and fill direction.

Chopped Fibres These may be subdivided as follows

Milled Fibres: These are finely ground or milled fibres. Lengths range from 30 to 3000 microns and the fibre (L/D) ratio is typically about 30. Fibres in this form are popular for closed mould manufacturing methods such as injection moulding.

Short Chopped Fibres: These are fibres with lengths up to about 6 mm. The fibre (L/D) ratio is typically about 800. They are more expensive than milled fibres but provide better strength and stiffness enhancement.

Long Chopped Fibres: These are chopped fibres with lengths up to 50 mm. They are used mainly in the manufacture of SMC and DMC (see Section 4.10.2).

Chopped Strand Mat This consists of strands of long chopped fibres deposited randomly in the form of a mat. The strands are held together by a resinous binder.

Manufacturing Methods

The methods used for manufacturing articles using fibre reinforced thermosets are almost as varied as the number of material variations that exist. They can, however, be divided into three main categories. These are manual, semi-automatic and automatic.

The *Manual* processes cover methods such as hand lay-up, spray-up, pressure bag and autoclave moulding.

The *Semi-Automatic* processes include processes such as cold pressing, hot pressing, compression moulding of SMC and DMC, resin injection.

The *Automatic* processes are those such as pultrusion, filament winding, centrifugal casting and injection moulding.

Typical market shares for the different methods are shown in Fig. 4.66.

Fig. 4.66   Typical market shares for composite moulding methods in Europe

4.10.1 Manual Processing Methods

(a) **Hand Lay-Up:** This method is by far the most widely used processing method for fibre reinforced materials. In the UK it takes up about 40% of the FRP market. Its major advantage is that it is a very simple process so that very little special equipment is needed and the moulds may be made from plaster, wood, sheet metal or even FRP. The first step is to coat the mould with a release agent to prevent the moulding sticking to it. This is followed by a thin layer (approximately 0.3–0.4 mm) of pure resin (called a gelcoat) which has a number of functions. Firstly it conceals the irregular mesh pattern of the fibres and this improves the appearance of the product when it is taken from the mould. Secondly, and probably most important, it protects the reinforcement from attack by moisture which would tend to break down the fibre/resin interface. A tissue mat may be used on occasions to back up the gelcoat. This improves the impact resistance of the surface and also conceals the coarse texture of the reinforcement. However, it is relatively expensive and is only used if considered absolutely necessary.

When the gelcoat has been given time to partially cure the main reinforcement is applied. Initially a coat of resin (unsaturated polyester is the most common) is brushed on and this is followed by layers of tailored glass mat positioned by hand. As shown in Fig. 4.67 a roller is then used to consolidate the mat and remove any trapped air. The advantage of this technique is that the strength and stiffness of the composite can be controlled by building up the thickness with further layers of mat and resin as desired. Curing takes place at room temperature but heat is sometimes applied to accelerate this. Ideally any trimming should be carried out before the curing is complete because the

(a) Basic hand lay-up method

(b) Vacuum bag   **(c) Pressure bag**

Fig. 4.67   Hand lay-up techniques

material will still be sufficiently soft for knives or shears to be used. After curing, special cutting wheels may be needed.

Variations on this basic process are (i) *vacuum bag moulding* and (ii) *pressure bag moulding*. In the former process a flexible bag (frequently rubber) is clamped over the lay-up in the mould and a vacuum is applied between the moulding and the bag. This sucks the bag on to the moulding to consolidate the layers of reinforcement and resin. It also squeezes out trapped air and excess resin. The latter process is similar in principle except that pressure is applied above the bag instead of a vacuum below it. The techniques are illustrated in Fig. 4.67(b) and (c).

(b) **Spray-Up:** In this process, the preparatory stages are similar to the previous method but instead of using glass mats the reinforcement is applied using a spray gun. Roving is fed to a chopper unit and the chopped strands are sprayed on to the mould simultaneously with the resin (see Fig. 4.68). The thickness of the moulding (and hence the strength) can easily be built up in sections likely to be highly stressed. However, the success of the method depends to a large extent on the skill of the operator since he controls the overall thickness of the composite and also the glass/resin ratio.

Fig. 4.68   Spray-up technique

(c) **Autoclave Moulding:** In order to produce high quality, high precision mouldings for the aerospace industries, for example, it is necessary to have strict control over fibre alignment and consolidation of the fibres in the matrix. To achieve this, fabric 'pre-pregs' (i.e. a fabric consisting of woven fibre yarns pre-impregnated with the matrix material) are carefully arranged in layers in an open mould. The arrangement of the layers will determine the degree of anisotropy in the moulded article. A typical layer arrangement is shown in Fig. 4.69(a). The pre-preg stack is then covered with a series of bleeder and breather sheets, as shown in Fig. 4.69(b) and finally with a flexible vacuum bag. When the air is extracted from between the flexible bag and the pre-preg stack, the latter will be squeezed tightly on to the mould. The whole assembly is then transferred to a very large oven (autoclave) for curing.

Fig. 4.69   Diagramatic cross section of a bagged lay-up

4.10.2 Semi-Automatic Processing Methods

(a) **Cold Press Moulding:** The basis of this process is to utilise pressure applied to two unheated halves of a mould to disperse resin throughout a prepared fabric stack placed in the mould. The typical procedure is as follows. Release agent and gelcoat are applied to the mould surfaces and the fibre fabric is laid into the lower part of the open mould. The activated resin is then poured on top of the mat and when the mould is closed the resin spreads throughout the reinforcement. High pressures are not necessary as the process relies on squeezing the resin throughout the reinforcement rather than forcing the composite into shape. A typical value of cycle time is about 10–15 minutes compared with several hours for hand lay-up methods. The process is illustrated in Fig. 4.70.

(a) Mould with gel coat and reinforced layers prepositioned

(b) Mould opened and moulding released

Fig. 4.70   Basic cold press moulding process

(b) **Hot Press Mouldings:** In this type of moulding the curing of the reinforced plastic is accelerated by the use of heat ($\simeq 180°C$) and pressure ($\simeq 15$ MN/m^2). The general heading of Hot Press Moulding includes both preform moulding and compression moulding.

(i) *Pre-form Moulding*: This technique is particularly suitable for mass production and/or more complex shapes. There are two distinct stages. In the first a preform is made by, for example, spraying chopped fibres on to a perforated metal screen which has the general shape of the article to be moulded. The fibres are held on the screen by suction applied behind it (see Fig. 4.71). A resin binder is then sprayed on the mat and the resulting preform is taken from the screen and cured in an oven at about 150°C for several minutes. Other methods by which the preform can be made include tailoring a continuous fibre fabric to shape and using tape to hold it together. The preform is then transferred to the lower half of the heated mould and the activated resin poured on top. The upper half of the mould is then brought into position to press the composite into shape. The cure time in the mould depends on the temperature, varying typically from 1 minute at 150°C to 10 minutes at 80°C. If the mould was suitably prepared with release agent the moulding can then

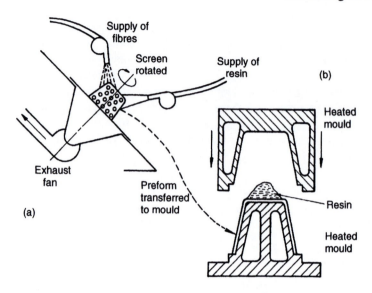

Fig. 4.71   Pre-form moulding of GFRP

be ejected easily. This method would not normally be considered for short production runs because the mould costs are high.

(ii) *Compression Moulding* (see also Section 4.7): Sheet Moulding Compounds: SMC is supplied as a pliable sheet which consists of a mixture of chopped strand mat or chopped fibres (25% by weight) pre-impregnated with resin, fillers, catalyst and pigment. It is ready for moulding and so is simply placed between the halves of the heated mould. The application of pressure then forces the sheet to take up the contours of the mould. The beauty of the method is that the moulding is done 'dry' i.e. it is not necessary to pour on resins. Fig. 4.72 illustrates a typical method used to manufacture SMC material.

Dough Moulding Compounds: DMC (also known as BMC – Bulk Moulding Compound) is supplied as a dough or rope and is a mixture of chopped strands (20% by weight) with resin, catalyst and pigment. It flows readily and so may be formed into shape by compression or transfer moulding techniques. In compression moulding the charge of dough may be placed in the lower half of the heated mould, in a similar fashion to that illustrated in Fig. 4.50(b) although it is generally wise to preform it to the approximate shape of the cavity. When the mould is closed, pressure is applied causing the DMC to flow in all sections of the cavity. Curing generally takes a couple of minutes for mould temperatures in the region of $120°-160°C$ although clearly this also depends on the section thickness.

In general, SMC moulds less well than DMC on intricate shapes but it is particularly suitable for large shell-like mouldings – automotive parts such as

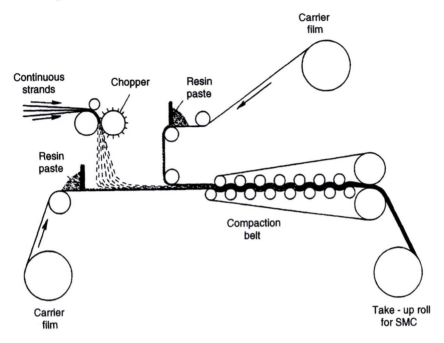

Fig. 4.72   Manufacture of SMC material

body panels and fascia panels are ideal application areas. An engine inlet manifold manufactured from SMC has recently been developed in the UK. DMC
finds its applications in the more complicated shapes such as business machine
housings, electric drill bodies, etc. In France, a special moulding method, called
ZMC, but based on DMC moulding concepts has been developed. Its most
famous application to date is the rear door of the Citroen BX saloon and
the process is currently under active consideration for the rear door of a VW
saloon car. Injection moulding of DMC is also becoming common for intri-
cately shaped articles (see Section 4.3.10).

Other types of compression moulding and stamp forming used for continuous
fibre reinforced composites are illustrated in Fig. 4.73.

(c) **Resin Injection:** This is a cold mould process using relatively low pressures
(approximately 450 kN/m^2). The mould surfaces are prepared with release
agent and gelcoat before the reinforcing mat is arranged in the lower half of
the mould. The upper half is then clamped in position and the activated resin
is injected under pressure into the mould cavity. The advantage of this type of
production method is that it reduces the level of skill needed by the operator
because the quality of the mould will determine the thickness distribution in the
moulded article (see Fig. 4.74). In recent times there has been a growing use of
pre-formed fabric shells in the resin injection process. The pre-form is produced

Fig. 4.73   Other types of compression moulding and stamp forming

using one of the methods described above and this is placed in the mould. This improves the quality and consistency of the product and reinforcements varying from chopped strand mat to close weave fabric in glass, aramid, carbon or hybrids of these may be used. It is possible, with care, to achieve reinforcement loadings in the order of 65%.

(d) **Vacuum Injection:** This is a development of resin injection in which a vacuum is used to draw resin throughout the reinforcement. It overcomes the

Fig. 4.74   Resin injection process

problem of voids in the resin/fibre laminate and offers faster cycle times with greater uniformity of product.

4.10.3 Automatic Processes

(a) **Filament Winding:** In this method, continuous strands of reinforcement are used to gain maximum benefit from the fibre strength. In a typical process rovings or single strands are passed through a resin bath and then wound on to a rotating mandrel. By arranging for the fibres to traverse the mandrel at a controlled and/or programmed manner, as illustrated in Fig. 4.75, it is possible to lay down the reinforcement in any desired fashion. This enables very high strengths to be achieved and is particularly suited to pressure vessels where reinforcement in the highly stressed hoop direction is important.

In the past a limitation on this process was that it tended to be restricted to shapes which were symmetrical about an axis of rotation and from which the mandrel could be easily extracted. However, in recent years there have been major advances through the use of collapsible or expendable cores and in particular through the development of computer-controlled winding equipment. The latter has opened the door to a whole new range of products which can be filament wound – for example, space-frame structures. Braiding machines for complex shapes are shown in Fig. 4.76.

(b) **Centrifugal Casting:** This method is used for cylindrical products which can be rotated about their longitudinal axis. Resin and fibres are introduced into the rotating mould/mandrel and are thrown out against the mould surface. The method is particularly suited to long tubular structures which can have a slight taper e.g. street light columns, telegraph poles, pylons, etc.

(c) **Pultrusion:** This is a continuous production method similar in concept to extrusion. Woven fibre mats and/or rovings are drawn through a resin bath and then through a die to form some desired shape (for example a 'plank'

Fig. 4.75   Filament winding of fibre composites

Fig. 4.76   Two types of filament winding

as illustrated in Fig. 4.77). The profiled shape emerges from the die and then passes through a tunnel oven to accelerate the curing of the resin. The pultruded composite is eventually cut to length for storage. A wide range of pultruded shapes may be produced – U channels, I beams, aerofoil shapes, etc.

(d) **Injection Moulding:** The injection moulding process can also be used for fibre reinforced thermoplastics and thermosets, for example DMC materials. This offers considerable advantages over compression moulding due to the higher production speeds, more accurate metering and lower product costs which can be achieved. The injection moulding process for thermosets has

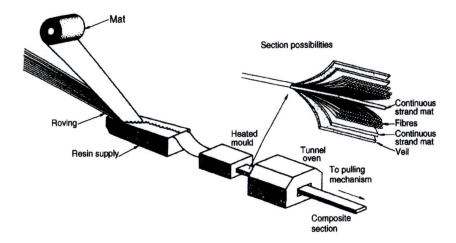

Fig. 4.77   Pultrusion process

already been dealt with in Section 4.3.8. See also the section on Reaction Injection Moulding (RIM) since this offers the opportunity to incorporate fibres.

Bibliography

Fisher, E.G, *Extrusion of Plastics*, Newnes-Butterworth, 1976.
Schenkel, G, *Plastics Extrusion Technology and Practice*, Iliffe, 1966.
Fisher, E.G, *Blow Moulding of Plastics*, Iliffe, 1971.
Rubin, I, *Injection Moulding-Theory and Practice*, Wiley, 1972.
Holmes-Walker, W.A, *Polymer Conversion*, Applied Science Publishers, 1975.
Bown, J, *Injection Moulding of Plastic Components*, McGraw-Hill, 1979.
Dym, J.B, *Injection Moulds and Moulding*, Van Nostrand Rheinhold, 1979.
Pye, R.G.W, *Injection Mould Design*, George Godwin, 1978.
Elden, R.A. and Swann, A.D, *Calendering of Plastics*, Plastics Institute Monograph, Iliffe, 1971.
Rosenzweig, N., Markis, M., and Tadmar, Z., *Wall Thickness Distribution in Thermoforming* Polym. Eng. Sci. October 1979 Vol 19 No. 13 pp 946–950.
Parker, F.J, *The Status of Thermoset Injection Moulding Today*, Progress in Rubber and Plastic Technology, October 1985, vol 1, No 4, pp 22–59.
Whelan, A and Brydson, J.A. *Developments with Thermosetting Plastics* App. Sci. Pub. London 1975.
Monk, J.F. *Thermosetting Plastics – Practical Moulding Technology*, George Godwin, London 1981.
Rosato, D.V. ed. Rosato, D.V. (ed.) *Blow Moulding Handbook*, Hanser, Munich (1989).
Hepburn, C. *Polyurethane Elastomers* (ch 6–RIM) Applied Science Publishers, London (1982).
Martin, J., *Pultrusion* Ch 3 in 'Plastics Product Design Handbook – B' ed by E. Miller, Dekker Inc, New York (1983).
Titow, W.V. and Lanham, B.J, *Reinforced Thermoplastics*, Applied Science Publisher, 1975.
Penn, W.S, *GRP Technology*, Maclaren, 1966.
Beck, R.D, *Plastic Product Design*, Van Nostrand Reinhold Co. 1980.
Schwartz, S.S. and Goodman, S.H. *Plastic Materials and Processes*, Van Nostrand, New York, 1982.
Crawford, R.J. *Rotational Moulding of Plastics*, Research Studies Press 2nd edition (1996).

Rosato, D.V. and Rosato, D.V. *Injection Molding Handbook*, 2nd edn, Chapman and Hall, New York (1995).

Stevens, M.J. and Covas, J.A. *Extruder Principles and Operation*, 2nd edn, Chapman and Hall, London (1995).

Michaeli, W. *Extrusion Dies*, Hanser, Munich (1984).

Baird, D.G. and Collias, D.I. *Polymer Processing*, Butterworth-Heinemann, Newton, USA (1995).

Michaeli, W. *Polymer Processing*, Hanser, Munich (1992).

Lee, N. (ed.), *Plastic Blow Moulding Handbook*, Van Nostrand Reinhold, New York (1990).

Throne, J.L. *Technology of Thermoforming*, Hanser, Munich (1996).

Florian, J. *Practical Thermoforming*, Marcel Dekker, New York (1987).

Boussinesq, M.J., *J. Math Pures et Appl.*, **2**, 13 (1868) pp. 377–424.

Meng Hou, Lin Ye and Yiu-Wing Mai, Advances in processing of continuous fibre reinforced composites *Plastics, Rubber and Composites Proc. and Appl.*, **23**, 5 (1995) pp. 279–292.

Mitchell, P. (ed.) *Tool and Manufacturing Engineers Handbook*, Vol 8, 4th edition, Soc. Man. Eng., Michigan (1996).

Questions

4.1 In a particular extruder screw the channel depth is 2.4 mm, the screw diameter is 50 mm, the screw speed is 100 rev/min, the flight angle is $17°$ $42'$ and the pressure varies linearly over the screw length of 1000 mm from zero at entry to 20 MN/m^2 at the die entry. Estimate

(a) the drag flow
(b) the pressure flow
(c) the total flow.

The plastic has a viscosity of 200 Ns/m^2. Calculate also the shear rate in the metering zone.

4.2 Find the operating point for the above extruder when it is combined with a die of length 40 mm and diameter 3 mm. What would be the effect on pressure and output if a plastic with viscosity 400 Ns/m^2 was used.

4.3 A single screw extruder has the following dimensions:

screw length = 500 mm
screw diameter = 25 mm
flight angle = $17°42'$
channel depth = 2 mm
channel width = 22 mm

If the extruder is coupled to a die which is used to produce two laces for subsequent granulation, calculate the output from the extruder/die combination when the screw speed is 100 rev/min. Each of the holes in the lace die is 1.5 mm diameter and 10 mm long and the viscosity of the melt may be taken as 400 Ns/m^2.

4.4 An extruder is coupled to a die, the output of which is given by (KP/η) where P is the pressure drop across the die, η is the viscosity of the plastic and K is a constant. What are the optimum values of screw helix angle and channel depth to give maximum output from the extruder.

4.5 A circular plate of diameter 0.5 m is to be moulded using a sprue gate in its centre. If the melt pressure is 50 MN/m^2 and the pressure loss coefficient is 0.6 estimate the clamping force required.

4.6 The container shown at the top of p. 341 is injection moulded using a gate at point A. If the injection pressure at the nozzle is 140 MN/m^2 and the pressure loss coefficient, m, is 0.5, estimate (i) the flow ratio and (ii) the clamping force needed.

4.7 Compare the efficiencies of the runners shown on p. 341.

4.8 A calender having rolls of diameter 0.3 m produces plastic sheet 1 m wide at the rate of 2000 kg/hour. If the roll speed is 5 rev/minute and the nip between the rolls is 4.5 mm, estimate the position and magnitude of the maximum pressure. The density of the material is 1400 kg/m^3 and its viscosity is 1.5×10^4 Ns/m^2.

4.9 A calender having rolls of 0.2 m diameter produces 2 mm thick plastic sheet at a linear velocity of 0.1 m/s. Investigate the effect of nips in the range 0.8 to 1.9 mm on the pressure profile. The viscosity is 10^3 Ns/m^2.

4.10 A hemispherical dome of 200 mm diameter has been vacuum formed from a flat sheet 4 mm thick. What is the thickness of the dome at the point furthest away from its diameter.

4.11 A disposable tumbler which has the shape of a frustrum of a cone is to be vacuum formed from a flat plastic sheet 3 mm thick. If the diameter of the mouth of the tumbler is 60 mm, the diameter of the base is 40 mm and the depth is 60 mm estimate the wall thickness at (a) a point 35 mm from the top and (b) in the centre of the base.

4.12 A blow moulding die which has an outside diameter of 40 mm and a die gap of 2 mm is used to produce a plastic bottle with a diameter of 70 mm. If the swelling ratio of the melt in the thickness direction is 1.8 estimate

(a) the parison dimensions
(b) the thickness of the bottle and
(c) a suitable inflation pressure if melt fracture occurs at a stress of 10 MN/m^2.

4.13 A plastic film, 0.1 mm thick, is required to have its orientation in the transverse direction twice that in the machine direction. If the film blowing die has an outer diameter of 100 mm and an inner diameter of 98 mm estimate the blow-up ratio which will be required and the lay flat film width. Neglect extrusion induced effects and assume there is no draw-down.

4.14 A molten polymer is to be coated on a cable at a speed of 0.5 m/s. The cable diameter is 15 mm and the coating thickness required is 0.3 mm. The die used has a length of 60 mm and an internal diameter of 16 mm. What pressure must be developed at the die entry if the viscosity of the polymer under these operating conditions is 100 Ns/m^2.

4.15 During a rotational moulding operation an aluminium mould with a uniform thickness of 3 mm is put into an oven at 300°C. If the initial temperature of the mould is 23°C, estimate the time taken for it to reach 250°C. The natural convection heat transfer coefficient is 28.4 J/m^{2}s.

K and the thermal diffusivity and conductivity of aluminium may be taken as 8.6×10^{-5} m^2/s and 230.1 J/m.s.K respectively.

4.16 A billet of PVC weighing 150 g is to be compression moulded into a long playing record of diameter 300 mm. If the maximum force which the press can apply is 100 kN estimate the time needed to fill the mould. The density and viscosity of the the PVC may be taken as 1200 kg/m^3 and 10 Ns/m^2 respectively.

CHAPTER 5 – Analysis of polymer melt flow

5.1 Introduction

In general, all polymer processing methods involve three stages – heating, shaping and cooling of a plastic. processing However, this apparent simplicity can be deceiving. Most plastic moulding methods are not straightforward and the practical know-how can only be gained by experience, often using trial and error methods. In most cases plastics processing has developed from other technologies (e.g. metal and glass) as an art rather than as a science. This is principally because in the early days the flow of polymeric materials was not understood and the rate of increase in the usage of the materials was much greater than the advances in the associated technology.

Nowadays the position is changing because, as ever increasing demands are being put on materials and moulding machines it is becoming essential to be able to make reliable quantitative predictions about performance. In Chapter 4 it was shown that a simple Newtonian approach gives a useful first approximation to many of the processes but unfortunately the assumption of constant viscosity can lead to serious errors in some cases. For this reason a more detailed analysis using a Non-Newtonian model is often necessary and this will now be illustrated.

Most processing methods involve flow in capillary or rectangular sections, which may be uniform or tapered. Therefore the approach taken here will be to develop first the theory for Newtonian flow in these channels and then when the Non-Newtonian case is considered it may be seen that the steps in the analysis are identical although the mathematics is a little more complex. At the end of the chapter a selection of processing situations are analysed quantitatively to illustrate the use of the theory. It must be stressed however, that even the more complex analysis introduced in this chapter will not give precisely accurate

343

solutions due to the highly complex nature of polymer melt flow. This chapter simply attempts to show how a quantitative approach may be taken to polymer processing and the methods illustrated are generally sufficiently accurate for most engineering design situations. Those wishing to take a more rigorous approach should refer to the work of Pearson, for example.

5.2 General Behaviour of Polymer Melts

In a fluid under stress, the ratio of the shear stress, τ. to the rate of strain, $\dot{\gamma}$, is called the shear viscosity, η, and is analogous to the modulus of a solid. In an ideal (Newtonian) fluid the viscosity is a material constant. However, for plastics the viscosity varies depending on the stress, strain rate, temperature etc. A typical relationship between shear stress and shear rate for a plastic is shown in Fig. 5.1.

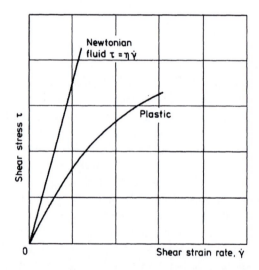

Fig. 5.1   Relations Between Shear Stress and Shear Rate

As a starting point it is useful to plot the relationship between shear stress and shear rate as shown in Fig. 5.1 since this is similar to the stress–strain characteristics for a solid. However, in practice it is often more convenient to rearrange the variables and plot viscosity against strain rate as shown in Fig. 5.2. Logarithmic scales are common so that several decades of stress and viscosity can be included. Fig. 5.2 also illustrates the effect of temperature on the viscosity of polymer melts.

When a fluid is flowing along a channel which has a uniform cross-section then the fluid will be subjected to shear stresses only. To define the flow behaviour we may express the fluid viscosity, η, as the ratio of shear stress, τ,

Fig. 5.2   Viscosity curves for polypropylene

to shear rate, $\dot{\gamma}$,

$$\eta = \frac{\tau}{\dot{\gamma}} \qquad (5.1)$$

If, on the other hand, the channel section changes then tensile stresses will also be set up in the fluid and it is often necessary to determine the tensile viscosity, λ, for use in flow calculations. If the tensile stress is σ and the tensile strain rate is ε then

$$\lambda = \frac{\sigma}{\dot{\varepsilon}} \qquad (5.2)$$

For many polymeric melts the tensile viscosity is fairly constant and at low stresses is approximately three times the shear viscosity.

To add to this picture it should be realised that so far only the viscous component of behaviour has been referred to. Since plastics are viscoelastic there will also be an elastic component which will influence the behaviour of the fluid. This means that there will be a shear modulus, G, and, if the channel section is not uniform, a tensile modulus, E, to consider. If γ_R and ε_R are the recoverable shear and tensile strains respectively then

$$G = \frac{\tau}{\gamma_R} \qquad (5.3)$$

$$E = \frac{\sigma}{\varepsilon_R} \qquad (5.4)$$

These two moduli are not material constants and typical variations are shown in Fig. 5.3. As with the viscous components, the tensile modulus tends to be about three times the shear modulus at low stresses. Fig. 5.3 has been included here as an introduction to the type of behaviour which can be expected from a polymer melt as it flows. The methods used to obtain this data will be described later, when the effects of temperature and pressure will also be discussed.

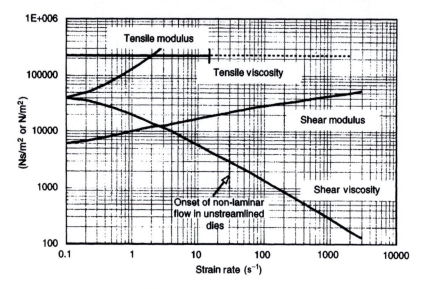

Fig. 5.3   Flow curves for polyethylene at 170°

5.3 Isothermal Flow in Channels: Newtonian Fluids

In the analysis of flow in channels the following assumptions are made:

1. There is no slip at the wall.
2. The melt is incompressible.
3. The flow is steady, laminar and time independent.
4. Fluid viscosity is not affected by pressure changes along the channel.
5. End effects are negligible.

The steady isothermal flow of incompresible fluids through straight horizontal tubes is of importance in a number of cases of practical interest.

(a) Flow of Newtonian Fluid along a Channel of Uniform Circular Cross-Section

Consider the forces acting on an element of fluid as shown in Fig. 5.4.

$$F_1 = \pi r^2 \left(P + \frac{\partial P}{\partial z} dz \right)$$

Fig. 5.4 Element of Fluid in a Capillary

$$F_2 = \pi r^2 P$$

$$F_3 = 2\pi r \, dz \, \tau_r$$

Since the flow is steady $\Sigma F_z = O$

$$\pi(r)^2 P = \pi r^2 \left(P + \frac{\partial P}{\partial z} dz \right) - 2\pi r \, dz \cdot \tau_r$$

$$\tau_r = \frac{r}{2} \left(\frac{dP}{dz} \right) \tag{5.5}$$

In many cases the pressure gradient is uniform, so that for a pressure drop, P, over a length, L, the maximum shear stress will be at the wall where $r = R$

$$\tau_w = \frac{PR}{2L} \tag{5.6}$$

Also since

$$\tau = \eta \dot{\gamma} = \eta \frac{\partial V}{\partial r}$$

So using equation (5.5)

$$\eta \cdot \frac{\partial V}{\partial r} = \frac{r}{2} \left(\frac{dP}{dz} \right)$$

Integrating this gives

$$\int_0^V dV = \int_R^r \frac{1}{2\eta} \frac{dP}{dz} \cdot r \, dr$$

$$V = \frac{1}{2\eta} \frac{dP}{dz} \left(\frac{r^2}{2} - \frac{R^2}{2} \right)$$

At $r = 0$, $V = V_0$

So

$$V_0 = -\frac{1}{4\eta} \cdot \frac{dP}{dz} \cdot R^2 \tag{5.7}$$

and

$$V = V_0 \left(1 - \left(\frac{r}{R}\right)^2\right) \tag{5.8}$$

The volume flow rate, Q, may be obtained from

$$Q = \int_0^R 2\pi r V \, dr$$

This can be rearranged using (5.7) to the form

$$Q = \int_0^R 2\pi r V_0 \left(1 - \left(\frac{r}{R}\right)^2\right) \cdot dr = \frac{1}{2}\pi V_0 R^2$$

$$Q = -\frac{\pi R^4}{8\eta} \cdot \frac{dP}{dz} \tag{5.9}$$

Once again if the pressure drop is uniform this may be expressed in the more common form

$$Q = \frac{\pi R^4 P}{8\eta L} \tag{5.10}$$

It is also convenient to derive an expression for the shear rate $\dot{\gamma}$.

$$\dot{\gamma} = \frac{\partial V}{\partial r}$$

then

$$\dot{\gamma} = \frac{\partial}{\partial r}\left[V_0\left(1 - \left(\frac{r}{R}\right)^2\right)\right] \tag{5.11}$$

This may be rearranged using the relation between flow rate and V_0 to give the shear rate at $r = R$ as

$$\dot{\gamma}_\omega = \frac{-4Q}{\pi R^3} \tag{5.12}$$

The negative signs for velocity and flow rate indicate that these are in the opposite direction to the chosen z-direction.

(b) Flow of Newtonian Fluid between Parallel Plates

Consider an element of fluid between parallel plates, T wide and spaced a distance H apart. For unit width of element the forces acting on it are:

$$F_1 = \left(P + \frac{\partial P}{\partial z} \cdot dz\right) 2y$$

Fig. 5.5   Element of Fluid Between Parallel Plates

$$F_2 = P \cdot \tau y$$

$$F_3 = \tau_y dz$$

For steady flow there must be equilibrium of forces so

$$P\tau y = \left(P + \frac{\partial P}{\partial z} \cdot dz \right) \tau y - 2\tau_y dz$$

$$y \frac{\partial P}{\partial z} = \tau_y \qquad\qquad (5.13)$$

In many cases the pressure gradient is uniform $(\partial P/\partial z = \Delta P/L)$ so the maximum shear stress will be at the wall where $y = \frac{1}{2}H$

$$\tau_w = \frac{PH}{2L} \qquad\qquad (5.14)$$

For a Newtonian Fluid the shear stress, τ, is also given by

$$\tau = \eta \dot{\gamma} = \eta \frac{\partial V}{\partial y}$$

So using (5.13)

$$\eta \frac{\partial V}{\partial y} = y \frac{\partial P}{\partial z}$$

Integrating this gives

$$\int_0^V dV = \int_{H/2}^y \frac{1}{\eta} \frac{dP}{dz} \cdot y \, dy$$

$$V = \frac{1}{2\eta}\frac{dP}{dz}\left(y^2 - \left(\frac{H}{2}\right)^2\right)$$

Now at $y = 0$, $V = V_0$

So
$$V_0 = -\frac{1}{2\eta}\frac{dP}{dz}\left(\frac{H}{2}\right)^2 \tag{5.15}$$

Substituting $\partial P/\partial z$ in the expression for V gives

$$V = V_0\left(1 - \left(\frac{2y}{H}\right)^2\right) \tag{5.16}$$

The volume flow rate, Q, is given by

$$Q = 2\int_0^{H/2} TV\,dy$$

Using equation (5.16) this may be expressed in the form

$$Q = 2\int_0^{H/2} TV_0\left(1 - \left(\frac{2y}{H}\right)^2\right)dy$$

$$= 2TV_0\left(\frac{H}{3}\right)$$

$$Q = -\frac{TH^3}{12\eta}\cdot\frac{dP}{dz} \tag{5.17}$$

which, for a uniform pressure gradient, reduces to

$$Q = \frac{TPH^3}{12\eta L} \tag{5.18}$$

An expression for the shear rate, $\dot{\gamma}$, may also be derived from

$$\dot{\gamma} = \frac{\partial V}{\partial y}$$

$$= \frac{\partial}{\partial y}\left[V_0\left(1 - \left(\frac{2y}{H}\right)^2\right)\right]$$

$$\dot{\gamma} = -8V_0 y/H^2 \tag{5.19}$$

but as shown above
$$V_0 = \frac{3Q}{2TH}$$

So
$$\dot{\gamma} = \frac{-12Qy}{TH^3}$$

At the wall
$$y = \frac{H}{2}$$

$$\dot{\gamma}_\omega = -\frac{6Q}{TH^2} \qquad (5.20)$$

It is worth noting that the equations for flow between parallel plates may also be used with acceptable accuracy for flow along a circular annular slot. The relevant terms are illustrated in Fig. 5.6.

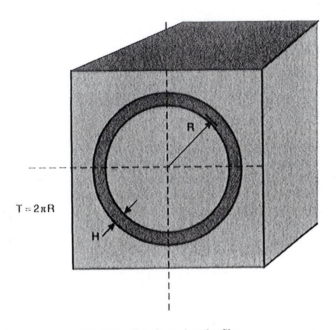

Fig. 5.6   Flow in an Annular Slot

5.4 Rheological Models for Polymer Melt Flow

When log (viscosity) is plotted against log (shear rate) or log (shear stress) for the range of shear rates encounterd in many polymer processing operations, the result is a straight line. This suggests a simple power law relation of the type

$$\tau = \tau_0 \left[\frac{\dot{\gamma}_a}{\dot{\gamma}_o} \right]^n \qquad (5.21)$$

where n is the Power Law index and τ_0 and γ_0 refer to the shear stress and shear rate at some reference state. This is often taken as $\dot{\gamma}_0 = 1 \text{ s}^{-1}$ and so the

Power Law may be expressed as

$$\tau = \tau_0 \, \dot{\gamma}_a^n \tag{5.22}$$

It is common practice nowadays to express the apparent viscosity as a function of the shear rate, γ_a. This may be obtained using the definition of apparent viscosity η_a as the ratio of shear stress to shear rate, ie

$$\frac{\tau}{\tau_0} = \frac{\eta_a}{\eta_0} \cdot \frac{\dot{\gamma}_a}{\dot{\gamma}_0}$$

So using

$$\frac{\eta_a}{\eta_0} \cdot \frac{\dot{\gamma}_a}{\dot{\gamma}_0} = \left[\frac{\dot{\gamma}_a}{\dot{\gamma}_0} \right]^n$$

$$\frac{\eta_a}{\eta_0} = \left[\frac{\dot{\gamma}_a}{\dot{\gamma}_0} \right]^{n-1} \tag{5.23}$$

or rearranging to use shear stress instead of shear rate

$$\frac{\eta_a}{\eta_0} = \left[\frac{\tau}{\tau_0} \right]^{\frac{n-1}{n}} \tag{5.24}$$

Equations (5.21), (5.22) and (5.23) are useful for the high strain rates experienced in injection moulding or extrusion but unfortunately they do not predict the low strain rate situation very well where plastic melts tend towards Newtonian behaviour (ie $n \to 1$). This is illustrated in Fig. 5.7.

Carreau proposed an alternative equation to overcome this problem. It has the form

$$\frac{\eta_a}{\eta_0} = \left[1 + (A_t \dot{\gamma}_a)^2 \right]^{\frac{n-1}{2}} \tag{5.25}$$

where A_t is a material constant.

It may be seen that when $\dot{\gamma} \gg 1$, this equation reduces to the Power Law (5.23) but as shown in Fig. 5.7, it also predicts the low shear rate situation very well.

In order to allow for the effect of temperature on viscosity a shift factor, a_T is often used. The Carreau equation then becomes

$$\frac{\eta_a}{\eta_0} = a_T \left[1 + a_T (A_t \dot{\gamma}_a)^2 \right]^{\frac{n-1}{2}} \tag{5.26}$$

where typically the shift factor, a_T, may be obtained from

$$\log a_T = \frac{C_1(T_1 - T_0)}{C_2 + (T_1 - T_0)} - \frac{C_1(T_2 - T_0)}{C_2 + (T_2 - T_0)} \tag{5.27}$$

Fig. 5.7   Comparison of Models for Flow Behaviour

where C_1 and C_2 are material constants, T_1 (°C) is the temperature at which the viscosity is known, T_2 (°C) is the temperature at which the viscosity is required and T_0 (°C) is a reference temperature.

When $T_0 = T_g$ then $C_1 = 17.4$ and $C_2 = 51.6$.

With the widespread use of software packages to assist with computational fluid dynamics (CFD) of polymer flow situations, other types of viscosity relationships are also used. For example, the regression equation of Klien takes the form

$$\ln \eta_a = a_0 + a_1 \ln \gamma_a + a_{11}(\ln \dot{\gamma})^2 + a_2 T + a_{22} T^2 + a_{12} \cdot T \cdot \ln \dot{\gamma}_a \qquad (5.28)$$

where a_{ij} are material constants and T is the temperature of the melt (°C).

It should be noted that if a_{11} is small (which is often the case) then this equation also reduces to the Power Law.

The polynomial of Muenstedt is also popular. It takes the form

$$\log \eta_a = \log a_T + A_0 + A_1 \log (a_T \dot{\gamma}_a) + A_2 [\log(a_T \dot{\gamma})]^2$$
$$+ A_3 [\log(a_T \dot{\gamma}_a)]^3 + A_4 [\log(a_T \dot{\gamma})]^4 \qquad (5.29)$$

However, for the high strain rates appropriate for the analysis of typical extrusion and injection moulding situations it is often found that the simple Power Law is perfectly adequate. Thus equations (5.22), (5.23) and (5.27) are important for most design situations relating to polymer melt flow.

Example 5.1 Applying the Carreau model to polypropylene, the following constants are known at 190°C.

$$\eta_0 = 2350 \text{ Ns/m}^2, \quad A_T = 0.05, \quad n = 0.33$$

Estimate the viscosity of polypropylene at 230°C and a shear rate of 1000 s^{-1}. The glass transition temperature for the polypropylene is -10°C.

Solution The temperature shift factor from 190°C to 230°C may be obtained from

$$\log a_T = \frac{C_1(T_1 - T_g)}{C_2 + (T_1 - T_g)} - \frac{C_1(T_2 - T_g)}{C_2 + (T_2 - T_g)}$$

$$= \frac{17.4(190 + 10)}{51.6 + (190 + 10)} - \frac{17.4(230 + 10)}{51.6 + (230 + 10)}$$

$$a_T = 0.324$$

Using the Carreau model

$$\eta_a = \eta_0 a_T \left(1 + a_T(A_T \cdot \dot{\gamma})^2\right)^{\frac{n-1}{n}}$$

$$= 2350(0.324) \left(1 + 0.324(0.05 \times 10^3)^2\right)^{\frac{0.33-1}{0.33}}$$

So, $\eta_a = 80.7$ Ns/m^2 at   230°C   and $\dot{\gamma} = 1000$

5.5 Isothermal Flow in Channels: Non-Newtonian Fluids

(a) Flow of Power Law Fluid Along a Channel of Uniform Circular Cross-section

Referring to Fig. 5.4 and remembering that equilibrium of forces was the only condition necessary to derive (5.5) then this equation may be used as the starting point for the analysis of any fluid.

$$\tau = \frac{r}{2} \left(\frac{\partial P}{\partial z}\right) \tag{5.5}$$

Now

$$\tau = \eta\dot{\gamma} = \eta\frac{\partial v}{\partial r}$$

so

$$\frac{\partial V}{\partial R} = \frac{r}{2\eta} \left(\frac{\partial P}{\partial z}\right)$$

but from (5.23)

$$\eta = \eta_0 \left[\frac{\dot{\gamma}}{\dot{\gamma}_0}\right]^{n-1} = \eta_0 \left[\frac{\partial V/\partial r}{\dot{\gamma}_0}\right]^{n-1}$$

So from above

$$\frac{\partial V}{\partial r} = \frac{r}{2\eta_0 \left[\dfrac{\partial V/\partial r}{\dot\gamma_0}\right]^{n-1}} \cdot \left(\frac{\partial P}{\partial z}\right)$$

$$\left(\frac{\partial V}{\partial r}\right)^n = \frac{r(\dot\gamma_0)^{n-1}}{2\eta_0} \cdot \left(\frac{\partial P}{\partial z}\right)$$

Integrating this between the limits

Velocity $= V$ at radius $= r$
and Velocity $= O$ at radius $= R$

The equation for velocity at any radius may then be expressed as

$$V = \left(\frac{n}{n+1}\right) \dot\gamma_0^{(n-1)/n} \left[\frac{(\partial P)/(\partial z)}{2\eta_0}\right]^{1/n} [r^{(n+1)/n} - R^{(n+1)/n}]$$

At $r = 0$, $V = V_0$

$$V_0 = -\left(\frac{n}{n+1}\right) \dot\gamma_0^{(n+1)/n} \left[\frac{(\partial P)/(\partial z)}{2\eta_0}\right]^{1/n} R^{(n+1)/n} \qquad (5.30)$$

$$V = V_0 \left[1 - \left(\frac{r}{R}\right)^{(n+1)/n}\right] \qquad (5.31)$$

The volume flow rate, Q, is given by

$$Q = \int_0^R 2\pi r V \, dr = 2\pi V_0 \int_0^R r \left(1 - \left(\frac{r}{R}\right)^{(n+1)/n}\right) dr$$

$$Q = \left(\frac{n+1}{3n+1}\right) \pi R^2 V_0 \qquad (5.32)$$

Also Shear Rate, $\dot\gamma = \dfrac{\partial V}{\partial r} = \dfrac{\partial}{\partial r}\left[V_0 \left(1 - \left(\frac{r}{R}\right)^{(n+1)/n}\right)\right]$

$$\dot\gamma = -\frac{\left(\dfrac{n+1}{n}\right) V_0}{R} \cdot \left(\frac{r}{R}\right)^{1/n} \qquad (5.33)$$

The shear rate at the wall is obtained by letting $r = R$ and the equation is more convenient if V_0 is replaced by Q from (5.33). In this case

Shear Rate at wall $\qquad \dot\gamma_\omega = -\left(\dfrac{3n+1}{4n}\right) \dfrac{4Q}{\pi R^3}$

It is worth noting that since a Newtonian Fluid is a special case of the Power Law fluid when $n = 1$, then the above equations all reduce to those derived in Section 5.3(a) if this substitution is made.

(b) Flow of Power Law Fluid between Parallel Plates

Referring to Fig. 5.5 and recognising that equation (5.13) is independent of the fluid then

$$\tau = y \frac{\partial P}{\partial z}$$

but

$$\tau = \eta \frac{\partial V}{\partial y} \quad \text{and} \quad \eta = \eta_0 \left[\frac{\dot{\gamma}}{\dot{\gamma}_0} \right]^{n-1}$$

So

$$\frac{\partial V}{\partial y} = \frac{y}{\eta_0 \left[\dfrac{\partial V / \partial y}{\dot{\gamma}_0} \right]^{n-1}} \cdot \left(\frac{\partial P}{\partial z} \right)$$

$$\frac{\partial V}{\partial y} = \dot{\gamma}_0^{(n-1)/n} y^{1n} \left(\frac{1}{\eta_0} \frac{\partial P}{\partial z} \right)^{1/n}$$

Integrating this equation between the limits

velocity $= V$ at distance $= y$
velocity $= O$ at distance $= H/2$

$$V = \left(\frac{n}{n+1} \right) \dot{\gamma}_0 \left(\frac{1}{\eta_0 \dot{\gamma}_0} \frac{\partial P}{\partial z} \right)^{1/n} \left[y^{(n+1)/n} - \left(\frac{H}{2} \right)^{(n+1)/n} \right]$$

At $y = 0 \, V = V_0$

$$V_0 = - \left(\frac{n}{n+1} \right) \dot{\gamma}_0 \left(\frac{1}{\eta_0 \dot{\gamma}_0} \frac{\partial P}{\partial z} \right)^{1/n} \left(\frac{H}{2} \right)^{(n+1)/n} \qquad (5.34)$$

And using this to simplify the expression for V

$$V = V_0 \left(1 - \left(\frac{2y}{H} \right)^{(n+1)/n} \right) \qquad (5.35)$$

The flow rate, Q, may then be obtained from the expression

$$Q = 2T \int_0^{H/2} V \, dy$$

$$= 2T \int_0^{H/2} V_0 \left(1 - \left(\frac{2y}{H} \right)^{(n+1)/n} \right) dy$$

$$Q = \left(\frac{n+1}{2n+1}\right) TV_0 H \qquad (5.36)$$

And as before $\quad \dot{\gamma} = \dfrac{\partial V}{\partial y} = \dfrac{\partial}{\partial y}\left[V_0\left(1 - \left(\dfrac{2y}{H}\right)^{(n+1)/n}\right)\right]$

$$\dot{\gamma} = -V_0\left(\frac{2}{H}\right)^{(n+1)/n}\left(\frac{n+1}{n}\right)y^{1/n} \qquad (5.37)$$

At the wall $y = H/2$ and substituting for V_0 from (5.36) then

$$\dot{\gamma}_\omega = -\left(\frac{2n+1}{n}\right)\frac{2Q}{TH^2}$$

It can be seen that in the special case of $n = 1$ these equations reduce to those for Newtonian Flow.

5.6 Isothermal Flow in Non-Uniform Channels

In many practical situations involving the flow of polymer melts through dies and along channels, the cross-sections are tapered. In these circumstances, tensile stresses will be set up in the fluid and their effects superimposed on the effects due to shear stresses as analysed above. Cogswell has analysed this problem for the flow of a power law fluid along coni-cylindrical and wedge channels. The flow in these sections is influenced by three factors:

(1) entry effects given suffix O
(2) shear effects given suffix S
(3) extensional effects given suffix E

Each of these will contribute to the behaviour of the fluid although since each results from a different deformation mode, one effect may dominate depending on the geometry of the situation. This will be seen more clearly later when specific design problems are tackled.

(a) Flow in Coni-Cylindrical Dies

Consider the channel section shown in Fig. 5.8(a) from which the element shown in Fig. 5.8(b) may be taken.

(i) Pressure Drop Due to Shear, P_s

Assuming that the different modes of deformation are separable then considering equilibrium of forces in regard to the shear stress only, gives

$$dP_s \cdot \pi r^2 = 2\pi r \, dl \sec\theta \cdot \tau \cos\theta$$

$$dP_s = \frac{2\tau}{r} \cdot dl \qquad (5.38)$$

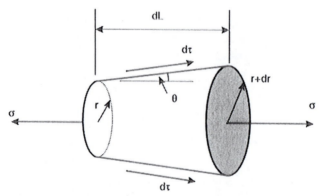

Fig. 5.8   Analysis of Coni-Cylindrical Die

but for a power law fluid

$$\tau = \tau_0 \dot{\gamma}^n \quad \text{and} \quad \dot{\gamma} = \left(\frac{3n+1}{n}\right)\frac{Q}{\pi r^3}$$

and using $dl = \dfrac{dr}{\tan\theta}$

then (5.38) becomes

$$dP_s = \frac{2\tau_0}{\tan\theta}\left(\left(\frac{3n+1}{n}\right)\frac{Q}{\pi}\right)^n \frac{dr}{r^{1+3n}}$$

Integrating

$$P_s = \frac{2\tau_0}{\tan\theta}\left(\left(\frac{3n+1}{n}\right)\frac{Q}{\pi}\right)^n \int_{R_1}^{R_0} \frac{dr}{r^{1+3n}}$$

$$P_s = \frac{2\tau_0}{3n\tan\theta}\left(\left(\frac{3n+1}{n}\right)\frac{Q}{\pi R_1^3}\right)^n\left[1-\left(\frac{R_1}{R_0}\right)^{3n}\right]$$

$$P_s = \frac{2\tau_1}{3n\tan\theta}\left[1-\left(\frac{R_1}{R_0}\right)^{3n}\right] \tag{5.39}$$

(ii) Pressure Drop Due to Extensional Flow P_E

$$dP_E \cdot \pi r^2 = \sigma[\pi(r+dr)^2 - \pi r^2]$$

$$dP_E = \frac{2\sigma\,dr}{r}$$

Integrating

$$P_E = \int_{R_1}^{R_0} \frac{2\sigma\,dr}{r} \tag{5.40}$$

In order to integrate the right hand side of this equation it is necessary to get an equation for σ in terms of r. This may be obtained as follows.

Consider a converging annulus of thickness, h, and radius, a, within the die. If the angle of convergence is ϕ then simple geometry indicates that for uniform convergence

$$\tan\phi = \frac{a}{r}\tan\theta \quad \text{and} \quad \frac{dh}{h} = \frac{da}{a}$$

Now if the simple tensile strain, ε, is given by

$$\varepsilon = \frac{dA}{A}$$

then the strain rate, $\dot{\varepsilon}$, will be given by

$$\dot{\varepsilon}_a = \frac{1}{area}\frac{d(area)}{dt} = \frac{1}{2\pi ah}\cdot\frac{d(2\pi ah)}{dt} = \frac{2}{a}\frac{da}{dt}$$

but

$$\frac{da}{dt} = \tan\phi\cdot\frac{dl}{dt} = \frac{a}{r}\tan\theta\cdot V$$

where V is the velocity of the fluid parallel to the centre line of the annulus. From (5.31) and (5.32) this velocity is given by

$$V = \left(\frac{3n+1}{n+1}\right)\frac{Q}{\pi r^3}\left[\left(\frac{a}{r}\right)^{(n+1)/n} - 1\right]$$

So

$$\dot{\varepsilon}_a = \frac{2}{r}\tan\theta\left(\frac{3n+1}{n+1}\right)\frac{Q}{\pi r^3}\left[\left(\frac{a}{r}\right)^{(n+1)/n} - 1\right]$$

The average extensional stress is obtained by integrating across the element so

$$\sigma\pi r^2 = \int_0^r \sigma_a 2\pi a \; da$$

and since by definition $\sigma_a = \lambda\dot{\varepsilon}_a$

$$\sigma\pi r^2 = \int_0^r \lambda\dot{\varepsilon}_a 2\pi a \; da = \int_0^r \lambda\cdot\frac{2}{r}\tan\theta\left(\frac{3n+1}{n+1}\right)\frac{Q}{\pi r^2}\left[\left(\frac{a}{r}\right)^{(n+1)/n} - 1\right]2\pi a \; da$$

$$\sigma = \lambda\left(\frac{2Q}{\pi r^3}\right)\tan\theta \tag{5.41}$$

So in (5.40)

$$P_E = \int_{R_1}^{R_0}\frac{2}{r}\lambda\left(\frac{2Q}{\pi r^3}\right)\tan\theta\cdot dr = \frac{\lambda\tan\theta}{3}\left(\frac{4Q}{\pi R_1^3}\right)\left[1 - \left(\frac{R_1}{R_0}\right)^3\right]$$

Using (5.41)

$$P_E = 2\sigma_1/3\left[1 - \left(\frac{R_1}{R_0}\right)^3\right] \tag{5.42}$$

(iii) Pressure Drop at Die Entry, P_0

When the fluid enters the die from a reservoir it will conform to a streamline shape such that the pressure drop is a minimum. This will tend to be of a coni-cylindrical geometry and the pressure drop, P_0, may be estimated by considering an infinite number of very short frustrums of a cone.

Consider a coni-cylindrical die with outlet radius, R_0, and inlet radius, R_i, and an included angle $2\alpha_0$.

From (5.40)

$$P_S = \frac{2\tau_0}{3n\tan\alpha_0}[1 - x^n]$$

From (5.42)

$$P_E = \frac{\lambda\tan\theta}{3}\left(\frac{4Q}{\pi R_0^3}\right)[1 - x]$$

where $x = \left(\dfrac{R_0}{R_1}\right)^3$ for convenience

The pressure drop, P_1, for such a die is

$$P_1 = P_S + P_E$$

$$= \frac{a}{\tan \alpha_0} + b \tan \alpha_0$$

where $\qquad a = \dfrac{2\tau_0}{3n}[1 - x^n], \quad b = \dfrac{\lambda}{3}\left(\dfrac{4Q}{\pi R_0^3}\right)[1 - x]$

For minimum pressure drop the differential of pressure drop with respect to the angle α_0 should be zero

$$\frac{dP_1}{d(\tan \alpha_0)} = \frac{-a}{\tan^2 \sigma_0} + b = 0$$

$$\tan \alpha_0 = \left(\frac{a}{b}\right)^{1/2}$$

So the minimum value of pressure is given by

$$P_1 = a^{1/2}b^{1/2} + a^{1/2}b^{1/2} = 2a^{1/2}b^{1/2}$$

If this procedure is repeated for other short coni-cylindrical dies then it is found that

$$P_2 = x^{(1+n)/2}P_1$$
$$P_3 = (x^{(1+n)/2})^2 P_1 \text{ and so on.}$$

So the total entry pressure loss, P_0, is given by

$$P_0 = \underset{x \to 1}{\text{Limit}} \sum_{i=1}^{i=\infty} P_i = \frac{2\sqrt{2}}{3}\left(\frac{4Q}{\pi R_0^3}\right)\left(\frac{\eta\lambda}{n}\right)^{1/2} \cdot \underset{x \to 1}{\text{Limit}} f(x)$$

where

$$f(x) = [(1 - x^n)(1 - x)]^{1/2} \cdot [1 + x^{(1+n)/2} + (x^{(1+n)/2})^2 + (x^{(1+n)/2})^3 + \cdots]$$

and η = viscosity corresponding to shear rate at die entry.

$$P_0 = \frac{4\sqrt{2}}{3(n+1)}\left(\frac{4Q}{\pi R_0^3}\right)(\eta\lambda)^{1/2} \qquad (5.43)$$

Also $\qquad \tan^2 \alpha_0 = \dfrac{2\tau_0}{3n} \cdot \dfrac{3}{\lambda}\left(\dfrac{\pi R_0^3}{4Q}\right)\underset{x \to 1}{\text{Limit}}\left[\dfrac{1 - x^n}{1 - x}\right]$

$$\tan \alpha_0 = \left(\frac{2\eta}{\lambda}\right)^{1/2}$$

So that from (5.44)

$$P_0 = \frac{8\sigma}{3(n + 1)} \tag{5.44}$$

(b) Flow in Wedge Shaped Die

Referring to the terminology in Fig. 5.9 and using analysis similar to that for the coni-cylindrical die, it may be shown that the shear, extensional and die entry pressure losses are given by

$$P_S = \frac{\tau_1}{2n \tan \phi}\left[1 - \left(\frac{H_1}{H_0}\right)^{2n}\right] \tag{5.45}$$

$$P_E = \frac{\sigma_1}{2}\left[1 - \left(\frac{H_1}{H_0}\right)^2\right] \tag{5.46}$$

$$P_0 = \frac{4}{(3n + 1)}\left(\frac{6Q}{TH_0^2}\right)(\eta\lambda)^{1/2} \tag{5.47}$$

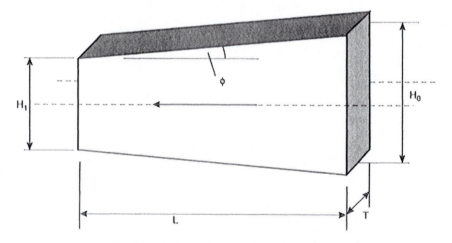

Fig. 5.9   Analysis of Flow in Tapered Wedge

where η corresponds to shear rate at die entry

Also

$$\dot{\gamma} = \left(\frac{2n + 1}{n}\right)\left(\frac{2Q}{TH^2}\right) \tag{5.48}$$

$$\dot{\varepsilon} = (\dot{\gamma}/3) \tan \phi \tag{5.49}$$

and
$$\tan \beta_0 = \frac{3}{2} \left(\frac{\eta}{\lambda} \right)^{1/2} \qquad (5.50)$$

where β_0 is the half angle of convergence to the die.

5.7 Elastic Behaviour of Polymer Melts

As discussed earlier, polymer melts can also exhibit elasticity. During flow they have the ability to store strain energy and when the stresses are removed, this strain is recoverable. A good example of elastic recovery is post extrusion swelling. After extrusion the dimensions of the extrudate are larger than those of the die, which may present problems if the dimensions of the extrudate are critical. In these circumstances some knowledge of the amount of swelling likely to occur is essential for die design. If the die is of a non-uniform section (tapered, for example) then there will be recoverable tensile and shear strains. If the die has a uniform cross-section and is long in relation to its transverse dimensions then any tensile stresses which were set up at the die entry for example, normally relax out so that only the shear component contributes to the swelling at the die exit. If the die is very short (ideally of zero length) then no shear stresses will be set up and the swelling at the die exit will be the result of recoverable tensile strains only. In order to analyse the phenomenon of post extrusion swelling it is usual to define the swelling ratio, B, as

$$B = \frac{\text{dimension of extrudate}}{\text{dimension of die}}$$

Swelling Ratios Due to Shear Stresses

(a) Long Capillary

Fig. 5.10 shows an annular element of fluid of radius r and thickness dr subjected to a shear stress in the capillary. When the element of fluid emerges from the die it will recover to the form shown by $ABCD$.

If the shear strain at radius r is γ_r then

$$\gamma_r = \tan \alpha = \frac{ED}{AE}$$

Also
$$\frac{\text{area of swollen annulus}}{\text{initial area of annulus}} = \frac{2\pi r \, dr^1}{2\pi r \, dr} = \frac{AD}{AE}$$

$$= \frac{(AE^2 + ED^2)^{1/2}}{AE}$$

$$= \left(1 + \left(\frac{ED}{AE} \right)^2 \right)^{1/2}$$

$$= (1 + \gamma_r^2)^{1/2}$$

Fig. 5.10 Polymer Melt Emerging from a Long Die

So from the definition of swelling ratio and using the subscripts **S** and **R** to denote **S**hear swelling in the **R**adial direction then

$$B_{SR}^2 = \frac{\text{area of swollen extrudate}}{\text{area of capillary}} = \frac{\displaystyle\int_0^R (1 + \gamma_r^2)^{1/2} 2\pi r \, dr}{\displaystyle\int_0^R 2\pi r \, dr}$$

Assuming that the shear strain, γ_r, varies linearly with radius, r, then

$$\gamma_r = \frac{r}{R}\gamma_R$$

where γ_R is the shear strain at the wall

So
$$B_{SR}^2 = \frac{\displaystyle\int_0^R \left(1 + \frac{r^2}{R^2}\gamma^2 R\right)^{1/2} 2\pi r \, dr}{\pi R^2}$$

$$B_{SR} = \left[\tfrac{2}{3}\gamma_R \left\{(1 + \gamma_R^{-2})^{3/2} - \gamma_R^{-3}\right\}\right]^{1/2} \tag{5.51}$$

(b) Long Rectangular Channel

When the polymer melt emerges from a die with a rectangular section there will be swelling in both the width (T) and thickness (H) directions. By a similar analysis to that given above; expressions may be derived for the swelling in these two directions. The resulting equations are

$$B_{ST} = \left[\frac{1}{2}(1 + \gamma_R^2)^{1/2} + \frac{1}{2\gamma_R} \ln\{\gamma_R + (1 + \gamma_R^2)^{1/2}\}\right]^{1/3} \tag{5.52}$$

$$B_{SH} = B_{ST}^2 \tag{5.53}$$

Equations (5.51), (5.52) and (5.53) can be cumbersome if they are to be used regularly so the relationships between swelling ratio and recoverable strain are often presented graphically as shown in Fig. 5.11.

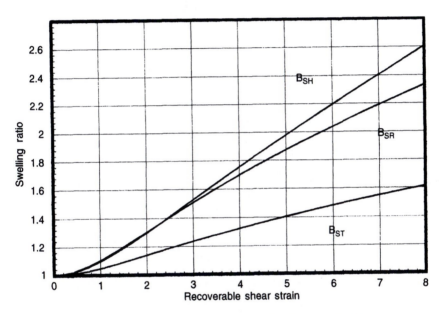

Fig. 5.11 Variation of Swelling Ratio for Capillary and Slit Dies

Swelling Ratio Due to Tensile Stresses

(a) Short Capillary (zero length)

Consider the annular element of fluid shown in Fig. 5.12. The true tensile strain ε_R in this element is given by

$$\varepsilon_R = \ln(1 + \varepsilon)$$

where ε is the nominal strain (extension ÷ original length)

$$\varepsilon_R = \ln\left(1 + \left(\frac{dr^1 - dr}{dr}\right)\right)$$

Fig. 5.12   Polymer Melt Emerging from a Short Die

$$e^{\varepsilon R} = 1 + \left(\frac{dr^1 - dr}{dr}\right)$$

$$(e^{\varepsilon R} - 1)dr + dr = dr^1$$

Now, $\dfrac{\text{area of swollen annulus}}{\text{original area of annulus}} = \dfrac{2\pi r \, dr^1}{2\pi r \, dr} = \dfrac{dr^1}{dr}$

$$= \frac{(e^{\varepsilon R} - 1)dr + dr}{dr}$$

$$= e^{\varepsilon R}$$

So from the definition of swelling ratio and using the subscript, E, to denote extensional stresses then

$$B_{ER}^2 = \frac{\text{area of swollen extrudate}}{\text{area of capillary}} = \frac{\displaystyle\int_0^R 2\pi r e^{\varepsilon R} \, dr}{\displaystyle\int_0^R 2\pi r \, dr}$$

$$B_{ER} = (e^{\varepsilon R})^{1/2} \qquad\qquad (5.54)$$

(b) Short Rectangular Channel

By similar analysis it may be shown that for a short rectangular slit the swelling ratios in the width (T) and thickness (H) directions are given by

$$B_{ET} = (e^{\varepsilon R})^{1/4} \qquad\qquad (5.55)$$

$$B_{EH} = (e^{\varepsilon_R})^{1/2} \tag{5.56}$$

Although these expressions are less difficult to use than the expressions for shear, it is often convenient to have the relationships available in graphical form as shown in Fig. 5.13.

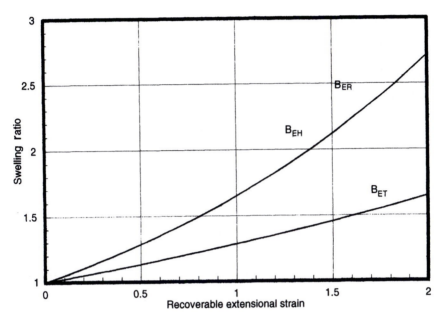

Fig. 5.13 Variation of Swelling Ratio for Capillary and Slit Dies

5.8 Residence and Relaxation Times

(a) Residence (or Dwell) Time

This refers to the time taken for the polymer melt to pass through the die or channel section. Mathematically it is given by the ratio

$$\text{Residence time, } t_R = \frac{\text{Volume of channel}}{\text{Volume flow rate}} \tag{5.57}$$

For the uniform channels analysed earlier, it is a simple matter to show that the residence times are:

Newtonian Flow

Circular Cross Section $\qquad\qquad t_R = \dfrac{8\eta L^2}{PR^2}$ $\qquad\qquad$ (5.58)

Rectangular Cross Section $\qquad t_R = \dfrac{12\eta L^2}{PH^2}$ $\qquad\qquad$ (5.59)

Power Law Fluid

Circular Cross Section $\qquad t_R = \left(\dfrac{3n+1}{n+1}\right)\dfrac{L}{V_0}$ $\qquad\qquad$ (5.60)

Rectangular Cross Section $\quad t_R = \left(\dfrac{2n+1}{n+1}\right)\dfrac{L}{V_0}$ $\qquad\qquad$ (5.61)

(b) Relaxation (or Natural) Time

In Chapter 2 when the Maxwell and Kelvin models were analysed, it was found that the time constant for the deformations was given by the ratio of viscosity to modulus. This ratio is sometimes referred to as the Relaxation or Natural time and is used to give an indication of whether the elastic or the viscous response dominates the flow of the melt.

To do this, a *Deborah Number* N_{deb}, has been defined as

$$N_{deb} = \frac{\text{Relaxation or Natural Time}}{\text{Timescale of the process}} \qquad\qquad (5.62)$$

If $N_{deb} > 1$ then the process is predominantly elastic whereas if $N_{deb} < 1$ then viscous effects dominate the flow.

5.9 Temperature Rise in Die

Power is the work done per unit time, where work in the simplest sense is defined as

$$\text{Work} = (\text{force}) \times (\text{distance})$$

Therefore

$$\text{Power} = (\text{force}) \times (\text{distance per unit time})$$

$$= (\text{Pressure drop} \times \text{area}) \times (\text{velocity})$$

But the volume flow rate, Q, is given by

$$Q = \text{area} \times \text{velocity} \qquad\qquad (5.63)$$

$$\text{So Power} = P \cdot Q$$

where P is the pressure drop across the die.

Using this expression it is possible to make an approximation for the temperature rise of the fluid during extrusion through a die. If it is assumed that all the work is changed into shear heating and that all the heat is taken up evenly by the polymer, then the work done may be equated to the temperature rise in the polymer.

$$\text{Power} = \text{Heat required to change temperature}$$

$$= \text{mass} \times \text{specific heat} \times \text{temperature rise}$$

So $$PQ = \rho Q \times C_p \times \Delta T \tag{5.64}$$

$$\Delta T = \frac{P}{\rho C_p}$$

where ρ is the density of the fluid and C_p is its specific heat.

5.10 Experimental Methods Used to Obtain Flow Data

In Section 5.11 design examples relating to polymer processing will be illustrated. In these examples the flow data supplied by material manufacturers will be referred to, so it is proposed in this section to show how this data may be obtained.

The equipment used to obtain flow data on polymer melts may be divided into two main groups.

(a) Rotational Viscometers – these include the cone and plate and the concentric cylinder.

(b) Capillary Viscometer – the main example of this is the ram extruder.

Cone and Plate Viscometer

In this apparatus the plastic to be analysed is placed between a heated cone and a heated plate. The cone is truncated and is placed above the plate in such a way that the imaginary apex of the cone is in the plane of the plate. The angle between the side of the cone and the plate is small (typically $<5°$).

The cone is rotated relative to the plate and the torque. T, necessary to do this is measured over a range of rotational rates, θ.

Referring to Fig. 5.14

Area of annulus $= 2\pi r \cdot dr$

Force $= 2\pi r \tau \cdot dr$

Torque $= 2\pi r^2 \tau \cdot dr$

So Total Torque, $T = \displaystyle\int_0^R 2\pi r^2 \tau \cdot dr$

$$T = \left(\tfrac{2}{3}\right) \pi R^3 \tau \tag{5.65}$$

Also shear strain, $$\gamma = \frac{x}{h} = \frac{r\theta}{r\alpha} = \frac{\theta}{\alpha}$$

Strain rate, $$\dot{\gamma} = \frac{\dot{\theta}}{\alpha} \tag{5.66}$$

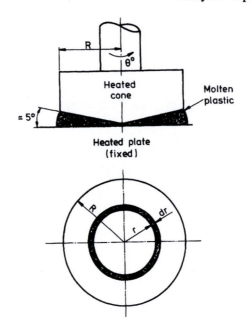

Fig. 5.14   Cone and Plate Viscometer

And since viscosity $\qquad \eta = \dfrac{\tau}{\dot\gamma}$

$$\eta = \frac{3T}{2\pi R^3} \cdot \frac{\alpha}{\theta} \qquad\qquad (5.67)$$

The disadvantage of this apparatus is that it is limited to strain rates in the region 10 to 1 s^{-1} whereas in plastics processing equipment the strain rates are in the order of $10 \ s^{3-1}$ to $10^4 \ s^{-1}$.

Concentric Cylinder Viscometer

In this apparatus the polymer melt is sheared between concentric cylinders. The torque required to rotate the inner cylinder over a range of speeds is recorded so that viscosity and strain rates may be calculated.

Referring to Fig. 5.15

Torque, $\qquad\qquad T = 2\pi RL \cdot R \cdot \tau = 2\pi R^2 L\tau \qquad\qquad (5.68)$

Strain rate, $\qquad\qquad \dot\gamma = \dfrac{du}{dy} = \dfrac{2\pi RN}{H} \qquad\qquad (5.69)$

Viscosity, $\qquad\qquad \eta = \dfrac{\tau}{\dot\gamma} = \dfrac{TH}{4\pi^2 R^3 LN} \qquad\qquad (5.70)$

Fig. 5.15   Concentric Cylinder Viscometer

As in the previous case, this apparatus is usually restricted to relatively low strain rates.

Ram Extruder

In this apparatus the plastic to be tested is heated in a barrel and then forced through a capillary die as shown in Fig. 5.16. Normally the ram moves at a constant velocity to give a constant volume flow rate, Q. From this it is conventional to calculate the shear rate from the Newtonian flow expression.

$$\dot{\gamma} = \frac{4Q}{\pi R^3}$$

Since it is recognised that the fluid is Non-Newtonian, this is often referred to as the *apparent* shear rate to differentiate it from the *true* shear rate. If the pressure drop, P, across the die is also measured then the shear stress, τ, may be calculated from

$$\tau = \frac{PR}{2L}$$

This leads to a definition of apparent viscosity as the ratio of shear stress to apparent shear rate

$$\eta = \frac{\pi P R^4}{8LQ}$$

A plot of apparent viscosity against shear rate produces a unique flow curve for the melt as shown in Fig. 5.3. Occasionally this information may be based on the true shear rate. As shown in Section 5.4(a) this is given by

$$\dot{\gamma} = \left(\frac{3n+1}{4n}\right)\frac{4Q}{\pi R^3}$$

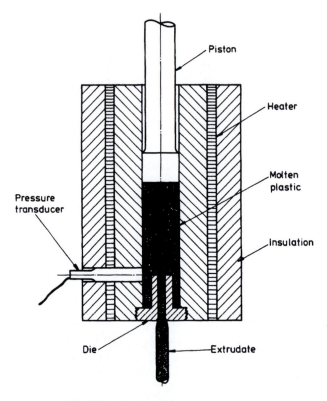

Fig. 5.16   Section Through Ram Extruder

However, this then means that the true shear rate must always be used in the flow situation being analysed. Experience has shown that this additional complexity is unnecessary because if the Newtonian shear rate is correlated with flow data which has been calculated using a Newtonian shear rate, then no error is involved.

Note that rotational viscometers give true shear rates and if this is to be used with Newtonian based flow curves then, from above, a correction factor of $(4n/3n + 1)$ needs to be applied to the true shear rate.

The ratio $(3n + 1)/4n$ is called the Rabinowitsch Correction Factor and it is used to convert Newtonian shear rates to true shear rates.

There are two other points worth noting about this test. Firstly the flow data is produced using a capillary die so that its use on channels of a different geometry would require a correction factor. However, in most cases of practical interest, the factor is not significantly different from 1 and so there is no justification for the additional complication caused by its inclusion.

Secondly, the pressure drop, P, in the above expression is the pressure drop due to shear flow along the die. If a pressure transducer is used to record the pressure drop as shown in Fig. 5.16, then it will also pick up the pressure losses at the die entry. This problem may be overcome by carrying out further tests using either a series of dies having different lengths or a die with a very short (theoretically zero) length. In the former case, the pressure drops for the various lengths of die may be extrapolated to give the pressure drop for entry into a die of zero length. In the second case this pressure is obtained directly by using the so-called zero length die. This is then subtracted from the measured pressure loss, P_L, on the long die being considered, so that

$$\tau = \frac{(P_L - P_0)R}{2L} \tag{5.71}$$

In addition, if the swelling of the extrudate is measured in each of these two tests then the swelling ratio using the long die will be B_{SR} and the swelling ratio using the short die will be B_{ER} (see Section 5.6). Using equation (5.44) and (5.47) this enables the shear and tensile components of the recoverable strains to be calculated and from them the shear and tensile moduli.

From this relatively simple test, therefore, it is possible to obtain complete flow data on the material as shown in Fig. 5.3. Note that shear rates similar to those experienced in processing equipment can be achieved. Variations in melt temperature and hydrostatic pressure also have an effect on the shear and tensile viscosities of the melt. An increase in temperature causes a decrease in viscosity and an increase in hydrostatic pressure causes an increase in viscosity. Typically, for low density polyethlyene an increase in temperature of 40°C causes a vertical shift of the viscosity curve by a factor of about 3. Since the plastic will be subjected to a temperature rise when it is forced through the die, it is usually worthwhile to check (by means of Equation 5.64) whether or not this is significant. Fig. 5.2 shows the effect of temperature on the viscosity of polypropylene.

A change in pressure from atmospheric ($= 0.1$ MN/m^2) to about 100 MN/m^2 (a pressure likely to be experienced during processing) causes a vertical shift of the viscosity curve by a factor of about 4 for LDPE. This effect may be important during processing because the material can be subjected to large changes in pressure in sections such as nozzles, gates, etc. However, it should be noted that in some cases the increase or decrease in pressure results in, or is associated with, an increase or decrease in temperatures so that the net effect on viscosity may be negligible.

Other factors such as the use of additives also have an effect on the shape of the flow curves. Flame retardants, if used, tend to decrease viscosity whereas pigments tend to increase viscosity. Fig. 5.17 shows flow curves for a range of plastics.

Melt Flow Rate (also known as Melt Flow Index)

The Melt Flow Rate Test is a method used to characterise polymer melts. It is, in effect, a single point ram extruder test using standard testing conditions (BS

Fig. 5.17   Viscosity Curves for a Range of Plastics (230°C unless otherwise shown)

Fig. 5.18   Diagram of Apparatus for Measuring Melt Flow Index

2782) as illustrated in Fig. 5.18. The polymer sample is heated in the barrel and then extruded through a standard die using a standard weight on the piston, and the weight (in gms) of polymer extruded in 10 minutes is quoted as the melt flow rate (MFR) of the polymer.

Flow Defects

When a molten plastic is forced through a die it is found that under certain conditions there will be defects in the extrudate. In the worst case this will take the form of gross distortion of the extrudate but it can be as slight as a dullness of the surface. In most cases flow defects are to be avoided since they affect the quality of the output and the efficiency of the processing operation. However, in some cases if the flow anomaly can be controlled and reproduced, it can be used to advantage – for example, in the production of sheets with matt surface finish. Flow defects result from a combination of melt flow properties, die design and processing conditions but the exact causes and mechanisms are not completely understood. The two most common defects are

(a) **Melt Fracture** When a polymer melt is flowing through a die, there is a critical shear rate above which the extrudate is no longer smooth. The defect may take the form of a spiralling extrudate or a completely random configuration. With most plastics it is found that increasing the melt temperature or the L/D ratio of the die will increase the critical value of shear rate. It is generally believed that the distortion of the extrudate is caused by slip-stick mechanism between the melt and the die wall due to the high shear rates. If there is an abrupt entry to the die then the tensile/shear stress history which the melt experiences is also considered to contribute to the problem.

(b) **Sharkskin** Although this defect is also a visual imperfection of the extrudate it is usually differentiated from melt fracture because the defects are perpendicular to the flow direction rather than helical or irregular. In addition, experience has shown that this defect is a function of the linear output rate rather than the shear rate or die dimensions. The most probable mechanism of sharkskin relates to the velocity of skin layers of the melt inside and outside the die. Inside the die the skin layers are almost stationary whereas when the extrudate emerges from the die there must be a rapid acceleration of the skin layers to bring the skin velocity up to that of the rest of the extrudate. This sets up tensile stresses in the melt which can be sufficient to cause fracture.

5.11 Analysis of Flow in Some Processing Operations

Design methods involving polymer melts are difficult because the flow behaviour of these materials is complex. In addition, flow properties of the melt are usually measured under well defined uniform conditions whereas unknown effects such as heating and cooling in processing equipment make service conditions less than ideal. However, sufficient experience has been gathered using the equations derived earlier that melt flow problems can be tackled quantitatively and with an accuracy which compares favourably with other engineering design situations. In this section a number of polymer processing methods will be analysed using the tools which have been assembled in this chapter. In some cases the heating and cooling of the melt may have an important effect on the flow behaviour and methods of allowing for this are developed.

Example 5.2 In a plunger-type injection moulding machine the torpedo has a length of 40 mm, a diameter of 23 mm and is supported by three spiders. If, during moulding of polythene at 170°C, the plunger moves forward at a speed of 15 mm/s, estimate the pressure drop along the torpedo and the shear force on the spiders. The barrel diameter is 25 mm.

Solution Assume that the flow is isothermal.

Volume flow rate, $Q = $ area $\times$ velocity

$$= \frac{\pi}{4}(25 \times 10^{-3})^2 \times 15 \times 10^{-3}$$

$$= 7.36 \times 10^{-6} \text{ m}^3/\text{s}.$$

The gap between the torpedo and the barrel may be considered as a rectangular slit with $T = (\pi \times 24 \times 10^{-3})$m and $H = 1 \times 10 \text{ m}^{-3}$.

So apparent strain rate, $\dot{\gamma} = \dfrac{6Q}{TH^2} = \dfrac{6 \times 7.36 \times 10^{-6}}{\pi \times 24 \times 10^{-3} \times 10^{-6}}$

$$= 585 \text{ s}^{-1}$$

From Fig. 5.3, at this strain rate, $\eta = 400 \text{ Ns/m}^2$.

So $\qquad \tau = \eta\dot{\gamma} = 400 \times 585 = 2.34 \times 10^5 \text{ N/m}^2$

$$P = \frac{2L\tau}{H} = \frac{2 \times 40 \times 10^{-3} \times 2.34 \times 10^5}{1 \times 10^{-3}} = 18.7 \text{ MN/m}^2$$

The force on the spider results from

(a) the force on the torpedo due to the pressure difference across it and
(b) the force on the torpedo due to viscous drag.

Cross-sectional area of torpedo $= \pi r^2$

So force due to pressure $= \pi r^2 P$

$$= \pi(11.5 \times 10^{-3})^2 \times 18.7 \times 10^6$$

$$= 7.8 \text{ kN}$$

Surface area of torpedo $= \pi DL$

Viscous drag force $= \pi DL\tau$

$$= \pi(23 \times 10^{-3})(40 \times 10^{-3})(2.34 \times 10^5)$$

$$= 0.68 \text{ kN}$$

So Total Force $= 8.48$ kN

Example 5.3 The output of polythene from an extruder is 30×10^{-6} m³/s. If the breaker plate in this extruder has 80 holes, each being 4 mm diameter and 12 mm long, estimate the pressure drop across the plate assuming the material temperature is 170°C at this point. The flow curves in Fig. 5.3 should be used.

Solution Assume the flow is isothermal.

Flow rate through each hole $\quad = \dfrac{30 \times 10^{-6}}{80} \text{ m}^3/\text{s}$

So Apparent shear rate, $\quad \dot{\gamma} = \dfrac{4Q}{\pi R^3}$

$$= \dfrac{4 \times 30 \times 10^{-6}}{\pi \times 80 \times 2^3 \times 10^{-9}}$$

$$= 59.6 \text{ s}^{-1}$$

From Fig. 5.3 the shear stress, τ, is 1.2×10^5 N/m² and since

$$\tau = \frac{PR}{2L}$$

$$P = \frac{2 \times 12 \times 10^{-3} \times 1.2 \times 10^5}{2 \times 10^{-3}}$$

$$= 1.44 \text{ MN/m}^2$$

Example 5.4 Eight polypropylene mouldings, each weighing 10 g are to be moulded using the runner layout shown in Fig. 5.19. If the injection time is 2 seconds and the melt temperature is 210°C, calculate the pressure at each cavity if the injection pressure at the sprue is 80 MN/m². The density of the polypropylene is 909 kg/m³ and the volume of the sprue is 5000 mm³. Assume that the flow is isothermal and ignore the pressure losses at corners.

Solution Firstly it is necessary to determine the volume flow rate through the runners and gates. The total volume of the runners, gates and cavities may be calculated as follows:

Volume of 8 cavities $\quad = 8 \times 10 \times 10^{-3}/909 = 8.8 \times 10^{-5}$ m³

$$= 88 \times 10^3 \text{ mm}^3$$

Fig. 5.19 Runner Layout, Example 5.4

Volume of long runners $= 4 \times \pi R^2 \times L$

$$= 4 \times \pi(3)^2 \times 100 = 11.3 \times 10^3 \text{ mm}^3$$

Volume of short runners $= 8 \times \pi(2.5)^2 \times 50 = 7.85 \times 10^3 \text{ mm}^3$

Volume of sprue $\quad = 5 \times 10^3 \text{ mm}^3$

Total volume $\quad = 112.2 \times 10^3 \text{ mm}^3$

The fill time is 2 seconds so the total volume flow rate, Q, is given by

$$Q = \frac{112.2 \times 10^3}{2} = 56.1 \times 10^3 \text{ mm}^3/\text{s} = 56.1 \times 10^{-6} \text{ m}^3/\text{s}$$

To get the pressure loss in the runners and gates. The flow rate in the long runner is $Q/4$, hence

(i) For long runners $\dot{\gamma} = \dfrac{4Q}{\pi R^3} = \dfrac{4 \times 56.1 \times 10^3}{4 \times \pi(3)^3} = 661 \text{ s}^{-1}$

From the flow curves for polypropylene at 210°C, at this shear rate, $\eta = 130$ Ns/m², so the shear stress, τ, is given by

$$\tau = 130 \times 661 = 85.9 \text{ kN/m}^2 = \frac{PR}{2L}$$

Hence, Pressure loss, $P = \dfrac{2L\tau}{R} = \dfrac{2 \times 100 \times 85.9}{3} = 5.73 \text{ MN/m}^2$

(ii) For the short runners, the flow rate is Q/8, and so

$$\dot{\gamma} = 571 \text{ s}^{-1}, \eta = 160 \text{ Ns/m}^2, \tau = 91.4 \text{ kN/m}^2$$

and these give the pressure loss = 3.66 MN/m².

(iii) For the gates, the flow rate is Q/8 and so

$$\dot{\gamma} = \frac{6Q}{TH^2} = \frac{6 \times 56.1 \times 10^3}{8 \times 3 \times 2^2} = 3506 \text{ s}^{-1}$$

From the flow curves, $\qquad\qquad \eta = 44 \text{ Ns/m}^2, \tau = 154.3 \text{ kN/m}^2$

So pressure loss, $\qquad\qquad p = \dfrac{2L\tau}{H} = \dfrac{2 \times 1.6 \times 154.3}{2} = 247 \text{ kN/m}^2$

Total pressure loss $\qquad\qquad = 5.73 + 3.66 + 0.247$

$$= 9.64 \text{ MN/m}^2$$

Thus, pressure available at cavity $\quad = 80 - 9.64$

$$= 70.4 \text{ MN/m}^2$$

Example 5.5 In a polyethylene film blowing die the geometry of the die lips is as shown in Fig. 5.20. If the output is 300 kg/hour and the density of the polythene is 760 kg/m³, estimate the pressure drop across the die lips.

Solution

$$\text{Volume flow rate} = \frac{300}{60 \times 60 \times 760} = 109.6 \times 10^{-6} \text{ m}^3/\text{s}$$

The die lips may be considered in three sections A, B and C and the total pressure drop is the sum of the losses in each.

Section A: Since this section is of a uniform section there will only be a pressure drop due to shear.

For a rectangular slot, as shown earlier, the apparent shear rate, $\dot{\gamma}$ at A_1 is given by

$$\dot{\gamma} = \frac{6Q}{TH_1^2} = \frac{6 \times 109.6 \times 10^{-6}}{\pi(260 \times 10^{-3})(0.7 \times 10^{-3})^2} = 1.64 \times 10^3 \text{ s}^{-1}$$

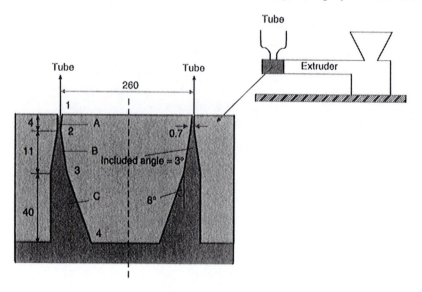

Fig. 5.20   Tube Extrusion Die

So from Fig. 5.3, $\tau = 3.3 \times 10^5$ N/m²

$$P_{S_{12}} = \frac{2L\tau_1}{H_1}$$

$$= \frac{2 \times 4 \times 3.3 \times 10^5}{0.7}$$

$$= 3.77 \text{ MN/m}^2$$

Section B: This section is tapered so there will be pressure losses due to both shear and extensional flows.

From simple geometry $H_3 = H_2 + (2L \tan \alpha)$ where α is half the included angle and $L = 11$ mm.

$$H_2 = 0.7 \text{ mm}, \quad H_3 = 1.28 \text{ mm}$$

Now at B_2, $\dot{\gamma} = 1.64 \times 10^3$ s⁻¹

$\dot{\varepsilon} = (\dot{\gamma}/3) \tan \alpha = 1/3 \times 1.64 \times 10^3 \tan (1.5°)$

$= 14$ s⁻¹

At B_3,

$$\dot{\gamma} = \frac{6Q}{TH_3^2} = \frac{6 \times 109.6 \times 10^{-6}}{\pi \times 260 \times 10^{-3}(1.276 \times 10^{-3})^2}$$

$$= 494.5 \text{ s}^{-1}$$

$$\dot\varepsilon = 1/3(494.5)\tan 1.5°$$

$$= 4.3 \text{ s}^{-1}$$

From Fig. 5.3 the flow curve can be taken as a straight line for the shear rate range 500 s^{-1} to 1600 s^{-1} and the power law index n may be taken as 0.33.

So from (5.45)

$$P_{S_{23}} = \frac{\tau_2}{2n\tan\alpha}\left[1 - \left(\frac{H_2}{H_3}\right)^{2n}\right]$$

$$= \frac{3.3\times10^5}{2(0.3)\tan 1.5}\cdot\left[1 - \left(\frac{0.7}{1.28}\right)^{0.66}\right]$$

$$= 6.9 \text{ MN/m}^2$$

Also from (5.46)

$$P_{E_{23}} = \frac{\sigma_2}{2}\cdot\left[1 - \left(\frac{H_2}{H_3}\right)^{2}\right]$$

Now $\dot\varepsilon = 14 \text{ s}^{-1}$ so from Fig. 5.3 $\sigma_2 = 3.15\times10^6 \text{ Nm}^{-2}$ (ie $2.25\times10^5 \times 14$)

$$P_{E_{23}} = \frac{3.15\times10^6}{2}\cdot\left[1 - \left(\frac{0.7}{1.28}\right)^{2}\right]$$

$$= 1.1 \text{ MN/m}^2$$

Section C: Although this section is tapered on one side only, since the angle is small the error in assuming it is tapered on both sides will be negligible. The included angle is taken as $8°$.

$$H_4 = H_3 + L\tan 8°$$

$$= 1.28 + 40\tan 8° = 6.9 \text{ mm}$$

At C_3

$$\dot\gamma = 494.5 \text{ s}^{-1}$$

$$\dot\varepsilon = 1/3(494.5)\tan 4°$$

$$= 11.52 \text{ s}^{-1}$$

At C_4

$$\dot\gamma = \frac{6Q}{TH_4^2} = \frac{6\times109.6\times10^{-6}}{\pi\times260\times10^{-3}\times(6.9\times10^{-3})^2} = 16.9 \text{ s}^{-1}$$

$$\dot\varepsilon = 1/3(16.9)\tan 4°$$

$$= 0.39 \text{ s}^{-1}$$

As before

$$P_{S_{34}} = \frac{\tau_3}{2n \tan \alpha} \cdot \left[1 - \left(\frac{H_3}{H_4} \right)^{2n} \right]$$

from Fig. 5.3, $\tau_3 = 2.1 \times 10^5$ Nm^{-2} at $\dot{\gamma} = 494.5$ s^{-1}

$$P_{S_{34}} = \frac{2.1 \times 10^5}{0.6 \tan 4°} \left[1 - \left(\frac{1.28}{6.9} \right)^{0.66} \right]$$

$$= 3.36 \text{ MN/m}^2$$

Also

$$P_{E_{34}} = \frac{\sigma_3}{2} \cdot \left[1 - \left(\frac{H_3}{H_4} \right)^2 \right]$$

$\dot{\varepsilon} = 11.52$ s^{-1} so from Fig. 5.3 $\sigma_3 = 2.6 \times 10^6$ Nm^{-2} (ie $2.25 \times 10^5 \times 11.52$)

$$P_{E_{34}} = \frac{2.6 \times 10^6}{2} \cdot \left[1 - \left(\frac{1.28}{6.9} \right)^2 \right]$$

$$= 1.26 \text{ MN/m}^2$$

In this case there will also be a pressure loss due to flow convergence at the die entry. This may be obtained from (5.47).

$$P_0 = \frac{4}{(3n + 1)} \cdot \left(\frac{6Q}{TH_4^2} \right) \cdot (\eta \lambda)^{1/2}$$

At C_4, $\dot{\gamma} = \frac{6Q}{TH_4^2} = 16.9$, so $\eta = 4 \times 10^3$ Nsm^{-2}, $\lambda = 2.25 \times 10^5$ Nsm^{-2}

$$P_0 = \frac{4}{1.9}(16.9)(4 \times 10^3 \times 2.25 \times 10^5)^{1/2}$$

$$= 1.07 \text{ MN/m}^2$$

So the total pressure drop through the die is given by

$$P = 3.77 + 6.9 + 1.1 + 3.36 + 1.26 + 1.07$$

$$= 17.5 \text{ MN/m}^2$$

Example 5.6 Estimate the dimensions of the tube which will be produced by the die in Example 5.5 assuming that there is no draw-down.

Solution Since the final section of the die lips is not tapered it may be assumed that the swelling at the die exit is due to shear effects only.

At exit, $\qquad\qquad \dot{\gamma} = \frac{6Q}{TH^2} = 1.64 \times 10^3$ s^{-1}

From Fig. 5.3, $\tau = 3.3 \times 10^5$ Nm^{-2} and $G = 5 \times 10^4$ Nm^{-2}

So recoverable strain, $\quad \gamma_R = \dfrac{\tau}{G} = \dfrac{3.3 \times 10^5}{5.0 \times 10^4} = 6.6$

Then using equations (5.52) and (5.53), or more conveniently Fig. 5.11, the swelling ratios may be obtained as

$$B_{SH} = 2.31 \quad \text{and} \quad B_{ST} = 1.52$$

Therefore swollen thickness of film $= H_1 \times B_{SH}$

$$= 0.7 \times 2.31 = 1.62 \text{ mm}$$

and swollen diameter of bubble $\quad = D \times B_{ST} = 260 \times 1.52 = 395$ mm

Example 5.7 In many practical situations it is the dimensions of the output which are known or specified and it is necessary to work back (often using an iterative procedure) to design a suitable die.

Design a die which will produce plastic film 0.52 mm thick at a linear velocity of 20 mm/s. The lay-flat width of the film is to be 450 mm and it is known that a blow-up ratio of 1.91 will give the necessary orientation in the film. Assume that there is no draw-down.

Solution Using the terminology from the analysis of blow moulding in Section 4.2.5.

$$2(LFW) = \pi D_b$$

$$D_b = \frac{2 \times 450}{\pi} = 286.5 \text{ mm}$$

Now it makes the solution simpler to assume that the blow up ratio is given by D_b/D_1 (ie rather than D_b/D_m). Also this seems practical because the change from D_1 to D_b is caused solely by inflation whereas the change from D_m to D_b includes die swell effects.

$$\therefore D_1 = 286.5/1.91 = 150 \text{ mm}$$

Then, for constant volume

$$\pi D_1 h_1 = \pi D_b h_b$$

$$h_1 = \frac{286.5 \times 0.52}{150} = 1 \text{ mm}$$

So having obtained the dimensions of the tube to be produced, the procedure is to start by assuming that there is no die swell. This means that the die dimensions will be the same as those of the tube.

$$\text{Apparent shear rate, } \dot{\gamma} = \frac{6Q}{TH^2}$$

but volume flow rate, $Q = VTH$

where V is the velocity of the plastic melt.

$$\dot{\gamma} = \frac{6V}{H}$$

$$= \frac{6 \times 20}{1} = 120 \text{ s}^{-1}$$

from Fig. 5.3, $\tau = 1.3 \times 10^5$ N/m^{-2} and $G = 3 \times 10^4$ N/m^{-2}

So recoverable shear strain, γ_R, is given by

$$\gamma_R = \frac{\tau}{G} = \frac{1.3 \times 10^5}{3 \times 10^4} = 4.33$$

Assuming that swelling results from shear effects only then from Fig. 5.13 at $\gamma_R = 4.33$, the swelling ratios are

$$B_{SH} = 1.81 \quad \text{and} \quad B_{ST} = 1.35$$

The initial die dimension must now be adjusted to allow for this swelling.

$$\text{New die gap} = H^1 = 1/1.81 = 0.55 \text{ mm}$$

$$\text{New die circumference} = T' = \frac{\pi \times 150}{1.35} = 349 \text{ mm}$$

The volume flow rate, Q, will be constant so the velocity of the melt in the die will also be adjusted.

$$Q = VTH = V'T'H'$$

$$V' = \frac{VTH}{T'H'} = V \cdot B_{SH} \cdot B_{ST}$$

New shear rate, $\dot{\gamma} = \frac{6V'}{H'} = \frac{6 \times 20 \times 2 \times 1.35}{0.55} = 589 \text{ s}^{-1}$

From Fig. 5.3, $\tau = 2.36 \times 10^5$ N/m^2 and $G = 4 \times 10^4$ N/m^2

So $\gamma_R = 5.9$

and from Fig. 5.13

$$B_{SH} = 2.18 \quad \text{and} \quad B_{ST} = 1.48$$

$$\text{So new die gap} = H'' = \frac{1}{2.18} = 0.459 \text{ mm}$$

$$\text{new die circumference} = T'' = \frac{\pi \times 150}{1.48} = 318 \text{ mm}$$

If this iterative procedure is repeated again the next values obtained are

$$\text{die gap, } H''' = 0.455 \text{ mm}$$

$$\text{die circumference, } T''' = \frac{\pi \times 150}{1.49} \text{ mm}$$

These values are sufficiently close to the previous values so the die exit dimensions are

$$\text{Gap} = 0.455 \text{ mm}$$

$$\text{Diameter} = \frac{\pi \times 150}{1.49 \times \pi} = 100.67 \text{ mm}$$

The shear rate, $\dot{\gamma}$, in the die land is

$$\dot{\gamma} = \frac{6 \times 20 \times 1.49 \times 2.2}{0.455}$$

$$= 864.5 \text{ s}^{-1}$$

From Fig. 5.3 it may be seen that this exceeds the shear rate for non-laminar flow (approximately 30 s^{-1}) so that the entry to this region would need to be streamlined. Fig. 5.3 also shows that the extensional strain rate, $\dot{\varepsilon}$, in the tapered entry region should not exceed about 15 s^{-1} if turbulence is to be avoided.

Therefore since $\dot{\varepsilon} = 1/3\dot{\gamma}\tan\alpha$

$$\alpha = \tan^{-1}\left(\frac{3\dot{\varepsilon}}{\dot{\gamma}}\right)$$

$$= \tan^{-1}\left(\frac{3 \times 15}{864.5}\right)$$

$$= 2.97°$$

So the full angle of convergence to the die land must be less than $(2 \times 2.97°) = 5.94°$.

To complete the die design it is likely that a third tapered section will be necessary as shown in Fig. 5.20. This will be designed so that the entry to the die is compatible with the outlet of the die body on the extruder. The maximum angle of convergence of this section can be estimated as shown earlier and the lengths of the two tapered sections selected to ensure the angle is not exceeded for the fixed die inlet dimension. If the length of the 5.94° tapered region is chosen as 10 mm it may be shown that the maximum angle of convergence for the entry to this taper is 58.5°. If the length of this section is fixed at 40 mm and the annular gap at the exit from the extruder is 16 mm, then a suitable angle of convergence would be about 40°.

Example 5.8 It is desired to blow mould a plastic bottle with a diameter of 60 mm and a wall thickness of 2 mm. If the extruder die has an annular slot of outside diameter = 32 mm and inside diameter = 28 mm, calculate the output rate needed from the extruder and recommend a suitable inflation pressure. Use the flow characteristics given in Fig. 5.3. Density of molten polythene = 760 kg/m^3.

Solution In the analysis of blow moulding in Section 4.2.7, it was shown that the thickness, h, of the inflated bottle is given by

$$h = B_{ST}^3 h_d \left(\frac{D_d}{D_m}\right)$$

where h_d = die gap = 2 mm

D_m = mould diameter = 60 mm

D_d = die diameter = $1/2(28 + 32)$ = 30 mm

So
$$B_{ST}^3 = \frac{2 \times 60}{30 \times 2}$$

$$B_{ST} = 1.26$$

From Fig. 5.11 at $B_{ST} = 1.26$, $\gamma_R = 3.15$

So $3.15 = \tau/G$

From Fig. 5.3 it is now necessary to determine the combination of τ and G to satisfy this equation.

For example, at $\dot{\gamma} = 5$ s^{-1}

$$G = 1.5 \times 10^4 \text{ N/m}^2, \quad \tau = 5 \times 8.8 \times 10^3 = 4.4 \times 10^4 \text{ N/m}^2$$

So $\tau/G = 2.93$

At $\dot{\gamma} = 50$ s^{-1}

$$G = 2.5 \times 10^4 \text{ N/m}^2, \quad \tau = 5.9 \times 10^4 \text{ N/m}^2, \quad \tau/G = 4.2$$

At $\dot{\gamma} = 10$ s^{-1}

$$G = 1.9 \times 10^4 \text{ N/m}^2, \quad \tau = 5.9 \times 10^4 \text{ N/m}^2, \quad \tau/G = 3.1$$

Therefore the latter combination is correct.

$$\dot{\gamma} = 10 = \frac{6Q}{TH^2}$$

and
$$Q = \frac{10 \times \pi \times 30 \times 2^2 \times 10^{-9}}{6} = 1.26 \times 10^{-6} \text{ m}^3/\text{s}$$

$$= 1.26 \times 10^{-6} \times 760 \times 3600 = 3.45 \text{ kg/hour}$$

To calculate the inflation pressure it is necessary to get the melt fracture stress. From Fig. 5.3 it may be seen that this is 4×10^6 N/m^2

Therefore since the maximum stress in the inflated bubble is the hoop stress, σ, the maximum inflation pressure, P, is given by

$$P = \frac{2h\sigma}{D_m} = \frac{2 \times 2 \times 4 \times 10^6}{60}$$

$$= 0.13 \text{ MN/m}^2$$

Fig. 5.21

Example 5.9 Calculate the time taken to inflate the bottle in the previous example if the inflation pressure is 50 kN/m^2. The flow curves given in Fig. 5.3 may be used.

Solution It is necessary to derive an expression for the inflation time. Referring to Fig. 5.22 where R is the radius of the inflated tube at any time, t, and h_t is the wall thickness at this time, then a force balance on the shaded element gives (for unit length).

$$(Rd\theta)P_1 - (R + h_t)d\theta P_2 - 2\sigma h_t \sin\left(\frac{d\theta}{2}\right) = 0$$

$$\text{Letting } \sin\left(\frac{d\theta}{2}\right) = \frac{d\theta}{2} \quad \text{and} \quad P_2 = 0, \quad \text{then}$$

$$RP_1 = \sigma h_t \qquad\qquad (5.72)$$

Now $\sigma = \lambda\dot{\varepsilon}$ where λ is the tensile viscosity.

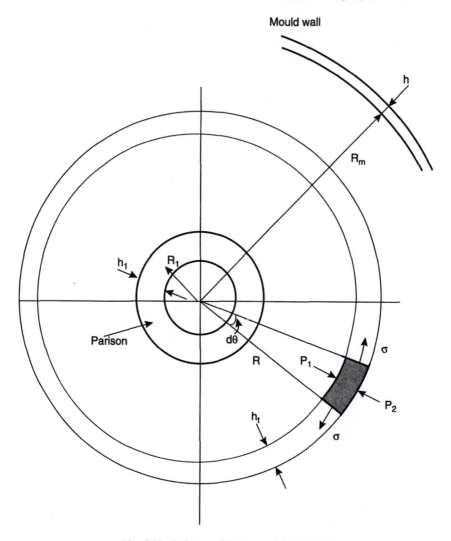

Fig. 5.22   Inflation of Parison to Mould Wall

$\dot{\varepsilon}$ is the extensional strain rate and is given by

$$\dot{\varepsilon} = \frac{\text{change in length per unit time}}{\text{original length}} = \frac{(R + dR)d\theta - Rd\theta}{Rd\theta dt} = \frac{1}{R}\frac{dR}{dt}$$

Substituting into (5.72) we get

$$RP_1 = \frac{\lambda}{R}\frac{dR}{dt}h_t \qquad\qquad (5.73)$$

Before this can be integrated, h_t needs to be expressed as a function of R or t. Assuming constant volume of the parison

$$Rh_t = R_1 h_1$$

Hence we can write

$$\int_0^t dt = \int_{R_1}^{R_m} \frac{\lambda R_1 h_1}{P_1} \cdot \frac{dR}{R^3}$$

$$t = \frac{\lambda h_1}{2P_1 R_1} \left\{ 1 - \left(\frac{R_1}{R_m} \right)^2 \right\}$$

or expressing this in terms of the original die dimensions (see Blow Moulding, Chapter 4),

$$\text{Inflation time, } t = \frac{\lambda B_{ST} h_d}{P_1 D_d} \left\{ 1 - B_{ST}^2 \left(\frac{D_d}{D_m} \right) \right\} \tag{5.74}$$

Using the information given and with $\lambda = 2.2 \times 10^5$ Ns/m^2 (from Fig. 5.3) then

$$t = \frac{2.2 \times 10^5 \times 1.26 \times 2}{50 \times 10^3 \times 30} \left\{ 1 - 1.26^2 \left(\frac{30}{60} \right) \right\}$$

$$t = 0.08 \text{ seconds}$$

Example 5.10 During the blow moulding of polythene bottles the parison is 0.3 m long and is left hanging for 5 seconds. Estimate the amount of sagging which occurs. The density of polythene is 760 kg/m^3

Solution Consider the small element of parison as shown as shown in Fig. 5.21.

(a) Elastic Strain

From the relationship between stress, strain and modulus

$$\delta L = \frac{F dx}{AE}$$

$$= \frac{\rho \pi D_1 h_1 x dx}{\pi D_1 h_1 E}$$

So total extension $= \dfrac{\rho}{E} \displaystyle\int_0^L x \, dx = \dfrac{\rho L^2}{2E}$

$$\text{So Elastic Strain} = \varepsilon_R = \frac{\rho L}{2E}$$

(b) Viscous Strain

$$\text{The stress on the element} = \frac{\text{force}}{\text{area}} = \frac{\text{weight below } YY}{\text{area}}$$

$$= \frac{\rho \pi D_1 h_1 \, dx}{\pi D_1 h_1} = \rho \, dx$$

$$\text{So total stress, } \sigma = \int_0^L \rho \, dx = \rho L$$

Now the viscous strain on the element is given by

$$\delta\varepsilon_v = \dot{\varepsilon} \, dt = \frac{\sigma}{\lambda} dt$$

$$\varepsilon_v = \int_0^t \frac{\sigma}{\lambda} dt = \int_0^t \frac{\rho L}{\lambda} dt = \frac{\rho L t}{\lambda}$$

Therefore the total strain, $\varepsilon = \varepsilon_R + \varepsilon_v$

$$= \frac{\rho L}{2E} + \frac{\rho L t}{\lambda}$$

$$\varepsilon = \rho L \left[\frac{1}{2E} + \frac{t}{\lambda} \right] \tag{5.75}$$

Note that this solution applies for values of t less than the relaxation time (λ/E) for the process. This is generally the case for blow moulding.
    For the situation given,

stress, $\sigma = \rho L = 760 \times 9.81 \times 0.3 = 2.24 \times 10^3 \text{ N/m}^2$

From Fig. 5.3, $\lambda = 2.25 \times 10^5 \text{ Ns/m}^2$.

$$\text{Also, since } \sigma = \lambda \, \dot{\varepsilon}, \quad \dot{\varepsilon} = \frac{2.24 \times 10^3}{2.25 \times 10^5} = 0.01$$

Extrapolating to the left on Fig. 5.3, it may be found that at a strain rate of 0.01, $E = 1.9 \times 10^4 \text{ N/m}^2$.
    So characteristic time $= \lambda/E = 11.6 \text{ s}^{-1}$
    Therefore since the parison sag time is less than this, the above expression may be used to calculate the amount of stretching

$$\varepsilon = 760 \times 9.81 \times 0.3 \left[\frac{1}{2 \times 1.9 \times 10^4} + \frac{5}{2.2 \times 10^5} \right] = 0.11$$

So extension $= 0.11 \times 0.3$ m

$$= 33 \text{ mm}$$

5.12 Analysis of Heat Transfer during Polymer Processing

Most polymer processing methods involve heating and cooling of the polymer melt. So far the effect of the surroundings on the melt has been assumed to be small and experience in the situations analysed has proved this to be a reasonable assumption. However, in most polymer flow studies it is preferable to consider the effect of heat transfer between the melt and its surroundings. It is not proposed to do a detailed analysis of heat transfer techniques here, since these are dealt with in many standard texts on this subject. Instead some simple methods which may be used for heat flow calculations involving plastics are demonstrated.

Fouriers equation for non-steady heat flow in one dimension, x, is

$$\frac{\partial^2 T}{\partial x^2} = \frac{1}{\alpha} \frac{\partial T}{\partial t}$$

where T is temperature and α is the thermal diffusivity defined as the ratio of thermal conductivity, K, to the heat capacity per unit volume

$$\alpha = \frac{K}{\rho C_p}$$

where ρ is density and C_p is specific heat.

Most materials manufacturers supply data on the thermal diffusivity of their plastics but in the absence of any information a value of 1×10^{-7} m^2/s may be used for most thermoplastics (see Table 1.8 and Table 5.1).

Solutions to Fourier's equation are in the form of infinite series but are often more conveniently expressed in graphical form. In the solution the following dimensionless groups are used.

(1) Fourier Number, $\qquad\qquad F_0 = \dfrac{\alpha t}{x^2}$ $\qquad\qquad$ (5.76)

where t is time
and $\quad x$ is the radius of the sphere or cylinder (or half thickness of the\sheet) considered.

In the case of a flat sheet, if the heating/cooling is from one side only then the dimension, x, is taken as the full thickness.

(2) Temperature Gradient, $\qquad \Delta T = \dfrac{T_t - T_m}{T_i - T_m}$ $\qquad\qquad$ (5.77)

where T_i = Initial uniform temperature of the melt
$\qquad T_m$ = Temperature of heating or cooling medium
$\qquad T_t$ = Temperature at time t.

Fig. 5.23 show the solution to Fourier's equation in terms of the temperature gradient at the centre line of section considered and the Fourier Number for

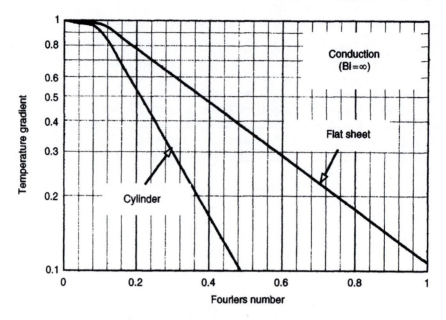

Fig. 5.23   Dimensionless Temperature Data for two Geometries

the cases of a flat sheet and a cylinder respectively. In each case it is assumed that there is no resistance to heat transfer at the boundary between the melt and the heating cooling medium.

The way in which this data may be used is illustrated by the following example.

Example 5.11 A polyethylene injection moulding is in the form of a flat sheet 100 mm square and 3 mm thick. If the melt temperature is 230°C, the mould temperature is 30°C and the plastic may be ejected at a centre-line temperature of 90°C, estimate

(a) the temperature of the material at the centre of the moulding after 7 seconds

(b) the time taken for the centre of the moulding to reach the ejection temperature of 90°C

(c) the time taken for the centre of the moulding to cool to 60°C in air at 20°C after ejection from the mould. Comment on the meaning of the answer. The heat transfer coefficient for the air is 20 W/m^2 K and the thermal conductivity of the polyethylene is 0.25 W/mK.

Solution (a) Fourier Number is

$$F_0 = \frac{\alpha t}{x^2}$$

$$F_0 = \frac{1 \times 10^{-7} \times 7}{(1.5 \times 10^{-3})^2}$$

$$= 0.31$$

From Fig. 5.23 the temperature gradient is 0.58

$$0.58 = \frac{T_t - 30}{230 - 30}$$

$$T_t = 146^\circ \text{ C}$$

(b) For freeze-off to occur the temperature gradient $\Delta T = \dfrac{T_f - T_M}{T_m - T_M}$

$$\Delta T = \frac{90 - 30}{230 - 30} = 0.3$$

From Fig. 5.23, $F_0 = 0.59$

$$0.59 = \frac{\alpha t}{x^2}$$

$$\text{or} \quad t = \frac{0.59 \times (1.5 \times 10^{-3})^2}{1 \times 10^{-7}}$$

$$= 13.3 \text{ s}$$

(c) The above analysis relates to the situation where there is little resistance to heat transfer between the plastic and the heating/cooling medium, ie conduction. In situations where the plastic is in contact with a fluid such as air, the main mode of heat transfer will be convection. The interrelationships between conduction and convection may be expressed by the Biot Number, B_i. This is given by

$$B_i = \frac{hx}{K} \tag{5.78}$$

where h = heat transfer coefficient (typically 5–10 W/m^2 K for still air and 10–100 W/m^2 K for forced air)

K = thermal conductivity of the plastic

x = full or half thickness of the plastic as described above.

A large Biot Number means that conduction controls the energy transfer to/from the plastic and large temperature gradients will exist in the plastic. A small Biot Number means that convection is the dominant factor. The above analysis was for conduction heat transfer ($B_i \rightarrow \infty$). When the plastic moulding is taken out of the mould we need to check the value of B_i. In this case

$$B_i = \frac{20 \times 1.5 \times 10^{-3}}{0.25} = 0.12$$

The dimensionless temperature gradient is

$$\Delta T = \frac{60 - 20}{80 - 20} = 0.57$$

Fig. 5.24 shows the modified $\Delta T/F_o$ graph for a flat sheet for different values of B_i. In this case $1/B_i = 8.33$. So from Fig. 5.24

$$F_o = 4 = \frac{\alpha t}{x^2}$$

$$t = \frac{4 \times (1.5 \times 10^{-3})^2}{1 \times 10^{-7}} = 90 \text{ s}$$

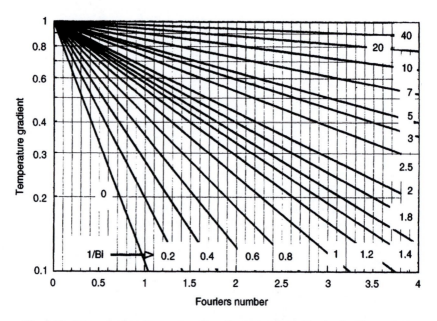

Fig. 5.24   Dimensionless Temperature Data for a Flat Sheet (Conduction/Convection)

Hence the plastic cools by 140°C (230°C to 90°C) in 13.3 seconds when it is in contact with the mould but it takes 90 seconds to cool a further 30°C when it emerges out into the air. This highlights the much faster cooling rate achieved in the mould.

Some typical moulding data for a range of plastics are given in Table 5.1. Note that the de-moulding temperature will be generally about 30°C below the Freeze-off temperature in order to ensure that the moulding is sufficiently solid for handling.

The above analysis illustrates how to estimate the cooling time using the relationship between the temperature gradient ΔT and the Fouriers Number

Table 5.1

Plastic	Moulding temperature (°C)	Mould temperature (°C)	Freeze-off temperature (°C)	Thermal diffusivity (m²/s)
LDPE	230	30	90	1.1×10^{-7}
Polypropylene	250	40	135	1.1×10^{-7}
Polystyrene	230	50	130	0.8×10^{-7}
PVC	180	40	140	1.2×10^{-7}
POM	210	60	130	0.9×10^{-7}
Acrylic	250	50	160	0.9×10^{-7}
Polycarbonate	300	90	200	1.0×10^{-7}
Nylon 66	290	80	240	1.3×10^{-7}
ABS	230	60	140	0.9×10^{-7}

F_0. A number of empirical relationships have also been proposed. Two of the more useful equations take the form

(a) for flat sheet

$$\text{time, } t = \frac{4x^2}{\pi^2 \alpha} \ln \left(\frac{4}{\pi \cdot \Delta T} \right) \qquad (5.79)$$

or (b) for cylinder

$$\text{time, } t = \frac{1.7r^2}{\pi^2 \alpha} \ln \left(\frac{1.7}{\Delta T} \right) \qquad (5.80)$$

The reader may wish to check the predictions of freeze-off time for the data in the above Example.

Example 5.12 Estimate the heat transfer from the die to the melt as it passes through the die land in Example 5.5.

Solution

$$\text{Volume of die land} = \pi DHL$$
$$= (\pi \times 260 \times 0.7 \times 4 \times 10^{-9}) \text{m}^3$$
$$\text{Residence time} = \frac{\pi \times 260 \times 0.7 \times 4 \times 10^{-9}}{109.6 \times 10^{-6}}$$
$$= 2.08 \times 10^{-2} \text{ s}$$
$$\text{Fourier number} = \frac{\alpha t}{x^2}$$
$$F_0 = \frac{1 \times 10^{-7} \times 2.08 \times 10^{-2}}{(0.35 \times 10^{-3})^2}$$
$$= 0.017$$

From Fig. 5.23 it may be seen that the centre line temperature gradient at this Fourier Number is almost 1.

$$\text{Since } \Delta T = \frac{T_3 - T_2}{T_1 - T_2} \simeq 1$$

$$\text{then } T_3 \simeq T_1$$

Thus the melt temperature after 2.08×10^{-2} seconds is the same as the initial melt temperature (T_1) so that as the melt passes through the die land it is relatively unaffected by the temperature of the die.

Any temperature rise which occurs in the melt as it passes through this section will be as a result of the work done on the melt and may be estimated using (5.64).

$$\text{Temperature rise} = \frac{P}{\rho C_p}$$

$$\text{So temperature rise} = \frac{3.66 \times 10^6}{760 \times 2.5 \times 10^3}$$

$$= 1.9°C$$

Example 5.13 Derive an expression for the flow length of a power law fluid when it is injected at constant pressure into a rectangular section channel assuming

(a) the flow is isothermal

(b) there is freezing off as the melt flows (ie non-isothermal flow).

Solution (a) As shown earlier the flow rate of a power law fluid in a rectangular section is given by

$$Q = \left(\frac{n+1}{2n+1}\right) TH \left(\frac{n}{n+1}\right) \left(\frac{P}{\eta_0 \ell}\right)^{1/n} \left(\frac{H}{2}\right)^{(n+1)/n} \tag{5.81}$$

During any increment of time, dt, the volume of the fluid injected into the channel is Qdt. This will be equal to the increase in volume of the fluid in the channel

$$Qdt = TH d\ell$$

So from above

$$TH\frac{d\ell}{dt} = \left(\frac{n}{2n+1}\right) TH \left(\frac{P}{\eta_0 \ell}\right)^{1/n} \left(\frac{H}{2}\right)^{(n+1)/n}$$

$$\therefore \quad \int_0^L \ell^{1/n} d\ell = \left(\frac{n}{2n+1}\right) \left(\frac{P}{\eta_0}\right)^{1/n} \left(\frac{H}{2}\right)^{(n+1)/n} \int_0^t dt$$

$$\left(\frac{n}{n+1}\right) L^{(n+1)/n} = \left(\frac{n}{2n+1}\right) \left(\frac{P}{\eta_0}\right)^{1/n} \left(\frac{H}{2}\right)^{(n+1)/n} \cdot t$$

$$L = \left(\frac{n+1}{2n+1}\right)^{n/(n+1)} \left(\frac{P}{\eta_0}\right)^{1/(n+1)} \left(\frac{H}{2}\right) \cdot t^{n/(n+1)} \quad (5.82)$$

The volume flow rate at any instant in time may be determined by substituting for L in equation (5.81).

(b) If the melt is freezing off as it flows then the effective channel depth will be h instead of H as shown in Fig. 5.23. Therefore the above expression may be written as

$$\ell^{1/n} d\ell = \left(\frac{n}{2n+1}\right) \left(\frac{P}{\eta_0}\right)^{1/n} \left(\frac{h}{2}\right)^{(n+1)/n} dt \quad (5.83)$$

This equation cannot be integrated as simply as before because h is now a function of time. Fig. 5.26 shows how the depth of the cavity changes with time as the melt flows. Barrie has investigated this situation and concluded that the freezing-off could be described by a relation of the form

$$\Delta y = Ct^s \quad (5.84)$$

Fig. 5.25   Flow of Molten Polymer into a Cold Mould

where C and s are constants and Δy is the thickness of the frozen layer as shown in Fig. 5.25. For most polymer melt flow situations, Barrie found that $s = 1/3$.

From (5.84) using the boundary condition that $t = t_f$ (freeze-off time) at

$$\Delta y = H/2$$

then $C = (H/2t_f^{1/3})$

Fig. 5.26   Variation of Cavity Depth with Time

also $2\Delta y = (H - h)$ so in (5.84)

$$\left(\frac{H - h}{2}\right) = \frac{H}{2}\left(\frac{t}{t_f}\right)^{1/3}$$

$$h = H\left(1 - \left(\frac{t}{t_f}\right)^{1/3}\right) \qquad (5.85)$$

So substituting this expression for h into (5.83), then

$$\ell^{1/n}d\ell = \left(\frac{n}{2n + 1}\right)\left(\frac{P}{\eta_0}\right)^{1/n}\left(\frac{1}{2}\right)^{(n+1)/n}\left[H\left(1 - \left(\frac{t}{t_f}\right)^{1/3}\right)\right]^{(n+1)/n} dt$$

This may then be integrated to give

$$\left(\frac{n}{n + 1}\right)L^{(n+1)/n} = \left(\frac{n}{2n + 1}\right)\left(\frac{P}{\eta_0}\right)^{1/n}\left(\frac{H}{2}\right)^{(n+1)/n}$$

$$\times \left[6t_f\left(\frac{n}{4n + 1}\right)\left(\frac{n}{3n + 1}\right)\left(\frac{n}{2n + 1}\right)\right]$$

So $L = \left(\frac{P}{\eta_0}\right)^{1/(n+1)}\frac{H}{2}\left[6t_f\left(\frac{n}{4n + 1}\right)\left(\frac{n}{3n + 1}\right)\left(\frac{n(n + 1)}{(2n + 1)^2}\right)\right]^{n/(n+1)}$

$$(5.86)$$

Similar expressions may also be derived for a circular section channel and for the situation where the injection rate is held constant rather than the pressure (see questions at the end of the chapter). In practical injection moulding situations the injection rate would probably be held constant until a pre-selected value of pressure is reached. After this point, the pressure would be held constant and the injection rate would decrease.

Note that for a Newtonian fluid, $n = 1$, so for the isothermal case, equation (5.82) becomes.

$$L = 0.408 \left(\frac{Pt}{\eta}\right)^{1/2} H$$

and for the non-isothermal case, equation (5.86) becomes

$$L = 0.13 \left(\frac{Pt_f}{\eta}\right)^{1/2} H$$

For the non-isothermal cases the freeze-off time, t_f, may be estimated by the method described in Example 5.11.

Example 5.14 A power law fluid with constants $\eta_0 = 1.2 \times 10^4$ Ns/m^2 and $n = 0.35$ is injected through a centre gate into a disc cavity which has a depth of 2 mm and a diameter of 200 mm. If the injection rate is constant at 6×10^{-5} m^3/s, estimate the time taken to fill the cavity and the minimum injection pressure necessary at the gate for (a) Isothermal and (b) Non-isothermal conditions.

Solution (a) *Isothermal Situation.* If the volume flow rate is Q, then for any increment of time, dt, the volume of material injected into the cavity will be given by (Qdt). During this time period the melt front will have moved from a radius, r, to a radius $(r + dr)$. Therefore a volume balance gives the relation

$$Qdt = 2\pi r H dr$$

where H is the depth of the cavity.

Since the volume flow rate, Q, is constant this expression may be integrated to give

$$\int_0^t dt = \frac{2\pi H}{Q} \int_0^R r \, dr$$

$$t = \frac{\pi R^2 H}{Q}$$

For the conditions given

$$t = \frac{\pi (100 \times 10^{-3})^2 \, (2 \times 10^{-3})}{6 \times 10^{-5}}$$

$$= 1.05 \text{ seconds}$$

It is now necessary to derive an expression for the pressure loss in the cavity. Since the mould fills very quickly it may be assumed that effects due to freezing-off of the melt may be ignored. In Section 5.4(b) it was shown that for the flow of a power law fluid between parallel plates

$$Q = \left(\frac{n+1}{2n+1}\right) T V_0 H$$

Now for the disc, $T = 2\pi r$ and substituting for V_0 from (5.27)

$$Q = \left(\frac{n+1}{2n+1}\right) \left(2\pi r H \left[-\left(\frac{n}{n+1}\right) \left(\frac{1}{\eta_0}\right)^{1/n} \left(\frac{dP}{dr}\right)^{1/n} \left(\frac{H}{2}\right)^{(n+1/n)}\right]\right.$$

$$dP = Q^n \left(\frac{2n+1}{2\pi n}\right)^n \left\{\frac{\eta_0(2)^{n+1}}{(H)^{2n+1}}\right\} \frac{dr}{r^n}$$

$$\int_{P_1}^{P_2} dP = \int_{R_1}^{R_2} -Q^n \left\{\frac{2n+1}{2\pi n}\right\}^n \frac{\eta_0(2)^{n+1}}{(H)^{2n+1}} \cdot \frac{dr}{r^n}$$

$$P_2 - P_1 = -Q^n \left\{\frac{2n+1}{2\pi n}\right\}^n \frac{\eta_0(2)}{H^{2n+1}} \left[\frac{r^{1+n}}{1-n}\right]_{R_1}^{R_2}$$

$$= -Q^n \left\{\frac{2n+1}{2\pi n}\right\} \frac{\eta_0(2)^{n+1}}{(1-n)H^{2n+1}} [R_2^{1-n} - R_1^{1-n}]$$

Now letting $P_1 = 0$ at $R_1 = R$
and $\qquad P_2 = P$ at $R_2 = r$

$$P = -Q^n \left\{\frac{2n+1}{2\pi n}\right\} \frac{\eta_0(2)^{n+1}}{(1-n)H^{2n+1}} [R^{1-n} - r^{1-n}]$$

$$= Q^n \left\{\frac{2n+1}{2\pi n}\right\} \frac{\eta_0(2)^{n+1}}{(1-n)H^{2n+1}} R^{1-n} \left[1 - \left(\frac{r}{R}\right)^{1-n}\right]$$

now $P = P_0$ at $r = 0$

$$P_0 = Q^n \left\{\frac{2n+1}{2\pi n}\right\} \frac{\eta_0(2)^{n+1}}{(1-n)H^{2n+1}} R^{1-n}$$

So for the conditions given

$$P_0 = (6 \times 10^{-5})^{0.35} \left(\frac{1.7}{0.7\pi}\right)^{0.35} \frac{(1.2 \times 10^4)(2)^{1.35}(100 \times 10^{-3})^{0.65}}{(2 \times 10^{-3})^{1.7}(0.65)}$$

$$= 12.42 \text{ MN/m}^2$$

Note that if P_0 is substituted back into the expression for P then the expression used earlier (Chapter 4) to calculate the mould clamping force is obtained. That is

$$P = P_0 \left\{ 1 - \left(\frac{r}{R} \right)^{1-n} \right\}$$

(b) *Non-isothermal Situation*. In this case the injection pressure will need to be higher because as the melt flows the cavity effectively gets thinner due to the freezing-off of the melt as it makes contact with the relatively cold mould.

$$\Delta P = Q^n \left\{ \frac{2n+1}{2\pi n} \right\} \frac{\eta_0 (2)^{n+1}}{(1-n)H^{2n+1}} R^{1-n}$$

but when non-isothermal conditions apply, $H = H(t)$, ie the cavity thickness varies with time due to the build up of a frozen layer at the mould.

As shown in the previous Example, the variation of cavity thickness with time is given by

$$H(t) = H \left(1 - \left(\frac{t}{t_f} \right)^s \right)$$

where 's' is often taken as 1/3.

So $\qquad \Delta P = Q^n \left\{ \frac{2n+1}{2\pi n} \right\} \frac{\eta_0 (2)^{n+1}}{(1-n)H^{2n+1}} \frac{R^{1-n}}{\left\{ 1 - \left(\dfrac{t}{t_f} \right)^{1/3} \right\}^{2n+1}}$

For the conditions given, $T_i = 230°C$, $T_t = 90°C$, $T_m = 30°C$, the freeze-off time may be calculated as shown in Example 5.11. Thus $t_f = 5.8$ seconds.

Thus $\qquad \left\{ 1 - \left(\frac{t}{t_f} \right)^{1/3} \right\}^{2n+1} = \left\{ 1 - \left(\frac{1.05}{5.8} \right)^{1/3} \right\}^{1.7} = 0.24$

So, for non-isothermal conditions

$$\Delta P = \frac{12.42}{0.24} = 51.3 \text{ MN/m}^2$$

5.13 Calculation of Clamping force

(a) Isothermal Situation

It may be seen from the above analysis and that in Chapter 4 for the calculation of clamping force on an injection moulding machine the Mean Effective pressure (MEP) across the cavity may be obtained from

$$(MEP) A = \pi R^2 P_0 \left\{ \frac{1-n}{3-n} \right\} \tag{5.87}$$

where A = projected area (πR^2, for a disc mould).

Hence, for the isothermal situation

$$MEP = \left\{\frac{1-n}{3-n}\right\} \{Q^n\} \left\{\frac{2n+1}{2\pi n}\right\} \frac{\eta_0(2)^{n+1}}{(1-n)H^{2n+1}} R^{1-n}$$

and for a constant injection rate, $Q = \pi R^2 H / t$

So, $$MEP = \left(\frac{1-n}{3-n}\right) \left[\left(\frac{\pi}{t}\right)^n \left\{\frac{2n+1}{2\pi n}\right\} \frac{\eta_0(2)^{n+1}}{(1-n)} \left(\frac{R}{H}\right)^{n+1}\right]$$

$$= \frac{\pi^{n-1}}{t^n} \left\{\frac{(2n+1)2^{n+1}}{2n(3-n)}\right\} \eta_0 \left(\frac{R}{H}\right)^{n+1}$$

$$MEP = \left\{\frac{(2n+1)\cdot 2^{n+1}\cdot \pi^{n-1}}{2n\cdot(3-n)}\right\} \frac{1}{t^n} \cdot \eta_0 \cdot \left(\frac{R}{H}\right)^{n+1} \tag{5.88}$$

Thus for any plastic where the Power Law constants are known, the clamping force can be calculated for a given radius, R, cavity depth, H, and fill time, t.

Fig. 5.27 shows the variation of MEP with flow ratio (R/H) for spreading flow in discs of different depths. The material is polypropylene and the constant injection rate is 3.4×10^{-3} m^3/s. This is a high injection rate but has been chosen because the clamp forces predicted by this diagram are representative of those occurring in real moulding situations (even though it is based

Fig. 5.27   Variation of Cavity Pressure with Flow Ratio

on isothermal conditions). Note that the clamp force is calculated simply by multiplying the mean effective pressure (MEP) by the projected area of the moulding, as illustrated in Chapter 4.

For plastics other than polypropylene it would be necessary to produce a similar set of curves using equation (5.88) and the Power Law data in Table 5.2.

Table 5.2

Typical Power Law Parameters for Plastics

Material	Temperature (°C)	$\eta_0 (\text{Ns}^n \text{ m}^{-2})$	n
LDPE	170	0.85×10^4	0.33
PVC	170	0.36×10^4	0.67
Polypropylene	230	0.85×10^4	0.33
Acetal (POM)	210	2.0×10^4	0.33
Polystyrene	230	1.4×10^4	0.33
ABS	230	1.95×10^4	0.33
PMMA	230	3.0×10^4	0.33
Polycarbonate	290	3.6×10^4	0.5
Nylon 66	290	0.8×10^4	0.67

(b) Non-isothermal Situation

Once again

$$MEP = \left\{ \frac{1-n}{3-n} \right\} Q^n \left\{ \frac{2n+1}{2\pi n} \right\} \frac{\eta_0 (2)^{n+1}}{(1-n)H^{2n+1}} R^{1-n}$$

but in this case, the thickness H is a function of time.

$$H(t) = H \left\{ 1 - \left(\frac{t}{t_f} \right)^{1/3} \right\}$$

and the constant volume flow rate $Q = \pi R^2 H / t$ where t is the fill time for the cavity.

$$MEP = \left(\frac{1-n}{3-n} \right) \left(\frac{\pi}{t} \right)^n \left\{ \frac{2n+1}{2\pi n} \right\} \frac{\eta_0 (2)^{n+1}}{(1-n)} \left(\frac{R}{H} \right)^{n+1} \frac{1}{\left\{ 1 - \left(\frac{t}{t_f} \right)^{1/3} \right\}^{2n+1}}$$

$$MEP = \left\{ \frac{(2n+1) \cdot 2^{n+1} \pi^{n-1}}{2n(3-n)} \right\} \eta_0 \left(\frac{R}{H} \right)^{n+1} \frac{1}{t^n \left\{ 1 - \left(\frac{t}{t_f} \right)^{1/3} \right\}^{2n+1}} \quad (5.89)$$

Fig. 5.28 shows a comparison of the isothermal and non-isothermal situations for polypropylene. When pressure is plotted as a function of injection rate, it

Fig. 5.28   Variation of Cavity Pressure Loss with Injection Rate

may be seen that for the isothermal situation the pressure increases as the injection rate increases. However, in the situation where the melt freezes off as it enters the mould, the relationship is quite different. In this case the pressure is high at high injection rates, in a similar manner to the isothermal situation. In fact at extremely high injection rates the two curves would meet because the melt would enter the mould so fast it would not have time to be affected by the melt temperature.

However, in the non-isothermal case the pressure is also high at low injection rates. This is because slow injection gives time for significant solidification of the melt and this leads to high pressures. It is clear therefore that in the non-isothermal case there is an optimum injection rate to give minimum pressure. In Fig. 5.28 this is seen to be about 3.0×10^{-4} m^3/s for the situation considered here. This will of course change with melt temperature and mould temperature since these affect the freeze-off time, t_f, in the above equations.

Some typical values for η_0 and n are given in Table 5.2.

The viscosity flow curves for these materials are shown in Fig. 5.17. To obtain similar data at other temperatures then a shift factor of the type given in equation (5.27) would have to be used. The temperature effect for polypropylene is shown in Fig. 5.2.

Example 5.15 During injection moulding of low density polyethylene, 15 kg of material are plasticised per hour. The temperature of the melt entering the mould is 190°C and the mould temperature is 40°C. If the energy input from the screw is equivalent to 1 kW, calculate

(a) the energy required from the heater bands
(b) the flow rate of the circulating water in the mould necessary to keep its temperature at $40 \pm 2°$

Solution The steady flow energy equation may be written as

$$q - W = \Delta h \qquad (5.90)$$

where q is the heat transfer per unit mass
       W is the work transfer per unit mass
       h is enthalpy

Enthalpy is defined as the amount of heat required to change the temperature of unit mass of material from one temperature to another. Thus the amount of heat required to change the temperature of a material between specified limits is the product of its mass and the enthalpy change.

The enthalpy of plastics is frequently given in graphical form. For a perfectly crystalline material there is a sharp change in enthalpy at the melting point due to the latent heat. However, for semi-crystalline plastics the rate of enthalpy change with temperature increases up to the melting point after which it varies linearly with temperature increases up to the melting point after which it varies linearly with temperature as shown in Fig. 5.29. For amorphous plastics there is only a change in slope of the enthalpy line at glass transition points. Fig. 5.29 shows that when LDPE is heated from 20°C to 190°C the change in enthalpy is 485 kJ/kg.

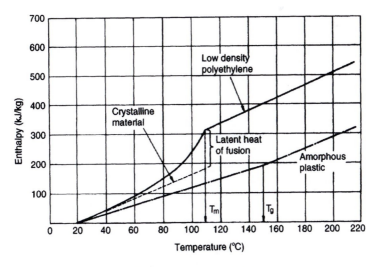

Fig. 5.29   Enthalpy Variation with Temperature

(a) In equation (5.90) the sign convention is important. Heat is usually taken as positive when it is applied to the system and work is positive when done by the system. Hence in this example where the work is done on the system by the screw it is regarded as negative work.

So using (5.90) for a mass of 15 kg per hour

$$q - (-4) = 15 \left(485 \times \frac{1}{60} \times \frac{1}{60} \right)$$

$$q = 1.02 \text{ kW}$$

The heater bands are expected to supply this power.

(b) At the mould there is no work done so in terms of the total heat absorbed in cooling the melt from 190°C to 40°C.

$$q = m\Delta h$$

$$= 15 \left[(485 - 40) \frac{1}{60} \times \frac{1}{60} \right]$$

$$= 1.85 \text{ kW}$$

This heat must be removed by the water circulating in the mould at a rate, Q

$$q = Q\Delta h$$

or by definition of enthalpy

$$q = QC_p\Delta T$$

where C_p is the specific heat (= 4.186 kJ/kg°C for water) and ΔT is the temperature change (= 4°C i.e. ±2°C)

$$1.85 = Q \times 4.186 \times 4$$

$$Q = 0.11 \text{ kg/s}$$

$$= 0.11 \text{ litres/s}$$

It is also possible to estimate the number of cooling channels required. If the thermal conductivity of the mould material is K then the heat removed through the mould per unit time will be given by

$$q = \frac{K \cdot A \, (\Delta T)}{Y}$$

where ΔT is the temperature between the melt and the circulating fluid

Y is the distance of the cooling channels from the mould

and A is the area through which the heat is conducted to the coolant. This is usually taken as half the circumference of the cooling channel multiplied by its length.

$$q = \frac{K\pi DL(\Delta T)}{2Y}$$

$$L = \frac{2Yq}{K\pi D(\Delta T)}$$

The K value for steel is 11.5 cal/m.s.°C, so assuming that the cooling channels have a diameter of 10 mm and they are placed 40 mm from the cavity, then

$$L = \frac{2 \times 40 \times 10^{-3} \times 1.85 \times 10^3}{11.5 \times 4.2 \times \pi \times 10 \times 10^{-3} \times (190 - 40)}$$

$$= 0.65 \text{ m}$$

If the length of the cavity is 130 mm then five cooling channels would be needed to provide the necessary heat removal.

Bibliography

Pearson, J.R.A. *Mechanics of Polymer Processing*, Elsevier Applied Science, London (1985)

Throne, J.L. *Plastics Process Engineering*, Marcel Dekker, New York (1979)

Fenner, R.T. *Principles of Polymer Processing*, Macmillan, London, (1979)

Tadmor, Z. and Gogos, C.G. *Principles of Polymer Processing*, Wiley Interscience, New York (1979)

Brydson, J.A. *Flow Properties of Polymer Melts*, George Godwin, London (1981)

Grober, H. *Fundamentals of Heat Transfer*, McGraw-Hill, New York (1961)

Cogswell, F.N. *Polymer Melt Rheology*, George Godwin, London (1981)

Carreau, P.J., De Kee, D. and Domoux, M. *Can. J. Chem. Eng.*, **57** (1979) p. 135.

Rao, N.S., *Design Formulas for Plastics Engineers*, Hanser, Munich (1991).

Kliene, I., Marshall, D.J. and Friehe, C.A. *J. Soc. Plast. Eng.*, **21** (1965) p. 1299.

Muenstedt, H. *Kunststoffe*, **68** (1978) p. 92.

Questions

5.1 In a particular type of cone and plate rheometer the torque is applied by means of a weight suspended on a piece of cord. The cord passes over a pulley and is wound around a drum which is on the same axis as the cone. There is a direct drive between the two. During a test on polythene at 190°C the following results were obtained by applying a weight and, when the steady state has been achieved, noting the angle of rotation of the cone in 40 seconds. If the diameter of the cone is 50 mm and its included angle is 170°, estimate the viscosity of the melt at a shear stress of 10^4 N/m².

Weight (g)	50	100	200	500	1000	2000
Angle ($\theta°$)	0.57	1.25	2.56	7.36	17.0	42.0

5.2 Derive expressions for the velocity profile, shear stress, shear rate and volume flow rate during the isothermal flow of a power law fluid in a rectangular section slit of width W, depth H and length L. During tests on such a section the following data was obtained.

Flow rate (kg/min)	0.21	0.4	0.58	0.8	1.3	2.3
Pressure drop (MN/m²)	1.8	3.0	4.0	5.2	7.6	12.0

If the channel has a length of 50 mm, a depth of 2 mm and a width of 6 mm, establish the applicability of the power law to this fluid and determine the relevant constants. The density of the fluid is 940 kg/m^3.

5.3 The viscosity characteristics of a polymer melt are measured using both a capillary rheometer and a cone and plate viscometer at the same temperature. The capillary is 2.0 mm diameter and 32.0 mm long. For volumetric flow rates of 70×10^{-9} m^3/s and 200×10^{-9} m^3/s, the pressures measured just before the entry to the capillary are 3.9 MN/m^2 and 5.7 MN/m^2, respectively.

The angle between the cone and the plate in the viscometer is 3° and the diameter of the base of the cone is 75 mm. When a torque of 1.18 Nm is applied to the cone, the steady rate of rotation reached is observed to be 0.062 rad/s.

Assuming that the melt viscosity is a power law function of the rate of shear, calculate the percentage difference in the shear stresses given by the two methods of measurement at the rate of shear obtained in the cone and plate experiment.

5.4 The correction factor for converting apparent shear rates at the wall of a circular cylindrical capillary to true shear rates is $(3n + 1)/4n$, where n is the power law index of the polymer melt being extruded.

Derive a similar expression for correcting apparent shear rates at the walls of a die whose cross-section is in the form of a very long narrow slit.

A slit die is designed on the assumption that the material is Newtonian, using apparent viscous properties derived from capillary rheometer measurements, at a particular wall shear stress, to calculate the volumetric flow rate through the slit for the same wall shear stress. Using the correction factors already derived, obtain an expression for the error involved in this procedure due to the melt being non-Newtonian. Also obtain an expression for the error in pressure drop calculated on the same basis. What is the magnitude of the error in each case for a typical power law index n = 0.37?

5.5 Polyethylene is extruded through a cylindrical die of radius 3 mm and length 37.5 mm at a rate of 2.12×10^{-6} m^3/s. Using the flow curves supplied, calculate the natural time of the process and comment on the meaning of the value obtained.

5.6 Polythene is passed through a rectangular slit die 5 mm wide, 1 mm deep at a rate of 0.7×10^{-9} m^3/s. If the time taken is 1 second, calculate the natural time and comment on its meaning.

5.7 In a plunger type injection moulding machine the torpedo has a length of 30 mm and a diameter of 23 mm. If, during the moulding of polythene at 170°C (flow curves given), the plunger moves forward at a speed of 50 mm/s estimate the pressure drop along the torpedo. The barrel diameter is 25 mm.

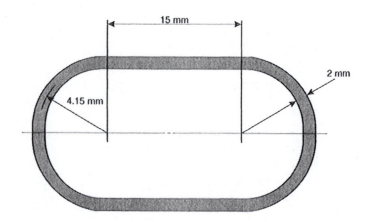

5.8 The exit region of a die used to extrude a plastic section is 10 mm long and has the cross-sectional dimensions shown below. If the channel is being extruded at the rate of 3 m/min calculate the power absorbed in the die exit and the melt temperature rise in the die. Flow curves for the polymer melt are given in Fig. 5.3. The product ρC_p for the melt is 3.3×10^6.

5.9 The exit region of a die used to extrude a plastic channel section is 10 mm long and has the dimensions shown below. If the channel is being extruded at the rate of 3 m/min. calculate the power absorbed in the die exit, and the dimensions of the extrudate as it emerges from the die. The flow curves in Fig. 5.3 may be used.

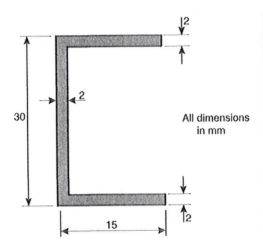

All dimensions in mm

5.10 During extrusion blow moulding of 60 mm diameter bottles the extruder output rate is 46×10^{-3} m³/s. If the die diameter is 30 mm and the die gap is 1.5 mm calculate the wall thickness of the bottles which are produced. The flow curves in Fig. 5.3 should be used.

5.11 Polyethylene is injected into a mould at a temperature of 170°C and a pressure of 100 MN/m². If the mould cavity has the form of a long channel with a rectangular cross-section 6 mm × 1 mm deep, estimate the length of the flow path after 1 second. The flow may be assumed to be isothermal and over the range of shear rates experienced ($10^3 - 10^5$ s⁻¹) the material may be considered to be a power law fluid.

5.12 Repeat the previous question for the situation in which the mould temperature is 60°C and the freeze-off temperature is 128°C. What difference would it make if it had been assumed that the material was Newtonian with a viscosity of 1.2×10^2 Ns/m².

5.13 During the blow moulding of polypropylene bottles, the parison is extruded at a temperature of 230°C and the mould temperature is 50°C. If the wall thickness of the bottle is 1 mm and the bottles can be ejected at a temperature of 120°C estimate the cooling time in the mould.

5.14 An injection moulding is in the form of a flat sheet 100 mm square and 4 mm thick. The melt temperature is 230°C, the mould temperature is 30°C and the plastic may be ejected from the mould at a centre-line temperature of 90°C. If the runner design criterion is that it should be ejectable at the same instant as the moulding, estimate the required runner diameter. The thermal diffusivity of the melt is 1×10^{-7} m²/s.

5.15 For a particular polymer melt the power law constants are = 40 kN.s^n/m and n = 0.35. If the polymer flows through an injection nozzle of diameter 3 mm and length 25 mm at a rate of 5×10^{-5} m³/s, estimate the pressure drop in the nozzle.

5.16 Polythene at 170°C is used to injection mould a disc with a diameter of 120 mm and thickness 3 mm. A sprue gate is used to feed the material into the centre of the disc. If the injection rate is constant and the cavity is to be filled in 1 second estimate the minimum injection pressure needed at the nozzle. The flow curves for this grade of polythene are given in Fig. 5.3.

5.17 During the injection moulding of a polythene container having a volume of 4×10^{-6} m^3, the melt temperature is 170°C, the mould temperature is 50°C and rectangular gates with a land length of 0.6 mm are to be used. If it is desired to have the melt enter the mould at a shear rate of 10^3 s^{-1} and freeze-off at the gate after 3 seconds, estimate the dimensions of the gate and the pressure drop across it. It may be assumed that freeze-off occurs at a temperature of 114°C. The flow curves in Fig. 5.3 should be used.

5.18 Polyethylene at 170°C passes through the annular die shown, at a rate of 10×10^{-6} m^3/s. Using the flow curves provided and assuming the power law index n = 0.33 over the working section of the curves, calculate the total pressure drop through the die. Also estimate the dimensions of the extruded tube.

5.19 A polythene tube of outside diameter 40 mm and wall thickness 0.75 mm is to be extruded at a linear speed of 15 mm/s. Using the 170°C polythene flow curves supplied, calculate suitable die exit dimensions.

5.20 The exit region of a die used to blow plastic film is shown below. If the extruder output is 100×10^{-6} m^3/s of polythene at 170°C estimate the total pressure drop in the die between points A and C. Also calculate the dimensions of the plastic bubble produced. It may be assumed that there is no inflation or draw-down of the bubble. Flow data for polythene is given in Fig. 5.3.

5.21 A polyethylene moulding material at 170°C passes along the channel shown at a rate of 4×10^{-6} m^3/s. Using the flow curves given and assuming n = 0.33 calculate the pressure drop along the channel.

5.22 A power law plastic is injected into a circular section channel using a constant pressure, P. Derive an expression for the flow length assuming that

(a) the flow is isothermal
(b) the melt is freezing off as it flows along the channel.

5.23 A polymer melt is injected into a circular section channel under constant pressure. What is the ratio of the maximum non-isothermal flow length to the isothermal flow length in the same time for (a) a Newtonian melt and (b) a power law melt with index. $n = 0.3$.

5.24 A power law fluid with the constants $\eta_0 = 10^4$ Ns/m^2 and $n = 0.3$ is injected into a circular section channel of diameter 10 mm. Show how the injection rate and injection pressure vary with time if.

(a) the injection pressure is held constant at 140 MN/m^2
(b) the injection rate is held constant at 10^{-3} m^3/s.

The flow in each case may be considered to be isothermal.

5.25 Polyethylene at 170°C is used to injection mould a flat plaque measuring 50 mm × 10 mm × 3 mm. A rectangular gate which is 4 mm × 2 mm with a land length of 0.6 mm is situated in the centre of the 50 mm side. The runners are 8 mm diameter and 20 mm long. The material passes from the barrel into the runners in 1 second and the pressure losses in the nozzle and sprue may be taken as the same as those in the runner. If the injection rate is fixed at 10^{-6} m^3/s, estimate (a) the pressure losses in the runner and gate and (b) the initial packing pressure on the moulded plaque. Flow curves are supplied.

5.26 A lace of polyethylene is extruded with a diameter of 3 mm and a temperature of 190°C. If its centre-line must be cooled to 70°C before it can be granulated effectively, calculate the required length of the water bath if the water temperature is 20°C. The haul-off speed is 0.4 m/s and it may be assumed that the heat transfer from the plastic to the water is by conduction only.

5.27 Using the data in Tables 5.1 and 5.2, calculate the flow lengths which would be expected if the following materials were injected at 100 MN/m^2 into a wide rectangular cross-section channel, 1 mm deep.
Materials – LDPE, polypropylene, polystyrene, PVC, POM, acrylic, polycarbonate, nylon 66 and ABS. Note that the answers will give an indication of the flow ratios for these materials. The flow should be assumed to be non-isothermal.

5.28 It is desired to blow mould a cylindrical plastic container of diameter 100 mm and wall thickness 2.5 mm. If the extruder die has an average diameter of 40 mm and a gap of 2 mm, calculate the output rate needed from the extruder. Comment on the suitability of an inflation pressure in the region of 0.4 MN/m^2. The density of the molten plastic may be taken as 790 kg/m^3. Use the flow curves in Fig. 5.3.

5.29 During the blow moulding of polyethylene at 170°C the parison is 0.4 m long and is left hanging for 1 second. Estimate the natural time for the process and the amount of sagging which occurs. The density of the melt may be taken at 730 kg/m^3.

5.30 The viscosity, η, of plastic melt is dependent on temperature, T, and pressure, P. The variations for some common plastics are given by equations of the form

$$\eta/\eta_R = 10^{B\Delta P} \quad \text{and} \quad \eta/\eta_R = 10^{A\Delta T}$$

where $\Delta T = T - T_R(^\circ C)$, $\Delta P = P - P_R$ (MN/m²), and the subscript R signifies a reference value. Typical values of the constants A and B are given below.

	Acrylic	Polypropylene	LDPE	Nylon	Acetal
A $(\times 10^{-3})$	−28.32	−7.53	−11.29	−12.97	−7.53
B $(\times 10^{-3})$	9.54	6.43	6.02	4.22	3.89

During flow along a particular channel the temperature drops by 40°C and the pressure drops by 50 MN/m². Estimate the overall change in viscosity of the melt in each case. Determine the ratio of the pressure change to the temperature change which would cause no change in viscosity for each of the above materials.

APPENDIX A – Structure of Plastics

A.1 Structure of Long Molecules

Polymeric materials consist of long chain-like molecules. Their unique structural configuration affects many of their properties and it is useful to consider in more detail the nature of the chains and how they are built up. The simplest polymer to consider for this purpose is polyethylene. During the polymerisation of the monomer ethylene, the double bond (see Fig. A.1) is opened out enabling the carbon single bonds to link up with neighbouring units to form a long chain of CH_2 groups as shown in Fig. A.2. This is a schematic representation and conceals the fact that the atoms are jointed to each other at an angle as shown in Fig. A.3.

Fig. A.1 Ethylene monomer

$$-C - C - C - C - C -$$

Fig. A.2 Polyethylene molecule

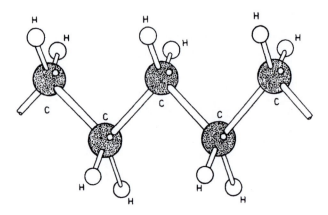

Fig. A.3   Polyethylene molecule

In all the groups along the chain, the bond angle is fixed. It is determined by considering a carbon atom at the centre of a regular tetrahedron and the four covalent bonds are in the directions of the four corners of the tetrahedron. This sets the bond angle at 109° 28′ as shown in Fig. A.4 and this is called the tetrahedral angle.

Fig. A.4   Tetrahedral angle

For a typical molecular weight of 300,000 there are about 21,000 carbon atoms along the backbone of the chain. Since the length of the C-C bond is 0.154×10^{-9} m the dimensions of an extended zig-zag chain would be about 2700 mm long and 0.3 mm diameter. This gives an idea of the long thread-like nature of the molecules. It must be remembered, however, that in any particular polymer, not all molecular chains have the same length. The length of each chain depends on a series of random events during the polymerisation process. One chain may grow rapidly in a region with an abundant supply of monomer whereas other chains stop growing prematurely as the supply of monomer dries up. This means that a particular sample of synthetic polymer will not have a unique value for its molecular weight. Instead statistical methods are used to determine an average molecular weight and the molecular weight distribution.

A.2 Conformation of the Molecular Chain

The picture presented so far of the polyethylene chain being of a linear zig-zag geometry is an idealised one. The conformation of a molecular chain is in fact random provided that the bond tetrahedral angle remains fixed. This is best illustrated by considering a piece of wire with one bend at an angle of 109° 28′ as shown in Fig. A.5a.

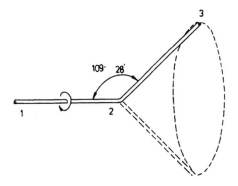

Fig. A.5(a)   Rigid Joint at Fixed Angle

If the horizontal arm is rotated about its axis, the other arm will form a cone of revolution. On the polyethylene molecule, the bent wire is similar to the carbon backbone of the chain with carbon atoms at positions 1, 2 and 3. Due to the rotation of the bond 2–3, atom 3 may be anywhere around the base of the cone of revolution. Similarly the next bond will form a cone of revolution with atom 3 as the apex and atom 4 anywhere around the base of this cone. Fig. A.5(b) illustrates how the random shape of the chain is built up. The hydrogen atoms have been omitted for clarity.

In practice the picture can take on a further degree of complexity if there is chain branching. This is where a secondary chain initiates from some point along the main chain as shown in Fig. A.6. In rubbers and thermosetting materials these branches link up to other chains to form a three dimensional network.

So far the structure of polymers has been described with reference to the material with the simplest molecular structure, i.e. polyethylene. The general principles described also apply to other polymers and the structures of several of the more common polymers are given below.

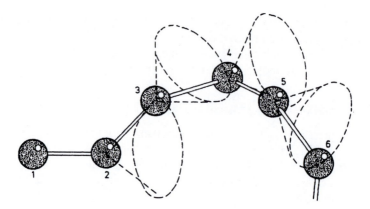

Fig. A.5(b)   Random conformation of carbon backbone on molecular chain

Fig. A.6   Chain branch in polyethylene

Polypropylene

Polyvinyl Chloride (PVC)

Polytetrafluoroethylene (PTFE)

Polymethylmethacrylate

Polystyrene

Nylon 66

Was Dupont Trade Mark

Polyoxymethylene (acetal)

It will be seen from this that a variety of atoms can be present along the carbon backbone and indeed carbon atoms may also be replaced, as in the cases of nylon and acetal. During polymerisation it is possible to direct the way in which monomers join on to a growing chain. This means that side groups (X) may be placed randomly (**atactic**) or symmetrically along one side of the chain (**isotactic**) or in regular alternating pattern along the chain (**syndiotactic**) as shown in Fig. A.7. A good example of this is polypropylene which in the atactic form is an amorphous material of little commercial value but in the isotactic form is an extremely versatile large tonnage plastic material.

```
     H   H   H   H   X   H   H
     |   |   |   |   |   |   |
  - C - C - C - C - C - C - C -
     |   |   |   |   |   |   |
     X   H   X   H   H   H   X
```

(a)   **Atactic**

```
     H   H   X   H   H   H   X
     |   |   |   |   |   |   |
  - C - C - C - C - C - C - C -
     |   |   |   |   |   |   |
     X   H   H   H   X   H   H
```

(b)   **Syndiotactic**

```
     H   H   H   H   H   H   H
     |   |   |   |   |   |   |
  - C - C - C - C - C - C - C -
     |   |   |   |   |   |   |
     X   H   X   H   X   H   X
```

(c)   **Isotactic**

Fig. A.7   Possible molecular structures

Polymers can also be produced by combining two or more different monomers in the polymerisation process. If two monomers are used the product is called a **copolymer** and the second monomer is usually included in the reaction to enhance the properties of the polymer produced by the first monomer alone. It is possible to control the way in which the monomers (A and B) link up and there are four main configurations which are considered useful. These are:

(1) Alternating $-A - B - A - B - A - B-$

(2) Random $A - A - B - A - A - A - B - B - A-$

(3) Block $-A - A - A - B - B - B - B - B - B - A - A-$

```
                 B
                 |
(4) Graft    - A - A - A - A - A - A - A -
                     |               |
                     B               B
                     |               |
                                     B
                                     |
                                     B
```

(a) Atactic Polypropylene

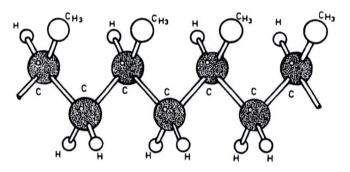

(b) Isotactic Polypropylene

Fig. A.8   Polypropylene structures

A.3 Arrangement of Molecular Chains

A picture of an individual molecular chain has been built up as a long randomly twisted thread-like molecule with a carbon backbone. It must be realised, however, that each chain must co-exist with other chains in the bulk material and the arrangement and interaction of the chains has a considerable effect on the properties of the material. Probably the most significant factor is whether the material is *crystalline* or *amorphous*. At first glance it may seem difficult to imagine how the long randomly twisted chains could exist in any uniform pattern. In fact X-ray diffraction studies of many polymers show sharp features associated with regions of three dimensional order (crystallinity) and diffuse features characteristic of disordered (amorphous) regions. By considering the polyethylene molecule again it is possible to see how the long chains can physically co-exist in an ordered crystalline fashion. This is illustrated in Fig. A.9.

During the 1940's it was proposed that partially crystalline polymers consisted of regions where the molecular chains where gathered in an ordered

O   Hydrogen   AB = 7·4 Å
                       BC = 4·93 Å
⬤   Carbon

Fig. A.9   Crystalline structure of polyethylene

fashion whereas adjacent regions had a random distribution of chains. The crystalline regions were considered to be so small that an individual chain could contribute to both crystalline and amorphous areas as shown in Fig. A.10. This was known as the **Fringed Micelle Model** and was generally accepted until the late 1950's when for the first time single polymer crystals were prepared from solution. These crystals took the form of thin platelets, their thickness being about 10 mm and their lateral dimensions as large as 0.01 mm. The most important discovery from the growth of these single crystals was that the chains were aligned perpendicular to the flat faces of the platelet. Since the length of an individual chain could be 1000 times greater than the thickness of the platelet the only conclusion was that the chains were folded.

Appropriately, this was called the **Folded Chain Theory** and is illustrated in Fig. A.11. There are several proposals to account for the co-existence of crystalline and amorphous regions in the latter theory. In one case, the structure is considered to be a totally crystalline phase with defects. These defects which include such features as dislocations, loose chain ends, imperfect folds, chain entanglements etc, are regarded as the diffuse (amorphous) regions viewed in X-ray diffraction studies. As an alternative it has been suggested that crystalline

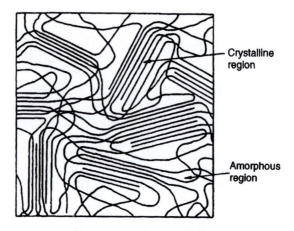

Fig. A.10   Fringed micelle model

Fig. A.11   Folded chain model

(folded chains) and amorphous (random chains) regions can exist in a similar manner to that proposed in the fringed micelle theory. In reality, time will probably show that in the complex structure of partially crystalline polymers, the crystalline regions consist of aligned and folded chains and the amorphous regions consist of crystal defects and randomly entangled chains.

Many crystalline polymers when viewed in cross polarised light, display characteristic Maltese crosses due to the presence of spherulites. These spherulites, shown in Fig. A.12, may vary in size from fractions of a micron to several millimetres in diameter, depending on the cooling rate from the melt. Slow cooling tends to produce larger spherulites than fast cooling. It is believed that the spherulites grow in all directions from a central nucleus, by the twisting of the folded chain platelets as shown in Fig. A.13. The size of a spherulite will be limited by the growth of adjacent spherulites. If the polymer melt is cooled very quickly it may undercool, i.e. remain molten at a temperature below its melting point. This results in a shower of nucleation sites becoming available and a mass of spherulites will start to grow. The solid polymer will then consist of a large number of small spherulites.

Fig. A.12   Typical illustration of spherulites

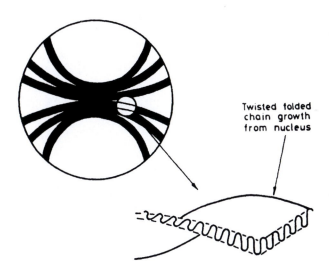

Twisted folded
chain growth
from nucleus

Fig. A.13   Structure of spherulite

The ease with which a polymer will form into crystalline regions depends on the structure of the molecular chain. It can be seen, for example, that if the polyethylene molecule has a high degree of branching then it makes it difficult to form into the ordered fashion shown in Fig. A.9. Also, if the side groups are large, it is not easy for a polymer with an atactic structure to form ordered regions. On the other hand isotactic and syndiotactic structures do have sufficient symmetry to be capable of crystallisation.

Another important feature of long chain molecules is the ease with which they can be rearranged by the application of stress. If a plastic is stretched the molecules will tend to align themselves in the direction of the stress and this is referred to as **orientation**. Molecular orientation leads to anisotropy of mechanical properties. This can be used to advantage in the production of fibres and film or may be the undesirable result of a moulding process. However, it is important that orientation should not be confused with crystallinity. It is possible to have an orientated polymer which shows no evidence of crystalline regions when X-ray diffraction studies are carried out. Equally, a polymer may be crystalline but optical measurements will show no signs of orientation. Orientation can be introduced into plastics such as polyethylene and polypropylene (both semi-crystalline) by cold drawing at room temperature. Other brittle plastics such as polymethyl methacrylate and polystyrene (both amorphous) cannot be cold drawn but can be drawn at elevated temperatures.

APPENDIX B – Solution of Differential Equations

The following type of differential equation is encountered in the text, for example, in the analysis of the models for viscoelastic behaviour:

$$A\frac{dy}{dx} + By - C = 0$$

where A, B and C are constants.

This has the solution

$$y = \frac{C}{B} - \frac{K}{B}e^{-Bx/A}$$

where the constant 'K' can be determined from the known boundary conditions in the specific case.

APPENDIX C – Stress/Strain Relationships

(i) Uniaxial Stresses

When a material is subjected to a uniaxial stress, there will be a strain in the direction of the stress and a strain of the opposite sense in the perpendicular directions. The latter is referred to as the Poisson's Ratio effect.

$$\varepsilon_x = \frac{1}{E}\sigma_x$$

where E = modulus of material.

$$\varepsilon_y = -v\left(\frac{1}{E} \cdot \sigma_x\right)$$

where v = Poisson's ratio (typical values are given in Chapter 2).

$$\varepsilon_z = -v\left(\frac{1}{E} \cdot \sigma_x\right)$$

The Z-direction is perpendicular to the page. For simplicity the material is assumed to be isotropic, ie same properties in all directions. However, in some cases for plastics and almost always for fibre composites, the properties will be anisotropic. Thus E and v will have different values in the x, y and z direction. Also, it should also be remembered that only at short times can E and v be assumed to be constants. They will both change with time and so for long-term loading, appropriate values should be used.

(ii) Biaxial stresses

If the material is subjected to biaxial stresses in both the x and y directions then the strains will be

426

$$\varepsilon_x = \frac{1}{E}\sigma_x - \nu\frac{1}{E}\sigma_y$$

That is, the total strain will be the sum of the tensile strain due to σ_x and the negative strain due to the Poisson's ratio effect caused by σ_y.
Similarly

$$\varepsilon_y = \frac{1}{E}\sigma_y - \nu\frac{1}{E}\sigma_x$$

$$\varepsilon_z = -\frac{\nu}{E}(\sigma_x + \sigma_y)$$

This situation is sometimes referred to as *plane stress* because there are only stresses in one plane. It is important to note, however, that there are strains in all three co-ordinate directions.

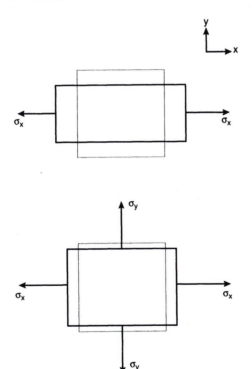

(iii) Triaxial stresses

If there are stresses in the x, y and z directions then the above approach is continued so that

$$\varepsilon_x = \frac{1}{E}\sigma_x - \nu\frac{1}{E}\sigma_y - \nu\frac{1}{E}\sigma_z = \frac{1}{E}\sigma_x - \frac{\nu}{E}(\sigma_y + \sigma_z)$$

$$\varepsilon_y = \frac{1}{E}\sigma_y - \frac{v}{E}(\sigma_x + \sigma_z)$$

$$\varepsilon_z = \frac{1}{E}\sigma_z - \frac{v}{E}(\sigma_y + \sigma_x)$$

In the above situation, if the stresses are such that $\varepsilon_z = 0$ then this condition is referred to as *plane strain*. This is because strains are experienced in only one plane even though there are stresses in all three co-ordinate directions.

APPENDIX D – Stresses in Cylindrical Shapes

(i) In a thin wall cylinder ($r > 10\,h$) subjected to internal pressure, p, the stresses are given by

$$\sigma_\theta = \frac{pr}{h}$$

$$\sigma_a = \frac{pr}{2h}$$

$$\sigma_r = -p \text{ (at bore)}$$

$$\sigma_r = 0 \text{ (at outside)}$$

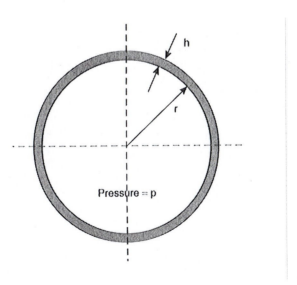

Pressure = p

Note that σ_r is small compared to σ_θ or σ_a and so it is often ignored.
Hoop strain,

$$\varepsilon_\theta = \frac{\Delta r}{r} = \frac{1}{E}\sigma_\theta - \frac{v}{E}\sigma_a$$

$$\varepsilon_\theta = \frac{pr}{Eh}\left\{1 - \frac{v}{2}\right\} = \frac{pr}{2Eh}\{2 - v\}$$

Axial strain,

$$\varepsilon_a = \frac{1}{E}\sigma_a - \frac{v}{E}\sigma_\theta$$

$$\varepsilon_a = \frac{pr}{2Eh}\{1 - 2v\}$$

(ii) In a thick wall cylinder subjected to both internal and external pressures
the stresses are given by

$$\sigma_\theta = \frac{1}{k^2 - 1}\left\{p_i\left(1 + \frac{r_0^2}{r^2}\right) - p_0 k\left(1 + \frac{r_i^2}{r^2}\right)\right\}$$

$$\sigma_r = \frac{1}{k^2 - 1}\left\{p_i\left(1 - \frac{r_0^2}{r^2}\right) - p_0 k\left(1 - \frac{r_i^2}{r^2}\right)\right\}$$

where $k = r_0/r_i$.

The axial stress will depend on the end constraints but if we consider the
most common situation where there are end caps and the cylinder is free to
change in length then for internal pressure only (ie $p_0 = 0$):

$$\sigma_a = \frac{p_i}{k^2 - 1}$$

For the zero external pressure situation, the axial strain will be given by

$$\varepsilon_a = \frac{1}{E}\sigma_a - \frac{v}{E}\sigma_r - \frac{v}{E}\sigma_\theta$$

$$\varepsilon_a = \frac{(1 - 2v)}{E(k^2 - 1)}(p_i)$$

APPENDIX E – Introduction to Matrix Algebra

E.1 Matrix definitions

A matrix is an array of terms as shown below:

$$[A] = \begin{bmatrix} a_{11} & a_{12} & a_{13} & \cdots & a_{1n} \\ a_{21} & a_{22} & a_{23} & \cdots & a_{2n} \\ a_{31} & a_{32} & a_{33} & \cdots & a_{3n} \\ \vdots & \vdots & \vdots & & \vdots \\ a_{m1} & a_{m2} & a_{m3} & \cdots & a_{mn} \end{bmatrix}$$

If $n = 1$ then we have a matrix consisting of a single column of terms and this referred to as a *column matrix*. If $m = 1$ then the matrix is called a *row matrix*.

If in the analysis of a problem there is a set of simultaneous equations then the use of matrices can be a very convenient shorthand way of expressing and solving the equations. For example, consider the following set of equations:

$$y_1 = a_{11}x_1 + a_{12}x_2 + a_{13}x_3 + \cdots + a_{1n}x_n$$

$$y_2 = a_{21}x_1 + a_{22}x_2 + a_{23}x_3 + \cdots + a_{2n}x_n$$

$$y_3 = a_{31}x_1 + a_{32}x_2 + a_{33}x_3 + \cdots + a_{3n}x_n$$

$$\vdots$$

$$y_m = a_{m1}x_1 + a_{m2}x_2 + a_{m3}x_3 + \cdots + a_{mn}x_n$$

These may be written in matrix form as follows:

$$\{y\} = [A]\{x\} \tag{E.1}$$

where $\{y\}$ and $\{x\}$ are column matrices.

E.2 Matrix multiplication

The matrix equation [E.1] involves the multiplication of the matrices $[A]$ and $\{x\}$. To do this one must apply the simple rules of *matrix multiplication*. These are:

- (a) two matrices may only be multiplied if the number of columns in the first is equal to the number of rows in the second;
- (b) the terms in the product matrix resulting from the multiplication of matrix $[A]$ with a matrix $[B]$ are given by

$$c_{ij} = \sum_{k=1}^{n} a_{ik}b_{kj} \qquad [E.2]$$

The use of these rules is illustrated in the following example:

$$\begin{bmatrix} a_{11} & a_{12} \\ a_{21} & a_{22} \end{bmatrix} \begin{bmatrix} b_{11} & b_{12} & b_{13} \\ b_{21} & b_{22} & b_{23} \end{bmatrix}$$

$$= \begin{bmatrix} (a_{11}b_{11} + a_{12}b_{21}) & (a_{11}b_{12} + a_{12}b_{22}) & (a_{11}b_{13} + a_{12}b_{23}) \\ (a_{21}b_{11} + a_{22}b_{21}) & (a_{21}b_{12} + a_{22}b_{22}) & (a_{21}b_{13} + a_{22}b_{23}) \end{bmatrix}$$

Suppose

$$[A] = \begin{bmatrix} 2 & 4 \\ 6 & 8 \end{bmatrix}, \quad [B] = \begin{bmatrix} 1 & 2 & 3 \\ 3 & 2 & -1 \end{bmatrix}$$

Then

$$[C] = [A][B] = \begin{bmatrix} 14 & 12 & 2 \\ 30 & 28 & 10 \end{bmatrix}$$

E.3 Matrix addition and subtraction

Matrix algebra also involves the *addition and subtraction* of matrices. The rules for this are as follows:

- (a) matrices may only be added or subtracted if they are of the same order, i.e. they each contain the same number of rows and columns;
- (b) the terms in the resulting matrix are given by

$$c_{ij} = a_{ij} \pm b_{ij} \qquad [E.3]$$

The following example illustrates the use of these rules:

$$\begin{bmatrix} d_{11} & d_{12} & d_{13} \\ d_{21} & d_{22} & d_{23} \end{bmatrix} \begin{bmatrix} b_{11} & b_{12} & b_{13} \\ b_{21} & b_{22} & b_{23} \end{bmatrix}$$

$$= \begin{bmatrix} (d_{11} \pm b_{11}) & (d_{12} \pm b_{12}) & (d_{13} \pm b_{13}) \\ (d_{21} \pm b_{21}) & (d_{22} \pm b_{22}) & (d_{23} \pm b_{23}) \end{bmatrix}$$

Suppose

$$[D] = \begin{bmatrix} -2 & 4 & 5 \\ 6 & 8 & -3 \end{bmatrix} \quad [B] = \begin{bmatrix} 1 & 2 & 3 \\ 3 & 2 & -1 \end{bmatrix}$$

Then

$$[E] = [D] + [B] = \begin{bmatrix} -1 & 6 & 8 \\ 9 & 10 & -4 \end{bmatrix}$$

$$[F] = [D] - [B] = \begin{bmatrix} -3 & 2 & 2 \\ 3 & 6 & -2 \end{bmatrix}$$

E.4 Inversion of a matrix

Referring back to the set of simultaneous equations at the beginning of this appendix, the objective is usually to solve these for the unknown x terms. This is where the use of matrices has a major advantage because referring to equation [E.1] we may rewrite this as

$$\{x\} = [A]^{-1}\{y\} \tag{E.4}$$

This equation expresses the solution to the set of simultaneous equations in that each of the unknown x terms is now given by a new matrix $[A]^{-1}$ multiplied by the known y terms. The new matrix is called the *inverse* of matrix $[A]$. The determination of the terms in the inverse matrix is beyond the scope of this brief introduction. Suffice to say that it may be obtained very quickly on a computer and hence the solution to a set of simultaneous equations is determined quickly using equation [E.4].

E.5 Symmetric matrix

A square matrix is one in which the number of columns is equal to the number of rows. An important type of square matrix which arises quite often in the finite element method is a *symmetric matrix*. Such matrices possess the property that $a_{ij} = a_{ji}$. An example of such a matrix is given below:

$$\begin{bmatrix} 2 & 4 & 7 & -3 \\ 4 & 5 & 1 & 9 \\ 7 & 1 & 6 & -5 \\ -3 & 9 & -5 & 4 \end{bmatrix} \quad \text{which is often written as}$$

$$\begin{bmatrix} 2 & 4 & 7 & -3 \\ & 5 & 1 & 9 \\ \text{sym.} & & 6 & -5 \\ & & & 4 \end{bmatrix}$$

APPENDIX F – Abbreviations for some Common Polymers

ABS	Acrylonitrile butadiene styrene	PET	Poly(ethylene terephtlhalate)
ASA	Acrylonitrile styrene acrylade	PF	Phenol-formaldehyde
CA	Cellulose acetate	PI	Polyimide
CAB	Cellulose acetate butyrate	PIB	Polyisobutylene
CAP	Cellulose acetate propionate	PMMA	Poly(methyl methacrylate), acrylic
CN	Cellulose nitrate		
CP	Cellulose propionate	PMP	Poly-4-methylpentene-1
EC	Ethyl cellulose	POM	Polyoxymethylene: polyformaldehyde, acetal
EP	Epoxide, epoxy		
EVA	Ethylene vinyl acetate	PP	Polypropylene
FEP	Fluorinated ethylene propylene	PPE	Polyphenylene ether
HDPE	High density PE	PPO	Polyphenylene oxide
HIPS	High impact PS	PPS	Poly(phenylene sulphide)
LDPE	Low density PE	PPSU	Poly(phenylene sulphone)
LLDPE	Linear low density PE	PS	Polystyrene
MDPE	Medium density PE	PSU	Polysulfone
MF	Melamine-formaldehyde	PTFE	Polytetrafluoroethylene
PA	Polyamide	PUR	Polyurethane
PAEK	Polyaryletherketone	PVAC	Poly(vinyl acetate)
PAI	Polyamide-imide	PVAL	Poly(vinyl alcohol)
PAN	Polyacrylonitrile	PVB	Poly(vinyl butyral)
PB	Polybutene-1	PVC	Poly(vinyl chloride)
PBA	Poly(butylene acrylate)	PVDC	Poly(vinylidene chloride)
PBT	Poly(butylene terephtlhalate)	PVDF	Poly(vinylidene fluoride)
PC	Polycarbonate	PVFM	Poly(vinyl formal)
PCTEE	Polychlorotrifluoroethylene	PVK	Polyvinylcarbazole
PE	Polyethylene	SAN	Styrene Acrylonitrile
PEEK	Polyetheretherketone	SBR	Styrene butadiene rubber
PEI	Polyether-imide	SBS	Styrene-butadiene-styrene
PEK	Polyetherketone	UF	Urea-formaldehyde
PES	Polyethersulfone	UP	Unsaturated polyester

Solutions to Questions

(2.1) Using the desirability factor D_f (Equation 1.16)

$$D_f = \left(\frac{E^{1/3}}{\rho C} \right)$$

Material	E	ρ	C	D_f
Polypropylene	0.3	905	1	7.4×10^{-4}
uPVC	2.1	1400	0.88	10.4×10^{-4}
ABS	1.2	1040	2.1	4.9×10^{-4}
Nylon 66	1.2	1140	3.9	2.4×10^{-4}
Polycarbonate	2.0	1150	4.2	2.6×10^{-4}
Acetal	1.0	1410	3.3	2.2×10^{-4}
Polysulphone	2.1	1240	11.0	4.9×10^{-4}

So uPVC would be the best choice on a cost basis.

(2.2)
$$L^3 = \frac{384 E I \delta}{W}$$

Weight, W = density × volume = $(2 \times 10^{-3} L)$N

From 1 week isochronous for polypropylene, $E = 427.5$ MN/m^2

So
$$L^4 = \frac{384 \times 427.5 \times 12.3 \times 10^3 \times 4}{2 \times 10^{-3}}$$

$$L = 1.417 \text{ m}$$

(2.3) $\bar{y} = 16.94$ mm

$$M = \frac{wL^2}{24} \text{ where } w = \frac{W}{L}$$

So, $\sigma = \dfrac{My}{I} = \dfrac{wL^2\bar{y}}{I} = \dfrac{2 \times 10^{-3} \times 1417^2 \times 16.94}{24 \times 12.3 \times 10^3}$

$$= 0.23 \text{ MN/m}^2$$

At this stress, $E = 550$ MN/m^2, so % error $= \dfrac{427.5 - 550}{427.5} = 28.7\%$

So $\delta = \dfrac{wL^3}{384EI} = \dfrac{2 \times 10^{-3} \times 1417 \times 1417^3}{384 \times 550 \times 12.3 \times 10^3} = 3.1$ mm

(2.4) From 3 year isochronous at 1.5%, $\sigma = 4.9$ MN/m^2

$$\sigma_0 = \frac{PR}{h} \quad \text{So} \quad h = \frac{PR}{\sigma_0} = \frac{0.5 \times 40}{4.9} = 4.08 \text{ mm}$$

If the density $= 905$ kg/m^3 ie a reduction of 4 kg/m^3 then there would be a $(4 \times 4)\%$ reduction in design stress. So new design stress $= (0.84 \times 4.9)$ MN/m^2

So   New $h = \dfrac{0.5 \times 40}{4.9 \times 0.84} = 4.86$ mm

Original weight $= 909 \times \pi \times 80 \times 4.08 \times 10^{-6}$

New weight $= 905 \times \pi \times 80 \times 4.86 \times 10^{-6}$

So   % change $= \dfrac{(905 \times 4.86) - (909 \times 4.08)}{(909 \times 4.08)} \times 100 = +18.6\%$

Note that this question involved a biaxial state of stress in the material and hence, strictly speaking, the creep curves used are not appropriate. However, creep curves for biaxial states of stress are rarely available, and one possible approach is to calculate an equivalent stress, σ_e, using a van Mises type criterion

$$\sigma_e = \frac{1}{\sqrt{2}} \sqrt{\{(\sigma_1 - \sigma_2)^2 + (\sigma_2 - \sigma_3)^2 + (\sigma_3 - \sigma_1)^2\}}$$

In the case of a cylinder under pressure, $\sigma_1 = pR/h$, $\sigma_2 = pR/2h$, $\sigma_3 = 0$, so

$$\sigma_e = \frac{\sqrt{3}}{2}\left(\frac{pR}{h}\right) = 0.866\sigma_\theta$$

In most cases it is probably sufficiently accurate to use σ_θ rather than σ_e and this approach will provide a built-in safety factor.

(2.5) Central Deflection,

$$\delta = \frac{WL^3}{48EI}$$

$$\text{Strain} = \frac{\sigma}{E} = \frac{My}{EI} = \left(\frac{W}{2}\right)\left(\frac{L}{2}\right)\left(\frac{d}{2}\right)\frac{1}{EI} = \frac{WLd}{8EI}$$

So,

$$\varepsilon = \frac{WL^3}{48EI}\left(\frac{6d}{L^2}\right) = \frac{6\delta d}{L^2}$$

This strain will occur in the outer fibres at mid-span.

(2.6) For rotating pipe, $\sigma_\theta = \rho\omega^2 r^2$

$$\sigma_\theta = 909\left(\frac{2\pi \times 3000}{60}\right)^2 (150 \times 10^{-3})^2 = 2 \text{ MN/m}^2$$

It is necessary to know how long the material can withstand this stress before it reaches a strain of $\varepsilon_\theta = \Delta D/D = 1.2/300 = 0.4\%$
From the creep curves $t = 2.1 \times 10^6$ seconds $= 24$ days

(2.7)

$$\varepsilon_\theta = \frac{\Delta D}{D} = \frac{\rho g H R}{Eh}\left(= \frac{12.5}{1250} = 1\%\right)$$

$$h = \frac{2\rho g H R^2}{E\Delta D}$$

From 1 year isochronous at 1% strain, $E = 360$ MN/m² Making the 20% correction for change in density from 909 kg/m³ to 904 kg/m³, $E = 288$ MN/m²

So

$$h = \frac{2 \times 1000 \times 9.81 \times 3 \times 0.625^2}{288 \times 10^6 \times 12.5 \times 10^{-3}} = 6.39 \text{ mm}$$

(2.8) From the 3 year isochronous curve for polypropylene, the initial modulus, $E = 466.7$ MN/m². Making the $(5 \times 4)\%$ allowance for the different density, $E = 0.8 \times 466.7 = 373.3$ MN/m².

$$h = \sqrt{\frac{P_c \times R^2}{0.365E}} = \sqrt{\frac{20 \times 10^{-3} \times 700^2}{0.365 \times 373.3}} = 8.48 \text{ mm}$$

Now

$$\sigma = \frac{PR}{2h} = \frac{20 \times 10^{-3} \times 700}{2 \times 8.48} = 0.825 \text{ MN/m}^2$$

At this stress the modulus has the same value as the initial modulus so no further iterations are necessary.

When the pipe is cooled it will contract by an amount given by $\varepsilon_T = \alpha(\Delta T)$ and when this equals the fixed strain, ε_θ, the pipe would leak

ie

$$\Delta T = \frac{0.00417}{9 \times 10^{-5}} = 46.3°\text{C}$$

So minimum temperature $= 20 - 46.3 = -26.3°\text{C}$

(2.9) The compressive stress in the bar is given by

$$\sigma = \frac{140}{10 \times 10} = 1.4 \text{ MN/m}^2$$

Now

$$I = \frac{bd^3}{12} = \frac{10 \times 10^3}{12} = 833.3 \text{ mm}^4$$

For buckling,

$$E = \frac{P_c L^2}{\pi^2 I} = \frac{140 \times 225^2}{\pi^2 \times 833.3} = 861.7 \text{ MN/m}^2$$

$$\therefore \quad \varepsilon = \frac{\sigma}{E} = \frac{1.4 \times 100}{861.7} = 0.16\%$$

From the creep curves this strain is reached at 10^4 seconds

$$\text{So time} = \frac{10^4}{3600} = 2.78 \text{ hours}$$

(Note: $\varepsilon < 0.5\%$ so no correction to tensile creep data is needed).

(2.10) Critical Stress,

$$\sigma_c = \frac{P_c}{A} = \frac{4\pi^2 EI}{AL^2}$$

and

$$I = \frac{Wh^3}{12}$$

So

$$\sigma_c = \frac{4\pi^2 EWh^3}{12WhL^2} = \frac{\pi^2 E}{3}\left(\frac{h}{L}\right)^2$$

$$\frac{h}{L} = \sqrt{\frac{\pi^2}{3}\left(\frac{E}{\sigma_c}\right)}$$

For many plastics $\sigma_c/E = 30 \times 10^{-3}$ to 35×10^{-3} for short-term loading

So

$$\frac{h}{L} = \sqrt{\frac{\pi^2}{3 \times 30 \times 10^{-3}}} = 10.5$$

For longer term loading, E will decrease more than σ_c so σ_c/E will get larger. This will cause the acceptable h/L to decrease.

(2.11)

$$P_c = \frac{\pi^2 EI}{L^2} \quad \text{So } \sigma_c = \frac{\pi^2 EI}{AL^2} \quad \text{and } \varepsilon_c = \frac{\pi^2 EI}{AL^2} = \frac{\pi^2 I}{AL^2}$$

Now for circular rod, $\varepsilon_c = \dfrac{\pi^2 d^2}{16L^2}$

So,

$$d = \sqrt{\frac{0.005 \times 16 \times 150^2}{\pi^2}} = 13.5 \text{ mm}$$

From 1 year isochronous at 0.5% strain, $E = 370$ MN/m^2

So
$$P_c = \frac{\pi^2 \times 370 \times \pi \times 13.5^4}{64 \times 150^2} = 227 \text{ N}$$

(2.12)
$$h^2 = \frac{3(3 + v)PR^2}{8\sigma} = \frac{3 \times 3.4 \times 0.04 \times 75^2}{8 \times 6} = 6.91 \text{ mm}$$

From 1 year isochronous curve at a stress of 6 MN/m^2, $E = 6/0.018 = 333.3$ MN/m^2

So
$$\delta = \frac{3 \times 0.6 \times 5.4 \times 0.04 \times 75^4}{16 \times 333.3 \times 6.91^2} = 7 \text{ mm}$$

(2.13) The maximum stress or strain is not specified so an iterative approach is needed. From the 1 year isochronous for PP the initial modulus is 370 MN/m^2

$$P = \frac{16Eh^3\delta}{3(1 - v)(5 + v)R^4} = \frac{16 \times 370 \times 2.5^3 \times 4}{3 \times 0.6 \times 5.4 \times 32^4} = 0.036 \text{ MN/m}^2$$

So
$$\sigma = \frac{3(3 + v)PR^2}{8h^2} = \frac{3 \times 3.4 \times 0.0363 \times 32^2}{8 \times 2.5^2} = 7.58 \text{ MN/m}^2$$

but at
$$\sigma = 7.58 \text{ MN/m}^2, E = \frac{7.58}{0.0248} = 305.6 \text{ MN/m}^2$$

So
$$P = \frac{16 \times 2.5^3 \times 4}{3 \times 0.6 \times 5.4 \times 32^4}(305.6) = 3 \times 10^{-2} \text{ MN/m}^2$$

$$\sigma = \frac{3 \times 3.4 \times 32^2}{8 \times 2.5^2}(3 \times 10^{-2}) = 6.27 \text{ MN/m}^2$$

but
$$E = \frac{6.27}{0.0196} = 319.6 \text{ MN/m}^2 \rightarrow P = 3.14 \times 10^{-2} \rightarrow \sigma = 6.55 \text{ MN/m}^2$$

$$E = 321.2 \text{ MN/m}^2 \rightarrow P = 3.15 \times 10^{-2} \text{ MN/m}^2 \rightarrow \sigma = 6.58 \text{ MN/m}^2$$

$$E = 321.2 \text{ MN/m}^2 \rightarrow P = 3.15 \times 10^{-2} \text{ MN/m}^2$$

$$\varepsilon_\theta = \frac{\Delta D}{D} = \frac{Pr}{2hE}(2 - v), \text{ So } \Delta D = \frac{Pr^2}{hE}(2 - v)$$

So
$$\Delta D = \frac{3.15 \times 10^{-2} \times 32^2}{2.5 \times 321.2}(1.6) = 6.43 \times 10^{-2} \text{ mm}$$

(2.14)
$$I = \frac{bd^3}{12} = \frac{12 \times d^3}{12} = d^3$$

From the 1 year isochronous curve, the initial modulus $= 370$ MN/m^2

Now
$$d^3 = \frac{5WL^3}{384E\delta} = \frac{5 \times 150 \times 200^3}{384 \times 370 \times 6} \text{ So } d = 19.16 \text{ mm, where } W = \omega L$$

$$\sigma = \frac{My}{I} = \frac{WL}{16d^2} = \frac{150 \times 200}{16d^2} = 5.1 \text{ MN/m}^2$$

At $\qquad \sigma = 5.1$ MN/m$^2 \rightarrow E = 342.3$ MN/m$^2 \rightarrow d = 19.67$ mm

$\qquad\qquad \sigma = 4.85$ MN/m$^2 \rightarrow E = 346.2$ MN/m$^2 \rightarrow d = 19.6$ mm

$\qquad\qquad \sigma = 4.88$ MN/m$^2 \rightarrow E = 346.4$ MN/m$^2 \rightarrow d = 19.6$ mm

(2.15) Once again an iterative type solution is required.

$$W = \frac{E(3)(1.5)}{\left\{0.48\left(\dfrac{1000}{40}\right)\left(\dfrac{40}{3}\right)^{1.22}\right\}} = 0.0795E$$

$$\sigma = \frac{2.4W}{3^2} = 0.267$$

From the 1 year isochronous curve, the initial modulus is 370 MN/m^2

So $\qquad W = 29.43$ N $\rightarrow \sigma = 7.86$ MN/m$^2 \rightarrow E = 302.3$ MN/m^2

$\qquad\qquad W = 24.03$ N $\rightarrow \sigma = 6.42$ MN/m$^2 \rightarrow E = 324.1$ MN/m^2

$\qquad\qquad W = 25.76$ N $\rightarrow \sigma = 6.88$ MN/m$^2 \rightarrow E = 317.0$ MN/m^2

$\qquad\qquad W = 25.2$ N $\rightarrow \sigma = 6.73$ MN/m$^2 \rightarrow E = 317.4$ MN/m^2

$\qquad\qquad W = 25.2$ N $\rightarrow \sigma = 6.73$ MN/m^2

So $W = 25.2$ N

(2.16) This is a stress relaxation problem but the isometric curves may be used.

From 2% isometric, after 10 seconds, $E = \dfrac{16.75}{0.02} = 837.5$ MN/m^2

$$W = \frac{Ed^4\delta}{128(1+v)R^3N} = \frac{837.5 \times 3^4 \times 10}{128(1.4)5^3 \times 10} = 3 \text{ N}$$

After 1 week, $\qquad E = \dfrac{8.55}{0.02} = 427.5$ MN/m^2

So $\qquad\qquad W = \dfrac{427.5 \times 3^4 \times 10}{128(1.4)5^3 \times 10} = 1.55$ N

(2.17) From the 1 day isochronous curve, the maximum stress at which the material is linear is 4 MN/m^2. This may be converted to an equivalent shear stress by the relation

$$\tau = \frac{\sigma}{2(1+v)} = \frac{4}{2(1.4)} = 1.43 \text{ MN/m}^2$$

Now $\qquad\qquad \tau = \dfrac{16WR}{\pi d^3}, \quad$ So $\quad d^3 = \dfrac{16 \times 3 \times 7.5}{\pi \times 1.43}$

$$d = 4.31 \text{ mm}$$

If $W = 4.5$ N, $\qquad \tau = \dfrac{16 \times 4.5 \times 7.5}{\pi(4.31)^3} = 2.15$ MN/m^2

Equivalent tensile stress $= \sigma = 2.15 \times 2 \times 1.4 = 6$ MN/m^2.

From the 1 day isochronous at this stress,

$$E = \frac{6}{0.0094} = 638.3 \text{ MN/m}^2$$

$$\delta = \frac{128(1 + v)R^3N}{Ed^4} = \frac{K}{E}$$

So

$$\left(\frac{\delta_2 - \delta_1}{\delta_1} \right) = \left\{ \frac{\dfrac{K}{E_2} - \dfrac{K}{E_1}}{\dfrac{K}{E_1}} \right\}$$

where $E_2 = 638.3$ MN/m^2 and

$$E_1 = \frac{1.4}{0.002} = 700 \text{ MN/m}^2$$

So % change $= +9.67\%$

(2.18) At short times, $E = \dfrac{5.6}{0.004} = 1400$ MN/m^2

Allowing for temperature, $E_{60} = 1400 \times 0.44 = 616$ MN/m^2

Overall strain, $\varepsilon = \varepsilon_T - \varepsilon_\sigma = 0$

$$\varepsilon_T = \varepsilon_\sigma$$
$$\alpha \Delta T = \sigma/E = P/AE$$

So $P = AE\alpha\Delta T = \dfrac{\pi(10)^2}{4} \times 616 \times 1.35 \times 10^{-4} \times 40 = 261$ N

(Note: $\sigma = (261 \times 4)/\pi(10)^2 = 3.3$ MN/m^2. Therefore since the strain is less than 0.5% no correction to the tensile data is needed for compressive loading).

After 1 year $E_{20} = 370$ MN/m^2, $E_{60} = 222$ MN/m^2

So $P = \dfrac{\pi(10)^2}{4} \times 222 \times 1.35 \times 10^{-4} \times 40 = 94$ N

(Note: Once again the strain is less than 0.5% so no correction is needed for compressive loading).

(2.19) $\varepsilon_\theta = \dfrac{\Delta D}{D} = \dfrac{0.05}{12} = 0.417\%$

From the 0.417% isometric curve, the stress after 1 year is 1.45 MN/m^2

The pipe would leak if the hoop stress caused by atmospheric pressure (0.1 MN/m^2) exceeded the stress in the pipe wall after 1 year.

For $P = 0.1$ MN/m^2, $\sigma_\theta = \dfrac{PR}{h} = \dfrac{0.1 \times 6}{1.5} = 0.4$ MN/m^2

Hence, leakage would not occur.

(2.20) This is a stress relaxation problem, but the question states that creep data may be used

$$\text{Strain} = \frac{\Delta D}{D} = \frac{0.16}{10} = 1.6\%$$

Then from a 1.6% isometric taken from the creep curves it may be determined that the stress after 10 seconds is 15.1 MN/m² and after 1 year it is 5.4 MN/m².

(a) Initial pressure at interface $= p = (h\sigma)/R = 3.02$ MN/m²

Thus normal force at interface, $F = p \times 2\pi RL = 3.02 \times 2\pi \times 5 \times 15$

$$= 1.423 \text{ kN}$$

So axial force, $W = \mu F = 0.3 \times 1.423 \text{ kN} = 0.427 \text{ kN}$

(b) Similarly, after 1 year for $\sigma = 5.4$ MN/m² the axial force $W = 0.153$ kN.

(2.21) (a) As illustrated in Example 2.7, the extraction force, F, is given by

$$F = \pi\mu p D W$$

So

$$p = \frac{1.2 \times 10^3}{\pi(0.24)(40)(10)} = 4 \text{ MN/m}^2$$

Treating the bush as a thick wall cylinder subjected to this value of external pressure, then Benham *et al.* show that at the outer surface of the bush, the stresses are:

$$\text{hoop stress, } \sigma_\theta = -p\left\{\frac{k^2 + 1}{k^2 - 1}\right\}$$

$$\text{radial stress, } \sigma_r = -p$$

Hence, the hoop strain, ε_θ, is given by

$$\varepsilon_\theta = \frac{\Delta R}{R} = \frac{\sigma_\theta}{E} - \frac{v\sigma_r}{E}$$

$$\Delta R = \frac{pR}{E}\left\{\left(\frac{k^2+1}{k^2-1}\right) - v\right\}$$

The negative sign simply means that this is a reduction.

So
$$\Delta R = \frac{4 \times 20}{2 \times 10^3}\left\{\left(\frac{1.14^2+1}{1.14^2-1}\right) - 0.4\right\}$$

$$\Delta R = 0.29 \text{ mm}$$

(b)
$$\Delta R = \alpha.R.\Delta T$$

$$\Delta T = \frac{0.29}{100 \times 10^{-6} \times 20} = 143°C$$

Hence the nylon bush would need to be cooled by 143°C to achieve the necessary contraction to have easy assembly. This suggests a cooled temperature of $-123°C$.

(c) At the bore of the bush, Benham *et al.* shows the stresses to be

$$\sigma_\theta = -\frac{2k^2 p}{k^2-1}, \qquad \sigma_r = 0$$

hence
$$\frac{\Delta r}{r} = \frac{\sigma_0}{E} - \frac{v\sigma_r}{E} = -\frac{2k^2 p}{E(k^2-1)}$$

$$\Delta_r = \frac{-2(1.14)^2 17.5(4)}{2 \times 10^3(1.14^2-1)} = -0.3 \text{ mm}$$

Therefore the internal diameter of the bush will be $35 - 2(0.3) = 34.4$ mm.

(d) If the long-term modulus of the nylon is 1 GN/m^2 then for the same interference conditions, the interface pressure would be reduced to half its initial value ie 2 MN/m^2. This means that the separation force would be half the design value ie 600 N.

(2.22) If the acetal ring is considered as a thick wall cylinder, then at the inner surface there will be hoop stresses and radial stresses if it is constrained in a uniform manner:

$$\text{hoop stress, } \sigma_\theta = p\left\{\frac{k^2+1}{k^2-1}\right\}$$

$$\text{radial stress, } \sigma_r = -p$$

where p is the effective internal pressure
k is the ratio of the outer to inner radii

Hence, the hoop strain, ε_θ, at the inner surface is given by

$$\varepsilon_\theta = \frac{\Delta r}{r} = \frac{\sigma_\theta}{E} - v\frac{\sigma_r}{E}$$

So,

$$\Delta r = \frac{pr}{E} \left\{ \left(\frac{k^2 + 1}{k^2 - 1} \right) + v \right\}$$

When the bobbin is cooled by 60°C the effective constrained contraction of the inner surface of the acetal will be

$$\Delta r = (\alpha_a - \alpha_s) r \Delta T$$

$$\Delta r = (80 - 11)10^{-6}(10)(60) = 0.0414 \text{ mm}$$

So,

$$0.0414 = \frac{p(10)}{3 \times 10^3} \left\{ \left(\frac{1.5^2 + 1}{1.5^2 - 1} \right) + 0.33 \right\}$$

$$p = 4.24 \text{ MN/m}^2$$

hence,

$$\sigma_\theta = p \left\{ \frac{k^2 + 1}{k^2 - 1} \right\} = 4.24 \left(\frac{3.25}{1.25} \right) = 11 \text{ MN/m}^2$$

(2.23) Plot $\log \dot{\varepsilon}$ vs. $\log \sigma$ and straight line confirms the Power Law with $A = 3 \times 10^{-11}$, $n = 0.774$

$$\varepsilon_t - \varepsilon_0 = \dot{\varepsilon} t = A \sigma^n t$$

So

$$\varepsilon_t = 0.95 \times 10^{-2} + (3 \times 10^{-11}(5)^{0.774})(9 \times 10^6 - 1 \times 10^6)$$

$$= 1.033\%$$

(2.24) A plot of $\log \varepsilon$ against $\log t$ is a straight line for $\sigma = 5.6$ MN/m² So $\varepsilon(t) = At^n$ where $A = 0.238$ and $N = 0.114$.

After 3 days ($= 2.59 \times 10^5$ seconds) strain $= 0.988\%$

$$F_r = \frac{\varepsilon(T) - \varepsilon_R(t)}{\varepsilon T} = \frac{0.988 - \varepsilon_R(t)}{0.988}$$

$$t_r = \frac{(2 \times 2.59 - 2.59) \times 10^5}{2.59 \times 10^5} = 1$$

So from (5) $F_r = 1 + t_R^n - (t_R + 1)^n$

$$1 - \frac{\varepsilon_R(t)}{0.988} = 1 + t_R^n - (t_R + 1)^n = 2 - (2)^{0.114}$$

$$0.988((2)^{0.114} - 1) = \varepsilon_r(t) = 0.0813\%$$

Alternatively: Strain after $2 \times 2.59 \times 10^5$ seconds $= 1.069\%$. Now recovery may be regarded as reversal of creep,

So residual strain $= 1.069 - 0.988 = 0.0812\%$

(2.25) $I = \dfrac{15 \times 15^3}{12} - \dfrac{10.5 \times 10.5^3}{12} + \dfrac{10.5}{12} \left(\dfrac{450}{909} \right)^2 (10.5)^3 = 3454.1 \text{ mm}^4$

For solid beam,

$$I = \frac{D^4}{12} = 3454.1, \quad D = 14.27 \text{ mm}$$

$$\text{Weight of solid beam} = 14.27^2 \times 10^{-6} \times 909 = 185.1 \text{ g}$$

Weight of foamed beam =

$$(4 \times 2.25 \times 12.75 \times 10^{-6} \times 909) + (450 \times 10.5^2 \times 10^{-6}) = 153.9 \text{ g}$$

% Saving = 16.9%

(2.26)

$$\sigma = \frac{My}{I} = \frac{WL^2 \times 7.5}{24 \times 3126.3}$$

$$W = \frac{7 \times 3126.3 \times 24}{250^2 \times 7.5} = 1.12 \text{ kN/mm}$$

From the 1 week isochronous, $E = \dfrac{7}{0.015} = 466.7 \text{ MN/m}^2$

$$\delta = \frac{WL^4}{384EI} = \frac{1.12 \times 250^4}{384 \times 466.7 \times 3126.3} = 7.8 \text{ mm}$$

(2.27) Weight of solid beam = $909 \times 12 \times 8 \times 300 \times 10^{-9} \times 10^3 = 26.18 \text{ g}$

Weight of composite = $(909 \times 2 \times 2 \times 12 \times 300 \times 10^{-6}) + (500 \times h12 \times 300 \times 10^{-6})$

$h = 7.27$ mm, so composite beam depth = 11.27 mm.

The ratio of stiffnesses will be equal to the ratio of second moment of area

$$I_{\text{solid}} = \frac{bd^3}{12} = \frac{12 \times 8^3}{12} = 512 \text{ mm}^4$$

$$I_{\text{composite}} = \frac{12 \times 11.27^3}{12} - \frac{12 \times 7.27^3}{12}$$

$$+ \frac{\left(\dfrac{500}{900}\right)^2 12 \times 7.27^3}{12} = 1163.45 \text{ mm}^4$$

$$I_c/I_s = \frac{1163.45}{512} = 2.27$$

(2.28) (a) *Solid*

Consider a flexural loading situation as above,

so

$$\frac{W}{\delta} = \frac{48EI}{L^3} \propto EI$$

So Solid
$$EI = E_s \left(\frac{bd^3}{12} \right) = E_s \left(\frac{1 \times 12^3}{12} \right)$$

$$\text{Weight} = \rho_s \times H \times 1 \times 1 = \rho_s H = 12\rho_s$$

$$\text{Ratio} = \frac{E_s}{\rho_s}(12)$$

(b) *Foamed*

$$EI = E_f \left(\frac{bd^3}{12} \right) = \left(\frac{\rho_f}{\rho_s} \right) E_s \left(\frac{1 \times 12^3}{12} \right)$$

$$\text{Weight} = \rho_f \times H \times 1 \times 1 = 12\rho_f = \frac{12}{1.5}\rho_s$$

$$\text{Ratio} = \frac{E_s}{\rho_s} \left(\frac{12 \times 1.5}{1.5^2} \right) = (8)\frac{E_s}{\rho_s}$$

(c) *Composite*

$$b' = \left(\frac{E_f}{E_s} \right)(I)$$

$$EI = E_s \left\{ \frac{1 \times 2^3}{12} + (1 \times 2)(5)^2 + \frac{b'(8)^3}{12} \right\}$$

$$= E_s \left\{ 2(0.67 + 50) + \left(\frac{\rho_f}{\rho_s} \right)^2 \frac{(8)^3}{12} \right\} = 120.3E_s$$

$$\text{Weight} = \rho_s(4 + (8 \times 0.44)) \times 1 \times 1 = 7.56\rho_s$$

$$\text{Ratio} = \frac{120.3E_s}{7.56\rho_s} = (15.9)\frac{E_s}{\rho_s}$$

So ratio foam: solid: composite = 8:12:15.9

(2.29) (a) Sandwich Beam of Minimum Weight for a given stiffness

$$\text{deflection, } \delta = \frac{KWL^3}{KI} \tag{1}$$

where K is a factor depending on loading and supports.

For $E_{\text{skin}} \gg E_{\text{core}}$, $\quad EI \approx Ebdh^2/2$

$$\delta = \frac{2KWL^3}{E_{\text{skin}}bdh^2}$$

Total mass

$$m = \rho_{core}bhL + \rho_{skin}bdL$$

From deflection equation (1)

$$d = \frac{2KWL^3}{E_{skin}bh^2\delta}$$

So

$$m = \rho_{core}bhL + \rho_{skin}\left(\frac{2KWL^4}{E_{skin}h^2\delta}\right)$$

To minimise the weight with respect to h, then $\partial m/\partial h = 0$

$$\frac{\partial m}{\partial h} = \rho_{core}bL - \frac{4\rho_{skin}KWL^4}{E_{skin}\delta h^3} = 0$$

∴

$$h^3 = \frac{4\rho_{skin}KWL^3}{E_{skin}\delta\rho_{core}b} = \frac{4\rho_{skin}th^2}{\rho_{core}}$$

∴

$$\rho_{core}h = 4_{skin}$$

or   weight of core = 2 × weight of skin

(b) Sandwich Beam of Minimum Weight for a Given Strength

Assuming the core does not fail in shear, failure occurs in bending when the stress in the skin reaches its yield strength σ_y, ie

$$\sigma_y = \frac{M}{bdh}$$

where M = bending moment.

Mass of beam

$$m = \rho_{core}bhL + \rho_{skin}bdL$$

Using

$$d = \frac{M}{bh\sigma_y}$$

$$m = \rho_{core}bhL + \rho_{skin}\left(\frac{LM}{h\sigma_y}\right)$$

To minimise the weight with respect to h, $\partial m/\partial h = 0$

$$\frac{\partial m}{\partial h} = \rho_{core}bL - \rho_{skin}\frac{LM}{h^2\sigma_y} = 0$$

∴

$$h = \frac{\rho_{skin}}{\rho_{core}}\frac{M}{b\sigma_y} = \frac{\rho_{skin}d}{\rho_{core}}$$

$\therefore$ $\rho_{core}h = \rho_{skin}d$

or   weight of core = weight of skin.

(2.30)

(a) *Maxwell*   (b) *Voigt*

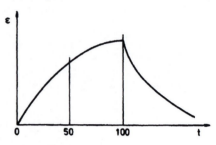

$$\varepsilon = \frac{\sigma_0}{\xi} + \dot{\varepsilon}t = \frac{\sigma_0}{\xi} + \frac{\sigma_0}{\eta}t \qquad\qquad \varepsilon = \frac{\sigma_0}{\xi}(1 - e^{-(\xi t)/\eta})$$

$$\varepsilon_{50} = \frac{12}{2000} + \frac{12 \times 10^6}{90 \times 10^9}(50) = 1.26\% \qquad \varepsilon_{50} = \frac{12}{2000}(1 - e^{(-2 \times 50)/90})$$
$$= 0.402\%$$

$$\varepsilon_{100} = \frac{12}{2000} + \frac{12 \times 10^6}{90 \times 10^9}(100) = 1.933\% \qquad \varepsilon_{100} = \frac{12}{2000}(1 - e^{(-2 \times 100)/90})$$
$$= 0.535\%$$

but $\dfrac{\sigma_0}{\xi} = 0.6\%$ $\varepsilon_{150} = \varepsilon_{100}e^{(\xi t)^\eta}$

$\varepsilon_{150} = 1.933 - 0.6 = 1.333\%$ $= 0.535e^{(-2 \times 50)/90} = 0.176\%$

(2.31) Maxwell Strain (50 s) = Kelvin Strain (50 s)

$$\frac{\sigma_0}{\xi_1} + \frac{\sigma_0 t}{\eta_1} = \frac{\sigma_0}{\xi_2}\left(1 - e^{(-\xi_2 t)/\eta_2}\right)$$

So $$\xi_1 = \left[\frac{1}{2}(1 - e^{(-\xi_2 t)/\eta_2}) - \frac{t}{\eta_1}\right]^{-1}$$

$$\xi_1 = \left[\frac{1}{2 \times 10^9}(1 - e^{(-2 \times 50)/100}) - \frac{50}{200 \times 10^9}\right]^{-1} = 15.1 \text{ GN/m}^2$$

(2.32)

From graph $\sigma_0/\xi = 0.01$

So $$\xi = \frac{2}{0.01} = 200 \text{ MN/m}^2$$

At $\varepsilon = 0.0089$, $t = 3000$ s

$$0.0089 = 0.01\left\{1 - e^{-3000(\xi/\eta)}\right\}$$

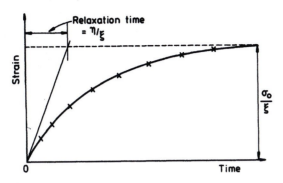

So
$$7.36 \times 10^{-4} = \xi/\eta$$

$$\eta = 270 \text{ GNs/m}^2$$

Relaxation time $= \eta/\xi = 1350$ seconds

for
$$\sigma_0 = 4.5 \text{ MN/m}^2$$

$$\varepsilon = \frac{4.5}{200} \left\{ 1 - e^{(-1500)/1350} \right\} = 0.0151$$

(2.33)

(a) Stress–Strain equations

$$\sigma_1 = \xi_1 \varepsilon_1$$

$$\sigma_2 = \xi_2 \varepsilon_2$$

$$\sigma_3 = \xi_3 \dot{\varepsilon}_3$$

(b) Equilibrium equations

$$\sigma = \sigma_1 = \sigma_2 + \sigma_3$$

(c) Geometry of Deformation

$$\varepsilon = \varepsilon_1 + \varepsilon_2 \text{ and } \varepsilon_2 = \varepsilon_3$$

From above

$$\dot{\varepsilon} = \dot{\varepsilon}_1 + \dot{\varepsilon}_2$$

but
$$\dot{\varepsilon}_1 = \dot{\sigma}_1/\xi_1 \quad \text{and} \quad \dot{\varepsilon}_2 = (\sigma_1/\eta) - (\xi_2/\eta)\varepsilon_2 \text{ (from Kelvin model)}$$

$$\dot{\varepsilon} = \frac{\dot{\sigma}}{\xi_1} + \frac{\sigma}{\eta} + \frac{\xi_2}{\eta}(\varepsilon - \varepsilon_1)$$

$$\dot{\varepsilon} = \frac{\dot{\sigma}}{\xi_1} + \frac{\sigma}{\eta} - \frac{\xi_2}{\eta}\left(\varepsilon - \frac{\sigma}{\xi_1}\right)$$

$$\dot{\varepsilon} + \frac{\xi_2}{\eta}\epsilon = \frac{\dot{\sigma}}{\xi_1} + \frac{\sigma}{\eta}\left(\frac{\xi_1 + \xi_2}{\xi_1}\right)$$

This is the governing equation for the Standard Model.

The solution to this differential equation will have the form

$$\varepsilon(t) = \frac{\sigma_0}{\xi_1} + \frac{\sigma_0}{\xi_2}\left(1 - e^{-t/T}\right)$$

where $T = \eta/\xi_2$ and σ_0/ξ_1 is the initial strain when the stress is applied.
The unrelaxed creep modulus is obtained by putting $t = 0$.

Hence, unrelaxed modulus $E_U = \dfrac{\sigma_0}{\varepsilon(0)} = \xi_1$

The relaxed modulus is obtained by putting $t = \infty$.

Hence, relaxed modulus $E_R = \dfrac{\sigma_0}{\varepsilon(\infty)} = \dfrac{\xi_1\xi_2}{\xi_1 + \xi_2}$.

(2.34) From figure below redrawn from the creep curves

$$\eta = \frac{\sigma}{\dot{\varepsilon}} \simeq 4.3 \times 10^{12} \text{ Ns/m}^2$$

$$\xi = \frac{\sigma_0}{\varepsilon_0} = 1400 \times 10^6 \text{ N/m}^2$$

So from text using the fact that for a strain of 0.4%, $\sigma_0 \simeq 5.5$ MN/m^2 (from creep data)

$$\sigma = \sigma_0 e^{(-\xi)/\eta t}$$

$$\sigma = 5.5 e^{(-1400 \times 10^6 \times 900)/4.3 \times 10^{12}} = 4.1 \text{ MN/m}^2$$

(2.35) From the graph below, and the theory of the 4-element model, $\xi_1 = \sigma_0/\varepsilon_1 = 4.2/0.003 = 1400$ MN/m^2

$$\xi_2 = \frac{\sigma_0}{\text{Retarded Creep}} = \frac{\sigma_0}{\varepsilon_2} = \frac{4.2}{(0.72 - 0.3)10^{-2}} = 1000 \text{ MN/m}^2$$

$$\eta_1 = \frac{\sigma_0}{d\varepsilon/dt} = \frac{4.2}{3.167 \times 10^{-6}} = 1.326 \times 10^6 \text{ (MN/m}^2\text{)hr}$$

$$\eta_1 = 4.775 \times 10^9 \text{ MNs/m}^2$$

Finally from the expression for the 4-element model

$$\varepsilon = \frac{\sigma_0}{\xi_1} + \frac{\sigma_0 t}{\eta_1} + \frac{\sigma_0}{\xi_2}(1 - e^{-(\xi_2 t)/\eta_2})$$

taking a value of strain from the *knee* of the curve

$$\varepsilon = 0.0055 \text{ at } t = 4 \times 10^4 \text{ seconds (11.1 hrs)}$$

So $\eta_2 = 4.525 \times 10^7$ MNs/m^2

For $\sigma = 5.6$ MN/m^2 at 3×10^5 seconds

$$\varepsilon = \frac{5.6}{1400} + \frac{5.6 \times 3 \times 10^5}{4.775 \times 10^9} + \frac{5.6}{1000}\left(1 - e^{-(1000 \times 3 \times 10^5)/4.525 \times 10^7}\right) = 0.99\%$$

(2.36)

$$E(t) = At^{-n}$$

$$\sigma(t) = Bt$$

$$\varepsilon(t) = \int_0^T \frac{1}{E(t)} \cdot \frac{d\sigma(t)}{dt} \cdot dt$$

$$= \int_0^T \frac{t^n}{A} \cdot B dt$$

$$= \left[\frac{t^{n+1}}{A(n+1)} \cdot B \right]_0^T$$

$$\varepsilon(t) = \frac{BT^{n+1}}{A(n+1)}$$

This will be a non-linear response for all non-zero values of n.

(2.37) For the Maxwell Model

$$\dot{\varepsilon} = \frac{1}{\xi}\dot{\sigma} + \frac{1}{\eta}\sigma = K$$

$$\frac{d\sigma}{dt} + \frac{\xi\sigma}{\eta} = K\xi$$

$$\int_0^t \xi \, dt = \int_0^a \left(\frac{1}{K - \dfrac{\sigma}{\eta}} \right) d\sigma$$

$$\xi t = -\eta \ln \left(K - \frac{\sigma}{\eta} \right) + \eta \ln K$$

$$\frac{\xi t}{\eta} = \ln \left(\frac{K}{K - \dfrac{\sigma}{\eta}} \right)$$

$$e^{(\xi t)/\eta} = \left(\frac{K}{K - \dfrac{\sigma}{\eta}} \right)$$

$$\sigma = K\eta(1 - e^{(-\xi t)/\eta})$$

Using

$$\frac{\xi}{\eta} = \frac{20}{100} = 2 \times 10^{-2}$$

and

$$K\eta = 10^{-5} \times 1000 \times 10^3 = 10$$

Time (s)	Strain (%)	Stress (MN/m²)
20	0.02	3.3
40	0.04	5.51
60	0.06	6.99
80	0.08	7.98
100	0.1	8.65

initial tangent modulus $= 7.00/0.0004 = 17.5$ GN/m²

0.1% secant modulus $= 8.65/0.001 = 8.65$ GN/m²

Note that in this question an alternative solution may be carried out using the *creep modulus* but this causes slight inaccuracies.

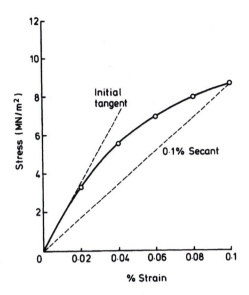

Alternative Solution:

$$\sigma(t) = \int\limits_0^t E(t-u)\frac{d\varepsilon(u)}{du} \cdot du$$

Now for Maxwell Model $E(t) = \dfrac{\xi\eta}{\eta + \xi t}$

So for case in question, the strain history is

$$0 < t < t_1 : \varepsilon(t) = Kt, d\varepsilon(u)/du = K$$

So $\qquad \sigma(t) = \displaystyle\int\limits_0^{t_1} \frac{\xi\eta}{\eta + \xi(t_1 - u)} \cdot K \cdot du = \int\limits_0^{t_1} \left(\frac{\xi\eta}{\eta + \xi t_1 - \xi u}\right) K\, du.$

$$= K\xi\eta \left[\log_e(\eta + \xi t_1 - \xi u) \cdot \left(-\frac{1}{\xi}\right)\right]_0^{t_1}$$

So $\qquad \sigma(t) = K\eta \left[\log_e \left(\dfrac{\eta + \xi t_1}{\eta}\right)\right]$

This equation predicts the following data

Time (s)	Strain (%)	Stress (MN/m^2)
20	0.02	3.36
40	0.04	5.88
60	0.06	7.88
80	0.08	9.56
100	0.1	10.98

From a plot of this data

$$\text{Initial tangent modulus} = \frac{10}{0.00057} = 17.54 \ \text{GN/m}^2$$

$$\text{0.1\% secant modulus} = \frac{10.98}{0.001} = 10.98 \ \text{GN/m}^2$$

(2.38) $\varepsilon(t_1) = K_1 t_1 \left(\dfrac{1}{\xi} + \dfrac{t_1}{2\eta} \right)$

$$\varepsilon(40) = 0.5 \times 40 \left(\frac{1}{3000} + \frac{40}{2 \times 45,000} \right) = 1.55\%$$

$$\varepsilon(t_2) = (K_1 T + K_2 T) \left(\frac{1}{\xi} + \frac{t_2}{\eta} - \frac{T}{2\eta} \right) - K_2 t_2 \left(\frac{1}{\xi} + \frac{t_2}{2\eta} \right)$$

$$\varepsilon(70) = (0.5 \times 60 + 1 \times 60) \left(\frac{1}{3000} + \frac{70}{45,000} - \frac{60}{90,000} \right)$$

$$- 70 \left(\frac{1}{3000} + \frac{70}{90,000} \right)$$

$$= 3.22\%$$

$$\varepsilon(t_3) = \frac{K_1 T}{\xi \eta} \{ \eta + \xi t_3 - 1/2 \xi T \} - \frac{K_2 (T' - T')}{\xi \eta} \{ \eta + \xi t_3 - 1/2 \xi (T' + T') \}$$

which, for $t_3 = 120$ s, $T = 60$ s and $T' = 90$ s, gives $\varepsilon(120) = 3\%$

(2.39) $\varepsilon(t_1) = K_1 t_1 \left(\dfrac{1}{\xi} + \dfrac{t_1}{2\eta} \right)$

$$\varepsilon(60) = 0.4 \times 60 \left(\frac{1}{3500} + \frac{60}{2 \times 50,000} \right) = 2.12\%$$

$$\varepsilon(t_2) = K_1 t' \left(\frac{1}{\xi} + \frac{t_2}{\eta} - \frac{t'}{2\eta} \right) - \frac{\Delta \sigma}{E(t_2 - t')}$$

$$= K_1 t^1 \left(\frac{1}{\xi} + \frac{t_2}{\eta} - \frac{t'}{2\eta} \right) - \frac{\Delta \sigma (\eta + (t_2 - t'))}{\xi \eta}$$

$$\varepsilon(130) = 0.4(100) \left(\frac{1}{3500} + \frac{130}{50,000} - \frac{100}{100,000} \right) - \frac{10(50,000 + 350(30))}{3500 \times 50,000}$$

$$= 6.61$$

(2.40)

(a)

$$\varepsilon(u_1) = \int\limits_0^{u_1} \frac{1}{E(t - t_1)} \cdot \frac{d\sigma(t)}{dt} \cdot dt$$

$$\sigma(t) = \frac{20}{800} \cdot t, \quad \frac{d\sigma(t)}{dt} = \frac{20}{800}$$

$$\varepsilon(u_1) = \int_0^{u_1} \frac{1}{E(t-0)} \cdot \frac{20}{800} dt$$

$$= \int_0^{u_1} \frac{t^{0.05}}{10^3} \frac{20}{800} dt$$

$$\varepsilon(500) = \left[\frac{t^{1.05}}{1.05 \times 1000} \cdot \frac{20}{800} \right]_0^{500}$$

$$= 0.0162 = 1.62\%$$

(b) After 800 s, the change in stress is $(-Kt)$ where $K = \dfrac{20}{800}$.

So
$$\varepsilon(u_2) = \int_0^{u_2} \frac{1}{E(t-t_1)} \cdot \frac{d\sigma(t)}{dt} \cdot dt + \int_{t_2}^{u_2} \frac{1}{E(t-t_2)} \cdot \frac{d\sigma(t)}{dt} \cdot dt$$

$$= \int_0^{u_2} \frac{t^{0.05}}{10^3} \cdot K \cdot dt + \int_{t_2}^{u_2} \frac{(t-t_2)^{0.05}}{10^3} \cdot (-K) dt$$

$$= \left[\frac{t^{1.05}}{1.05 \times 10^3} \cdot K \right]_0^{u_2} - \left[\frac{(t-t_2)^{1.05}}{1.05 \times 10^3} \cdot K \right]_{t_2}^{u_2}$$

then for $u_2 = 1000$ s, $t_2 = 800$ a

$$\varepsilon(1000) = 0.0336 - 6.21 \times 10^{-3} = 0.0274 = 2.74\%$$

(2.41)

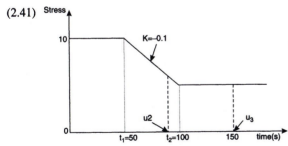

For a Kelvin–Voigt model

$$E(t) = \xi \left[1 - e^{-t/T} \right]^{-1}$$

where $T = \eta/\xi = 30$

(a) Using Boltzmanns Superposition Principle

$$\varepsilon(u_2) = \frac{\sigma_0}{\xi}(1 - e^{-u_2/T}) + \int_{t_1}^{u_2} \frac{1}{E(t-t_1)} \cdot \frac{d\sigma(t)}{dt} \cdot dt$$

where
$$\sigma(t) = -Kt \quad \text{so} \quad \frac{d\sigma(t)}{dt} = -K$$

$$\varepsilon(u_2) = \frac{\sigma_0}{\xi}(1 - e^{-u_2/T}) - K \int_{t_1}^{u_2} \frac{[1 - e^{-t/T} \cdot e^{-t_1/T}]}{\xi} dt$$

$$= \frac{\sigma_0}{\xi}(1 - e^{-u_2/T}) - K \left[\frac{t + T \cdot e^{t_1/T} e^{-t/T}}{\xi} \right]_{t_1}^{u_2}$$

$$\varepsilon(90) = \frac{10}{12 \times 10^3}(1 - e^{-90/30}) - 0.1 \left\{ \left(\frac{90 + 30e^{-40/30}}{12 \times 10^3} \right) - \left(\frac{50 + 30}{12 \times 10^3} \right) \right\}$$

$$\varepsilon(90) = 6.43 \times 10^{-4} = 0.064\%$$

(b) $$\varepsilon(u_3) = \frac{\sigma_0}{\xi}(1 - e^{-u_3/T}) + \int_{t_1}^{u_3} \frac{1}{E(t - t_1)} \frac{d\sigma(t)}{dt} dt + \int_{t_2}^{u_3} \frac{1}{E(t - t_2)} \frac{d\sigma(t)}{dt} dt$$

For the period after t_2 ($= 100$ s) the change in stress is Kt.

So $$\frac{d\sigma(t)}{dt} = K \text{ for this third period}$$

$$\varepsilon(u_3) = \frac{\sigma}{\xi}(1 - e^{-u_3/T}) + \int_{t_1}^{u_3} \frac{(1 - e^{-t/T} \cdot e^{t_1/T})}{\xi} \times (-K) \, dt + \int_{t_2}^{u_3} \frac{(1 - e^{-t_2/T})K \, dt}{\xi}$$

$$\varepsilon(u_3) = \frac{\sigma_0}{\xi}(1 - e^{-u_3/T}) - K \left[\frac{t + T \cdot e^{t_1/T} \cdot e^{-t/T}}{\xi} \right]_{t_1}^{u_3} + K \left[\frac{t + T \cdot e^{t_2/T} \cdot e^{-t/T}}{\xi} \right]_{t_2}^{u_3}$$

Substituting $u_3 = 150$ s, $K = 0.1$, $t_1 = 50$ s, $t_2 = 100$ s

$$\varepsilon(150) = 4.49 \times 10^{-4} = 0.045\%$$

(2.42) $$\varepsilon(t) = \sum_{i=0}^{i=N} \sigma_i \left\{ \frac{1}{E(t - u_i)} \right\} \text{ where } E(t - u_i) = \frac{\xi\eta}{\eta + \xi(t - u_i)}$$

So, $$\varepsilon(4500) = 10 \left\{ \frac{\eta + \xi(4500 - 0)}{\xi\eta} \right\} + 10 \left\{ \frac{\eta + \xi(4500 - 1000)}{\xi\eta} \right\}$$

$$- 15 \left\{ \frac{\eta + \xi(4500 - 2000)}{\xi\eta} \right\} + 20 \left\{ \frac{\eta + \xi(4500 - 3000)}{\xi\eta} \right\}$$

$$- 25 \left\{ \frac{\eta + \xi(4500 - 4000)}{\xi\eta} \right\}$$

$$= \frac{10}{\xi\eta} \{\eta + 4500\xi + \eta + 3500\xi - 1.5\eta - 3750\xi + 2\eta$$

$$+ 3000\xi - 2.5\eta - 1250\xi\}$$

$$= \frac{10}{\xi\eta} \{6000\xi\} = \frac{60,000}{4 \times 10^6} = 1.5\%$$

(2.43) From the information provided

$$F_r = \frac{\varepsilon_c(T) - \varepsilon_r(t)}{\varepsilon_c(T)} = \frac{0.8 - 0.058}{0.8} = 0.9275$$

also

$$t_R = \frac{t - T}{T} = \frac{200 - 100}{100} = 1$$

but,

$$F_r = 1 + t_R^n - (t_R + 1)^n$$

$$0.9275 = 1 + 1 - (2)^n \rightarrow n = 0.1$$

Also, since $\varepsilon_c(100) = 0.8\%$ and $\varepsilon_c(100) = A(100)^{0.1}$

then $A = 0.504766$ (for $\sigma = 10$ MN/m^2)

Therefore after 2400 seconds at 10 MN/m^2

$$\varepsilon(2400) = 0.504766(2400)^{0.1} = 1.1\%$$

For 2400 seconds on and 7200 seconds off, $t' = 9600$ seconds.

So

$$\varepsilon_r(9.6 \times 10^4) = \varepsilon_c(2400) \sum_{x=1}^{x=10} \left[\left(\frac{t'x}{T}\right)^n - \left(\frac{t'x}{T} - 1\right)^n \right]$$

$$\varepsilon_r(9.6 \times 10^4) = 1.1 \sum_{x=1}^{x=10} \left[(4x)^n - (4x - 1)^n \right] = 0.108\%.$$

(2.44) From question (2.20) at 5.6 MN/m^2 the grade of PP may be represented by
$\varepsilon(t) = 0.238(t)^{0.114}$
   So, after 1000 seconds, $\varepsilon(t) = 0.523\%$

After 10 cycles in the given sequence ($t' = 1500$ seconds)

$$\varepsilon(1.5 \times 10^4) = (0.523) + 0.523 \sum_{x=1}^{x=10} \left[(1.5x)^{0.114} - (1.5x - 1)^{0.114} \right] = 0.691\%$$

Then if a straight line is drawn from the point 0.523, 1000 to 0.691, 10,000 on Fig. 2.4 then this may be extrapolated to 1% strain which occurs at approximately $t = 9 \times 10^5$ seconds. This is the total creep time (ignoring recovery) and so the number of cycles for this time is

$$(9 = 10^5)/1000 = 900 \text{ cycles}$$

Notes

1000
cycles
$$\begin{cases} \varepsilon_c(1.5 \times 10^6) = 0.495\% \\ \varepsilon_r(1.5 \times 10^6) = 1.018\% \end{cases}$$

1500
cycles
$$\begin{cases} \varepsilon_c(2.25 \times 10^6) = 0.5329\% \\ \varepsilon_r(2.25 \times 10^6) = 1.056\% \end{cases}$$

Equivalent modulus after 900 cycles $= 5/0.01 = 500$ MN/m^2.

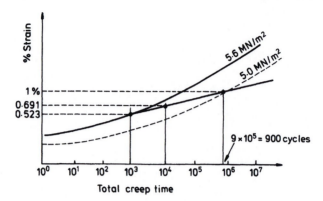

(2.45) For the PP creep curves in Fig. 2.4, $n = 0.114$ at $\sigma = 7$ MN/m^2, $\varepsilon_c(2.16 \times 10^4) = 0.99\%$ after 11 cycles of creep (10 cycles of load removal), $t = 66$ hrs $= 2.37 \times 10^5$ s. (using equation in text or from computer program) $\varepsilon_c(2.37 \times 10^5) = 1.105\%$

A line joining 0.99, 2.16×10^4 to 1.105, 2.375×10^6 on Fig. 2.4 allows strain at 365 days (ie 365 days @ 6 hrs per day $= 7.88 \times 10^6$ seconds) to be extrapolated to 1.29%. The computer program predicts

$$\varepsilon_c(7.88 \times 10^6) = 1.294\%$$

at $\sigma = 8.4$ MN/m^2, $\varepsilon_c(2.37 \times 10^5) = 1.406\%$ and $\varepsilon_c(2.16 \times 10^4) = 1.26\%$

Extrapolation on Fig. 2.4 to 7.88×10^6 seconds gives $\varepsilon_c(7.88 \times 10^6) = 1.62\%$. Computer program gives $\varepsilon_c(7.88) \times 10^6) = 1.64\%$

Extrapolation between these two values to get $\varepsilon_c(7.88 \times 10^6) = 1.5\%$ gives $\sigma = 7.8$ MN/m^2.

So since $\sigma = \dfrac{pD}{2h}$, $h = \dfrac{0.5 \times 150}{2 \times 7.8} = 4.8$ mm

For continuous loading for 1 year ($= 3.15 \times 10^7$ s) the design stress would be 5.08 MN/m^2 which gives $h = 7.38$ mm

So material saving $= \dfrac{7.38 - 4.8}{7.38} = 35\%$

(2.46) $\sigma = \sigma_0 e^{i(\omega t + \delta)}$

$$\varepsilon = \varepsilon_0 e^{i\omega t}$$

As shown in the previous question, the governing equation for this type of Standard Linear Solid is given by:

$$\dot{\varepsilon} + \frac{\xi_2}{\eta}\varepsilon = \frac{\dot{\sigma}}{\xi_1} + \frac{\sigma}{\eta}\left(\frac{\xi_1 + \xi_2}{\xi_1}\right)$$

$$\varepsilon_0 i\omega e^{i\omega t} + \frac{\xi_2}{\eta}\varepsilon_0 e^{i\omega t} = \frac{1}{\xi_1}\sigma_0 i\omega e^{i(\omega t + \delta)} + \left(\frac{\xi_1 + \xi_2}{\xi_1 \eta}\right)\sigma_0 e^{i(\omega t + \delta)}$$

Dividing across by $\varepsilon_0 e^{i\omega t}$ and letting $E^* = \dfrac{\sigma_0 e^{i(\omega t + \delta)}}{\varepsilon_0 e^{i\omega t}}$

then E^* is given by

$$E^* = \frac{\xi_1 \eta \omega i + \xi_1 \xi_2}{(\xi_1 + \xi_2) + \eta \omega i}$$

Multiplying top and bottom by the conjugate of the denominator,

$$E^* = \frac{\xi_1 \xi_2(\xi_1 + \xi_2) + \eta^2 \omega^2 \xi_1 + [(\xi_1 + \xi_2)\omega \xi_1 \eta - \xi_1 \xi_2 \eta \omega]i}{(\xi_1 + \xi_2)^2 + \eta^2 \omega^2}$$

Hence

$$E_1 = \frac{\xi_1 \xi_2(\xi_1 + \xi_2) + \eta^2 \omega^2 \xi_1}{(\xi_1 + \xi_2)^2 + \eta^2 \omega^2}$$

$$E_2 = \frac{(\xi_1 + \xi_2)\omega \xi_1 \eta - \xi_1 \xi_2 \eta \omega}{(\xi_1 + \xi_2)^2 + \eta^2 \omega^2}$$

$$\tan \delta = \frac{E_2}{E_1} = \frac{(\xi_1 + \xi_2)\omega \xi_1 \eta - \xi_1 \xi_2 \eta \omega}{\xi_1 \xi_2(\xi_1 + \xi_2) + \eta^2 \omega^2 \xi_1}$$

(2.47) The straight line graph of stress against log (time) confirms the relationship

At $t = 800, \sigma = 60$ ⎫ So, $B = 0.467 (MN/m^2)^{-1}$
At $t = 8.9 \times 10^5, \sigma = 45$ ⎭ $A = 1.225 \times 10^5$ s

(2.48) Zhurkov-Bueche equation gives

$$t = t_0 e^{(U_0 - \gamma \sigma)/RT}$$

comparing this with $t = A^{-B\sigma}$

$$t = t_0 e^{U_0/RT} \text{ and } B = \gamma/RT$$

So $\quad t_0 = \dfrac{A}{\exp\left(\dfrac{U_0}{RT}\right)} = \dfrac{1.225 \times 10^{15}}{\exp\left(\dfrac{150 \times 10^3}{8.314 \times 293}\right)} = 2.22 \times 10^{-12}$ s

Also $\gamma = BRT = 0.467 \times 8.314 \times 293 \times 10^{-6} = 1137.6 \times 10^{-6}$ m^3/mol

So at $\sigma = 44$ MN/m^2, $T = 40°$C

$$t = 2.22 \times 10^{-12} \exp\left(\frac{150 \times 10^3 - 1137.6 \times 10^{-6} \times 44 \times 10^6}{8.314 \times 313}\right) = 1.06 \times 10^5 \text{ s}$$

(2.49) 3 years $= 0.95 \times 10^8$ seconds

From Fig. 3.10 creep rupture strength $= 8$ MN/m^2

Using a safety factor of 1.5 the design stress $= 8/1.5 = 5.33$ MN/m^2 for the pipe, hoop stress $= pR/h$

$$5.33 = \frac{0.5(100)}{h}$$

$$h = 9.4 \text{ mm}$$

(2.50) The critical defect size in the material may be calculated from

$$K_{1c} = \sigma(\pi a_c)^{1/2}$$

$$a_c = \left(\frac{2}{20}\right)^2 \frac{1}{\pi} = 3.18 \text{ mm}$$

$$t_f = \frac{2}{C(\sigma)^m \pi^{(1/2)m}(m-2)} \left\{a_i^{1-(1/2)m} - a_c^{1-(1/2)m}\right\}$$

$$3.15 \times 10^7 = \frac{2}{3 \times 10^{-11}(20)^{3.2}(1.2)\pi^{1.6}} \left\{a_i^{-0.6} - (3.18 \times 10^{-3})^{-0.6}\right\}$$

$$a_i = 0.631 \text{ mm}$$

(2.51) acrylic, $\quad r_p = \dfrac{1}{2\pi}\left(\dfrac{K_{1c}}{\sigma_y}\right)^2 = \dfrac{(0.023)^2}{2\pi} = 0.084$ mm

ABS, $\quad r_p = \dfrac{(0.13)^2}{2\pi} = 2.69$ mm

polypropylene, $\quad r_p = \dfrac{(0.2)^2}{2\pi} = 6.37$ mm

(2.52) Using fracture mechanics

$$K_{1c} = \sigma(\pi a_c)^{1/2}$$

$$a_c = \left(\frac{K_{1c}}{\sigma}\right)^2 \frac{1}{\pi}$$

Table 3.1 gives K_{1c} for acrylic as $0.9 - 1.6$ MN/m$^{-(3/2)}$

$$a_c = \left(\frac{0.9}{57}\right)^2 \frac{1}{\pi} = 0.08 \text{ mm (up to 0.25 mm)}$$

So inherent flaw sizes are probably in the range 0.16 mm–0.5 mm.

(2.53) Stress range $= \dfrac{FL}{60}$

Stress amplitude = mean stress $= \dfrac{FL}{120}$

To allow for mean stress use

$$\sigma_a = \sigma_f \left\{1 - \frac{\sigma_m}{\sigma_c}\right\} \cdots\cdots (1)$$

Now at 1×10^7 cycles $\sigma_f = 16$ MN/m^2

also 1×10^7 cycles at $5H_f$ represents 2×10^6 seconds

So $\sigma_c = 42$ MN/m^2

In (1) $\dfrac{FL}{120} = \dfrac{16}{2}\left\{1 - \dfrac{FL}{120 \times 42}\right\}$

So $F = 16.13$ N

(2.54) Bending stress $= \dfrac{My}{I}$

So $\sigma_m = \dfrac{1 \times d/2}{\pi d^4/64} = \dfrac{32}{\pi d^3}$ N/m^2

$$\sigma_a = \frac{32 \times 0.75}{\pi d^3} \text{ N/m}^2$$

but $\sigma_a = \sigma_f\left(1 - \dfrac{\sigma_m}{\sigma_c}\right)$

$$2.5\left(\frac{32 \times 0.75}{\pi d^3}\right) = 25 \times 10^6 \left(1 - \frac{32 \times 2.5}{\pi d^3 \times 35 \times 10^6}\right)$$

So $d = 11.43$ mm

if $K_f = 2$

$$2.5 \times \left(\frac{32 \times 0.75}{\pi d^3}\right) = \frac{25 \times 10^6}{2}\left(1 - \frac{32 \times 2.5}{\pi d^3 \times 35 \times 10^6}\right)$$

$$d = 13.1 \text{ mm}$$

(2.55) $\sigma_f = 43.4 - 3.8\log 10^7 = 16.8$ MN/m^2

At 5 Hz, 10^7 cycles would take 2×10^6 seconds

from (3.1) $t = 1.225 \times 10^{15} e^{-0.467\sigma} c$ So $\sigma_c = 43.3$ MN/m²

Now $\qquad \sigma_m = \dfrac{500 \times 4}{\pi(10)^2} = 6.37$ MN/m²

$$\sigma_a = \frac{My}{I} = \frac{64M(5 \times 10^{-3})}{\pi(10 \times 10^{-3})^4} = 10.2 \times 10^6 M \ (\text{N/m}^2)$$

So $\qquad \sigma_a = \sigma_f \left(1 - \dfrac{\sigma_m}{\sigma_c}\right)$

$$2 \times 10.2 \times 10^6 M = \frac{16.8 \times 10^6}{1.8}\left(1 - \frac{6.37 \times 2}{43.3}\right)$$

$$M = 0.32 \text{ Nm}$$

(2.56) $\qquad$ Maximum stress $= \dfrac{P}{\pi r^2} + \dfrac{4(Pe)r}{\pi T^4}$

$\qquad\qquad$ Minimum stress $= -\dfrac{P}{2\pi r^2} - \dfrac{4(Pe)r}{\pi r^4}$

$$\sigma_m = \frac{\sigma_{max} + \sigma_{min}}{2} = \frac{P}{4\pi r^2} + \frac{Per}{\pi r^4}$$

$$\sigma_a = \frac{\sigma_{max} - \sigma_{min}}{2} = \frac{1}{2}\left(\frac{3P}{2\pi r^2} + \frac{12Per}{2\pi r^4}\right)$$

$$\sigma_a = \sigma_f \left(1 - \frac{\sigma_m}{\sigma_c}\right)$$

$$\sigma_f = 43.4 - 3.8\log(1 \times 10^8) = 13 \text{ MN/m}^2$$

At 10 Hz, 10^8 cycles would take 10^7 seconds so from $t = 1.225 \times 10^{15} \exp(-0.467\sigma_c)$, $\sigma_c = 39.9$ MN/m²

So $\qquad 2.5 \times \dfrac{1}{2}\left(\dfrac{3P}{2\pi r^2} + \dfrac{12Per}{2\pi r^4}\right) = 13 \times 10^6 \left(1 - \dfrac{2.5 \times 10^6}{39.9}\left(\dfrac{P}{4\pi r^2} + \dfrac{Per}{4\pi r^2}\right)\right)$

from which $P = 235.7$ N

(2.57) $\qquad\qquad K_t = 1 + 2(c/r)^{1/2}$

$$3.5 = 1 + 2(c/0.25)^{1/2} \rightarrow c = 0.39 \text{ mm}$$

(2.58) 4.5 kg = 44.145 N

   (i) loss of energy due to friction, etc = (44.145)(0.3 − 0.29) = 0.44 J
  (ii) energy absorbed due to specimen fracture = (44.145)(0.29 − 0.2) = 3.973 J

$$\text{impact strength} = \frac{3.973}{(12 \times 2)10^{-6}} = 165.5 \text{ kJ/m}^2$$

 (iii) Initial pendulum energy = 0.25 × 44.145 = 11.036 J

Loss of energy due to friction + specimen fracture = $0.44 + 3.947 = 4.414$ J

Remaining energy = $11.036 - 4.414 = 6.6218$ J

So height of swing $= \dfrac{6.6218}{44.145} = 0.15$ m

(2.59)

$$K = \frac{P}{Wh}(\pi a)^{1/2}\left[1.12 - 0.23\left(\frac{a}{w}\right) + 10.6\left(\frac{a}{w}\right)^2 - 21.7\left(\frac{a}{w}\right)^3 + 30.4\left(\frac{a}{w}\right)^4\right]$$

So

$$1.75 \times 10^6 = \frac{P}{100 \times 5 \times 10^{-6}}(\pi \times 10 \times 10^{-3})^{1/2}\left[f\left(\frac{a}{w}\right)\right]$$

where

$$\left[f\left(\frac{a}{w}\right)\right] \simeq 1$$

So

$$P = 4.89 \text{ kN}$$

(2.60)

$$K_c = \sigma\sqrt{\pi a_c} \rightarrow a_c = \left(\frac{1.6}{10}\right)^2 \frac{1}{\pi} = 8.15 \text{ mm}$$

from equation (3.28)

$$N = \frac{2((8.15 \times 10^{-3})^{-0.66} - (25 \times 10^{-6})^{-0.66})}{2 \times 10^{-6}(10)^{3.22}\pi^{1.66}(-1.32)}$$

$$= 57,794 \text{ cycles} = 16.05 \text{ hours}$$

(2.61)

$$N = \frac{2\left[\left(\left(\frac{K_c}{\sigma}\right)^2 \frac{1}{\pi}\right)^{1-(1/2)n} - a_i^{1-(1/2)n}\right]}{co^n\pi^{n/2}(2 - n)}$$

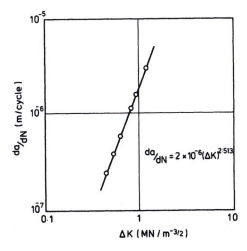

$$\frac{da}{dN} = 2 \times 10^{-6}(\Delta K)^{2.513}$$

from the graph $\dfrac{da}{dN} = 2 \times 10^{-6} \, (\Delta K)^{2.513}$

$$\sigma^{2.513} = \frac{2\left(\left(\dfrac{1.8^2}{\pi\sigma^2}\right)^{-0.256} - (20 \times 10^{-6})^{-0.256}\right)}{2 \times 10^{-6} \times 10^6 \times \pi^{1.256}(-0.513)}$$

$$\sigma = 2.13 \ \text{MN/m}^2$$

(2.62) For a sheet, $K = \sigma_f (\pi a)^{1/2}$

where σ_f = fatigue limit. Knowing K this would give the critical defect size, a_c, in the material. Equating this to the critical version of the expression for the pressure vessel

$$\sigma_f (\pi a_c)^{1/2} = 2\sigma_\theta (2a_c)^{1/2}$$

$$\sigma_\theta = \sigma_f \left(\frac{\pi^{1/2}}{2\sqrt{2}}\right) = 10 \left(\frac{\pi^{1/2}}{2\sqrt{2}}\right) = 0.627 \ \text{MN/m}^2$$

Now,

$$\sigma_\theta = \frac{PD}{2h} \rightarrow P = \frac{2h\sigma_\theta}{D^{1/8}}$$

$$P = \frac{2 \times 4 \times 0.627}{120} = 41.8 \ \text{kN/m}^2$$

(2.63)

$$K_c = \sigma(\pi a_c)^{1/2} \left(\frac{W}{\pi a} \tan\left(\frac{\pi a}{W}\right)\right)^{1/2}$$

$$= \frac{P}{Wb}(\pi a_c)^{1/2} \left(\frac{W}{\pi a} \tan\left(\frac{\pi a}{W}\right)\right)^{1/2}$$

$$43 = \frac{P}{30 \times 5} \left(\pi \times 5 \times 10^{-3}\right)^{1/2} (1.05)$$

$$P = 49 \ \text{kN}$$

(3.1) The energy absorbing ability of a material is given by

$$U = \frac{1}{2}\sigma\varepsilon = \frac{1}{2}\frac{\sigma^2}{E}$$

Hence the following table may be drawn up

	σ	E	$\sigma^2/2E$
Carbon (HS)	2.9	230	1.83×10^{-2}
Carbon (HM)	2.2	380	0.64×10^{-2}
Kevlar	3.0	130	3.46×10^{-2}
E-Glass	2.0	80	2.5×10^{-2}
S-Glass	3.3	91	6.0×10^{-2}

From this table it may be seen that carbon fibre has low energy absorbing capability compared with the other fibres. However, the other fibres are not as stiff. Hence it is quite common to use hybrid fibre composites, eg glass and carbon fibres in order to get a better combination of properties.

(3.2) It is necessary to calculate the volume fraction for each of the fibres. Firstly for the carbon

$$V_{fc} = \frac{\dfrac{20}{1800}}{\dfrac{20}{1800} + \dfrac{30}{2540} + \dfrac{50}{1300}} = 0.181$$

Similarly for the glass fibres

$$V_{fg} = \frac{\dfrac{30}{2540}}{\dfrac{20}{1800} + \dfrac{30}{2540} + \dfrac{50}{1300}} = 0.192$$

Hence, using the rule of mixtures

$$\rho_c = 0.181(1800) + 0.192(2540) + 0.627(1300)$$

$$= 1629 \text{ kg/m}^3$$

(3.3) $\rho_f = 1800$ kg/m^3, $\rho_m = 1250$ kg/m^3, $\rho_c = 1600$ kg/m^3

Using rule of mixtures

$$\rho_c = \rho_f V_f + \rho_m(1 - V_f)$$

$$1600 = 1800 V_f + 1250(1 - V_f)$$

$$V_f = 0.636$$

to get weight fraction $\quad W_f = \dfrac{\rho_f}{\rho_c} V_f = \dfrac{1800}{1600}(0.636) = 0.716$

hence $\qquad\qquad\qquad W_m = 0.284$

Weight ratio $W_m : W_f = 0.284 : 0.716$

Hence for 1 kg epoxy, the weight of carbon $= 2.52$ kg

(3.4) Using rule of mixtures

$$\rho_c = \rho_f V_f + \rho_m V_m = 2540(0.5) + 1250(0.4) = 2024 \text{ kg/m}^3$$

$$E_c = E_f V_f + E_m V_m = 80(0.6) + 6.1(0.4) = 50.44 \text{ GN/m}^2$$

$$K_c = K_f V_f = 1.05(0.6) + 0.25(0.4) = 0.73 \text{ W/m}^\circ\text{K}$$

(3.5) Using the rule of mixtures

$$E_c = E_f V_f + E_m(1 - V_f) = 120(0.4) + 6(0.6) = 51.6 \text{ GN/m}^2$$

Also, $\qquad\qquad \sigma_c = \sigma_f V_f + \sigma_m(1 - V_f)$

but $\qquad\qquad\quad \varepsilon_c = \varepsilon_m = \varepsilon_f$

$$\frac{\sigma_c}{E_c} = \frac{\sigma_m}{E_m} = \frac{\sigma_f}{E_f} \rightarrow \left(\frac{E_f}{E_m}\right)\sigma_m = \sigma_f$$

So,
$$\sigma_c = \frac{E_f}{E_m}(\sigma_m)V_f + \sigma_m(1 - V_f)$$

$$50 = 20(\sigma_m)0.4 + \sigma_m(0.6)$$

$$\sigma_m = 5.8 \text{ MN/m}^2, \sigma_f = 20(5.8) = 116.3 \text{ MN/m}^2$$

(3.6) In the composite

$$\varepsilon_f = \varepsilon_c$$

$$\frac{\sigma_f}{E_f} = \frac{\sigma_c}{E_c}$$

$$\frac{\sigma_f}{\sigma_c} = \frac{E_f}{E_c}$$

$$\frac{F_f A_c}{F_c A_f} = \frac{E_f}{E_f V_f + E_m V_m}$$

$$\frac{F_f}{F_c} = \frac{E_f/E_m}{E_f/E_m + V_m/V_f} = \frac{40}{40 + 1}$$

$$F_f = 97.6\% F_c$$

(3.7) $$\sigma_c = \sigma_f V_f + \sigma_m(1 - V_f)$$

Since $$\varepsilon_c = \varepsilon_m = \varepsilon_f \text{ then } \frac{\sigma_c}{E_c} = \frac{\sigma_m}{E_m} = \frac{\sigma_f}{E_f}$$

ie $$\sigma_f = \sigma_m \left(\frac{E_f}{E_m}\right)$$

So $$\sigma_c = \sigma_m \left(\frac{E_f}{E_m}\right) V_f + \sigma_m(1 - V_f)$$

$$= 60(25)0.3 + 60(0.7) = 492 \text{ MN/m}^2$$

Also $$E_c = E_f V_f + E_m(1 - V_f)$$

$$= (76 \times 0.3) + \frac{76}{25}(0.7) = 24.93 \text{ GN/m}^2$$

(3.8)

(i) $$\nu_{12} = \nu_f V_f + \nu_m(1 - V_f)$$

$$= (0.23)(0.58) + 0.35(0.42)$$

$$= 0.28$$

(ii) $$E_1 = E_f V_f + E_m V_m = 230(0.58) + 3.8(0.42) = 135 \text{ GN/m}^2$$

$$E_2 = \frac{E_f E_m}{V_f E_m + V_m E_f} = \frac{(230)(3.8)}{(0.58)(3.8) + (0.42)(230)} = 8.8 \text{ GN/m}^2$$

$$\nu_{21} = \nu_{12} = 0.28 \left(\frac{8.8}{135}\right) = 0.02$$

(3.9) The local compliance matrix is

$$
S = \begin{bmatrix} \dfrac{1}{E_1} & \dfrac{v_{12}}{E_1} & 0 \\[2mm] \dfrac{v_{12}}{E_1} & \dfrac{1}{E_2} & 0 \\[2mm] 0 & 0 & \dfrac{1}{G_{12}} \end{bmatrix} \quad \text{and} \quad Q = S^{-1}
$$

$$
\overline{Q} = T_\sigma^{-1} \cdot Q \cdot T_\varepsilon \quad \text{and} \quad \overline{S} = \overline{Q}^{-1}
$$

Then

$$
\begin{bmatrix} \varepsilon_x \\ \varepsilon_y \\ \gamma_{xy} \end{bmatrix} = \overline{S} \begin{bmatrix} \sigma_x \\ \sigma_y \\ \tau_{xy} \end{bmatrix}
$$

$$
\overline{S} = \begin{bmatrix} 8.81 \times 10^{-5} & -2.72 \times 10^{-5} & -5.44 \times 10^{-5} \\ -2.72 \times 10^{-5} & 1.04 \times 10^{-4} & -4.05 \times 10^{-5} \\ -5.44 \times 10^{-5} & -4.05 \times 10^{-5} & 1.70 \times 10^{-4} \end{bmatrix} \quad \text{m}^2/\text{MN}
$$

Directly by matrix manipulation

$$
\varepsilon_x = 2.238 \times 10^{-3} \qquad \varepsilon_y = 7.578 \times 10^{-4} \qquad \gamma_{xy} = -2.241 \times 10^{-3}
$$

and

$$
E_x = \frac{1}{\overline{S}_{11}}, \quad E_y = \frac{1}{\overline{S}_{22}}, \quad G_{xy} = \frac{1}{\overline{S}_{66}}, \quad v_{xy} = -E_x \cdot \overline{S}_{21}, \quad v_{yx} = -E_y \cdot \overline{S}_{12}
$$

Alternatively using the Constitutive Equation Approach

$$
A = \overline{Q} \cdot h
$$

$$
A = \begin{bmatrix} 2.78 \times 10^4 & 1.17 \times 10^4 & 1.16 \times 10^4 \\ 1.17 \times 10^4 & 2.06 \times 10^4 & 8.624 \times 10^3 \\ 1.16 \times 10^4 & 8.62 \times 10^3 & 1.45 \times 10^4 \end{bmatrix}
$$

Also, $a = A^{-1}$ (since $[B] = 0$)

$$
a = \begin{bmatrix} 5.87 \times 10^{-5} & -1.81 \times 10^{-5} & -3.63 \times 10^{-5} \\ -1.81 \times 10^{-5} & 6.99 \times 10^{-5} & -2.7 \times 10^{-5} \\ -3.63 \times 10^{-5} & -2.7 \times 10^{-5} & 1.13 \times 10^{-4} \end{bmatrix} \quad \text{mm/N}
$$

$$
E_x = \frac{1}{a_{11} \cdot h}, \quad E_y = \frac{1}{a_{22} \cdot h}, \quad G_{xy} = \frac{1}{a_{66} \cdot h}, \quad v_{xy} = \frac{-a_{12}}{a_{11}}, \quad v_{yx} = \frac{-a_{12}}{a_{22}}
$$

$$
E_x = 11.3 \text{ GMN/m}^2, \quad E_y = 9.53 \text{ GN/m}^2, \quad G_{xy} = 5.8 \text{ GN/m}^2
$$

$$
v_{xy} = 0.308 \qquad v_{yx} = 0.259
$$

For the applied Forces N_x and N_y

$$
\begin{bmatrix} \varepsilon_x \\ \varepsilon_y \\ \gamma_{xy} \end{bmatrix} = a. \begin{bmatrix} N_x \\ N_y \\ N_{xy} \end{bmatrix} \qquad \left(= a \cdot \begin{bmatrix} \sigma_x \\ \sigma_y \\ \tau_{xy} \end{bmatrix} \cdot h \right)
$$

$$
\varepsilon_x = 2.238 \times 10^{-3} \qquad \varepsilon_y = 7.578 \times 10^{-4} \qquad \gamma_{xy} = -2.241 \times 10^{-3}
$$

(3.10) The properties of the ply in the fibre and transverse directions are

$$E_1 = 200 \text{ GN/m}^2$$
$$E_2 = 11 \text{ GN/m}^2$$
$$G_{12} = 8 \text{ GN/m}^2$$
$$\nu_{12} = 0.32$$

The local compliance matrix is

$$S = \begin{bmatrix} \dfrac{1}{E_1} & -\dfrac{\nu_{12}}{E_1} & 0 \\ -\dfrac{\nu_{12}}{E_1} & \dfrac{1}{E_2} & 0 \\ 0 & 0 & \dfrac{1}{G_{12}} \end{bmatrix} \quad \text{and} \quad Q = S^{-1}$$

So, $\overline{Q} = T_\sigma^{-1} \cdot Q \cdot T_\varepsilon$ and $\overline{S} = Q^{-1}$

Then

$$\begin{bmatrix} \varepsilon_x \\ \varepsilon_y \\ \gamma_{xy} \end{bmatrix} = \overline{S} \cdot \begin{bmatrix} \sigma_x \\ \sigma_y \\ \tau_{xy} \end{bmatrix}$$

$$\overline{S} = \begin{bmatrix} 3.13 \times 10^{-5} & -6.45 \times 10^{-6} & -4.28 \times 10^{-5} \\ -6.45 \times 10^{-6} & 7.42 \times 10^{-5} & -3.15 \times 10^{-5} \\ -4.28 \times 10^{-5} & -3.15 \times 10^{-5} & 1.05 \times 10^{-4} \end{bmatrix} \quad \text{m}^2/\text{MN}$$

Directly by matrix manipulation

$$\varepsilon_x = 2.507 \times 10^{-3}, \qquad \varepsilon_y = 5.164 \times 10^{-4}, \qquad \gamma_{xy} = -3.424 \times 10^{-3}$$

and

$$E_x = \frac{1}{\overline{S}_{11}}, \quad E_y = \frac{1}{\overline{S}_{22}}, \quad G_{xy} = \frac{1}{\overline{S}_{66}}, \quad \nu_{xy} = -E_x \cdot \overline{S}_{21}, \quad \nu_{yx} = -E_y \cdot \overline{S}_{12}$$

$$E_x = 31.9 \text{ GN/m}^2, \quad E_y = 13.5 \text{ GN/m}^2, \quad G_{xy} = 9.5 \text{ GN/m}^2,$$

$$\nu_{xy} = 0.206, \qquad \nu_{yx} = 0.087$$

Alternatively using the Constitutive Equation Approach

$$A = \overline{Q} \cdot h$$

$$A = \begin{bmatrix} 2.42 \times 10^5 & 7.2 \times 10^4 & 1.19 \times 10^5 \\ 7.2 \times 10^4 & 5.22 \times 10^4 & 4.48 \times 10^4 \\ 1.19 \times 10^5 & 4.48 \times 10^4 & 8.09 \times 10^4 \end{bmatrix} \quad \text{N/mm}$$

$$D = \overline{Q} \cdot \frac{h^3}{12}$$

$$D = \begin{bmatrix} 8.07 \times 10^4 & 2.4 \times 10^4 & 3.99 \times 10^4 \\ 2.4 \times 10^4 & 1.74 \times 10^4 & 1.49 \times 10^4 \\ 3.99 \times 10^4 & 1.49 \times 10^4 & 2.69 \times 10^4 \end{bmatrix} \quad \text{Nmm}$$

$$a = A^{-1}$$

$$a = \begin{bmatrix} 1.56 \times 10^{-5} & -3.22 \times 10^{-6} & -2.14 \times 10^{-5} \\ -3.22 \times 10^{-6} & 3.71 \times 10^{-5} & -1.58 \times 10^{-5} \\ -2.14 \times 10^{-5} & 1.58 \times 10^{-5} & 5.27 \times 10^{-5} \end{bmatrix} \text{ mm/N}$$

$$d := D^{-1}$$

$$d = \begin{bmatrix} 4.7 \times 10^{-5} & -9.68 \times 10^{-6} & -6.42 \times 10^{-5} \\ -9.68 \times 10^{-6} & 1.11 \times 10^{-4} & -4.73 \times 10^{-5} \\ -6.42 \times 10^{-5} & -4.73 \times 10^{-5} & 1.58 \times 10^{-4} \end{bmatrix} \text{ (Nmm)}^{-1}$$

$$E_x = \frac{1}{a_{11} \cdot h}, \quad E_y = \frac{1}{a_{22} \cdot h}, \quad G_{xy} = \frac{1}{a_{66} \cdot h} \quad v_{xy} = \frac{-a_{12}}{a_{11}} v_{yx} = \frac{-a_{12}}{a_{22}}$$

$$E_x = 31.9 \text{ GN/m}^2, \quad E_y = 13.5 \text{ GN/m}^2, \quad G_{xy} = 9.5 \text{ GN/m}^2$$

$$v_{xy} = 0.206, \quad v_{yx} = 0.087$$

It may be seen that these agree with the previous values.

For the applied Force N_x

$$\begin{bmatrix} \varepsilon_x \\ \varepsilon_y \\ \gamma_{xy} \end{bmatrix} = a \cdot \begin{bmatrix} N_x \\ N_y \\ N_{xy} \end{bmatrix} \quad \left(= a \cdot \begin{bmatrix} \sigma_x \\ \sigma_y \\ \tau_{xy} \end{bmatrix} \cdot h \right)$$

$$\varepsilon_x = 2.507 \times 10^{-3} \quad \varepsilon_y = -5.164 \times 10^{-4} \quad \gamma_{xy} = -3.424 \times 10^{-3}$$

For the applied Moment $M_x = 180$ Nm/m

$$\begin{bmatrix} \kappa_x \\ \kappa_y \\ \kappa_{xy} \end{bmatrix} = d \cdot \begin{bmatrix} M_x \\ M_y \\ M_{xy} \end{bmatrix} \quad \text{and} \quad v_{yx} = \frac{-\kappa_x}{\kappa_y}, \quad v_{xy} = \frac{-\kappa_y}{\kappa_x}$$

$$\kappa_x = 8.46 \text{ m}^{-1}, \quad \kappa_y = -1.743 \text{ m}^{-1}, \quad \kappa_{xy} = -12 \text{ m}^{-1},$$

$$v_{yx} = 4.854, \quad v_{xy} = 0.206$$

It may be seen that when the moment is applied, the major Poisson's ratio v_{xy} corresponds as it should to the value when the in-plane stress, σ_x, is applied.

(ii) If an additional moment M_y is applied, then the D and d matrices do not change and so once again

$$\begin{bmatrix} \kappa_x \\ \kappa_y \\ \kappa_{xy} \end{bmatrix} = d \cdot \begin{bmatrix} M_x \\ M_y \\ M_{xy} \end{bmatrix} \quad \text{and} \quad v_{xy} = \frac{-\kappa_y}{\kappa_x}, \quad v_{yx} = \frac{-\kappa_x}{\kappa_y}$$

$$\kappa_x = 6.04 \text{ m}^{-1}, \quad \kappa_y = 26 \text{ m}^{-1}, \quad \kappa_{xy} = -23 \text{ m}^{-1}$$

$$v_{yx} = 0.23, \quad v_{xy} = -4.32$$

(3.11)

(i) Symmetric

(ii) Non-symmetric (there is an uneven number of plies)

(iii) Symmetric (could also be written as $[0/90/45/-45_3/-45_3/45/90/0]_T$

(3.12) This is treated as a 4 layer situation with

$$h_0 = -0.7, \quad h_1 = -0.2, \quad h_2 = 0, \quad h_3 = 0.2, \quad h_4 = 0.7 \text{ mm}$$

For   *Material A*   *Material B*

$$\overline{Q}_1 = \begin{bmatrix} 4.16 \times 10^3 & 1.66 \times 10^3 & 0 \\ 1.66 \times 10^3 & 4.16 \times 10^3 & 0 \\ 0 & 0 & 1.25 \times 10^3 \end{bmatrix}, \overline{Q}_2 = \begin{bmatrix} 683.5 & 239.5 & 0 \\ 239.5 & 683.5 & 0 \\ 0 & 0 & 222.2 \end{bmatrix}$$

$$\overline{Q}_4 = \overline{Q}_1 \quad \text{and} \quad \overline{Q}_3 = \overline{Q}_2$$

As the laminate is symmetrical, $B = 0$ and

$$A = \sum_{f=1}^{4} \overline{Q}_f \cdot (h_f - h_{f-1}) \quad \text{and} \quad a = A^{-1}$$

$$A = \begin{bmatrix} 4.44 \times 10^3 & 1.76 \times 10^3 & 0 \\ 1.76 \times 10^3 & 4.44 \times 10^3 & 0 \\ 0 & 0 & 1.33 \times 10^3 \end{bmatrix} \text{N/mm}$$

$$a = \begin{bmatrix} 2.67 \times 10^{-4} & -1.06 \times 10^{-4} & 0 \\ -1.06 \times 10^{-4} & 2.67 \times 10^{-4} & 0 \\ 0 & 0 & 7.46 \times 10^{-4} \end{bmatrix} \text{mm/N}$$

The strains are then obtained from

$$\begin{bmatrix} \varepsilon_x \\ \varepsilon_y \\ \gamma_{xy} \end{bmatrix} = a \cdot \begin{bmatrix} N_x \\ N_y \\ N_{xy} \end{bmatrix}$$

$$\varepsilon_x = 3.028 \times 10^{-3}, \quad \varepsilon_y = 8.257 \times 10^{-3}, \quad \gamma_{xy} = 0.016$$

and the stresses are given by

Stresses in Material A

$$\begin{bmatrix} \sigma_{ax} \\ \sigma_{ay} \\ \tau_{axy} \end{bmatrix} = \overline{Q}_1 \cdot \begin{bmatrix} \varepsilon_x \\ \varepsilon_y \\ \gamma_{xy} \end{bmatrix} \qquad \sigma_{ax} = 26.38 \text{ MN/m}^2 \qquad \sigma_{ay} = 39.45 \text{ MN/m}^2$$

$$\tau_{axy} = 19.6 \text{ MN/m}^2$$

Stresses in Material B

$$\begin{bmatrix} \sigma_{bx} \\ \sigma_{by} \\ \tau_{bxy} \end{bmatrix} = \overline{Q}_2 \cdot \begin{bmatrix} \varepsilon_x \\ \varepsilon_y \\ \gamma_{xy} \end{bmatrix} \qquad \sigma_{bx} = 4.04 \text{ MN/m}^2 \qquad \sigma_{by} = 6.36 \text{ MN/m}^2$$

$$\tau_{bxy} = 3.48 \text{ MN/m}^2$$

(3.13)

Compliance Matrix   *Stiffness Matrix*

$$S = \begin{bmatrix} \dfrac{1}{E_1} & -\dfrac{\nu_{12}}{E_1} & 0 \\ -\dfrac{\nu_{12}}{E_1} & \dfrac{1}{E_2} & 0 \\ 0 & 0 & \dfrac{1}{G_{12}} \end{bmatrix} \qquad Q = S^{-1}$$

Overall Stiffness Matrix   Overall Compliance Matrix

$$Q = T_\sigma^{-1} \cdot Q \cdot T_\varepsilon \qquad\qquad \overline{S} = \overline{Q}^{-1}$$

The Extension Stiffness Matrix is then obtained from

$$[A] = 0.4\overline{Q}(-60) + 0.4\overline{Q}(-30) + 0.4\overline{Q}(0) + 0.4\overline{Q}(30) + 0.4\overline{Q}(60) + 0.4\overline{Q}(90)$$

which gives

$$A = \begin{bmatrix} 1.36 \times 10^5 & 4.19 \times 10^4 & 0 \\ 4.19 \times 10^4 & 1.36 \times 10^5 & 0 \\ 0 & 0 & 4.71 \times 10^4 \end{bmatrix} \text{ N/mm}$$

As $[B] = 0$ in this symmetric laminate, the compliance matrix is obtained by inverting $[A]$

$$a = A^{-1}$$

$$a = \begin{bmatrix} 8.10 \times 10^{-6} & -2.49 \times 10^{-6} & 0 \\ -2.49 \times 10^{-6} & 8.10 \times 10^{-6} & 0 \\ 0 & 0 & 2.11 \times 10^{-5} \end{bmatrix} \text{ mm/N}$$

The stiffness terms in the global directions are (for $h = 2.4$ mm)

$$E_x = \frac{1}{a_{11} \cdot h}, \qquad E_x = 5.14 \times 10^4 \text{ MN/m}^2$$

$$E_y = \frac{1}{a_{22} \cdot h}, \qquad E_y = 5.14.10^4 \text{ MN/m}^2$$

$$G_{xy} = \frac{1}{a_{66} \cdot h}, \qquad G_{xy} = 1.96.10^4 \text{ MN/m}^2$$

$$v_x = \frac{a_{12}}{a_{11}}, \qquad v_x = 0.308$$

$$v_y = \frac{a_{12}}{a_{22}}, \qquad v_y = 0.308$$

and the strains may be obtained as

$$\begin{bmatrix} \varepsilon_x \\ \varepsilon_y \\ \gamma_{xy} \end{bmatrix} = a \cdot \begin{bmatrix} \sigma_x \\ \sigma_y \\ \tau_{xy} \end{bmatrix} \cdot h$$

$$\varepsilon_x = 1.94 \times 10^{-3}, \quad \varepsilon_y = -5.98 \times 10^{-4}, \quad \gamma_{xy} = -1.12 \times 10^{-7}$$

(3.14) Using the S and Q matrices to calculate $\overline{S}$ and $\overline{Q}$ for each layer, then

$$A = \sum_{f=1}^{8} \overline{Q}_f \cdot (h_f - h_{f-1}), \qquad a = A^{-1}$$

$$B = \frac{1}{2} \cdot \sum_{f=1}^{8} \overline{Q}_f \cdot [(h_f)^2 - (h_{f-1})^2], \, D = \frac{1}{3} \cdot \sum_{f=1}^{8} \overline{Q}_f \cdot [(h_f)^3 - (h_{f-1})^3], \, d = D^{-1}$$

$$\begin{bmatrix} \varepsilon_x \\ \varepsilon_y \\ \varepsilon_{xy} \end{bmatrix} = a \cdot \begin{bmatrix} N_x \\ N_y \\ N_{xy} \end{bmatrix} = a \cdot \begin{bmatrix} \sigma_x \\ \sigma_y \\ N_{xy} \end{bmatrix} \cdot H \quad \text{where } H = \text{full thickness of laminate}$$

$$\varepsilon_x = 8.989 \times 10^{-4}, \qquad \varepsilon_y = -1.286 \times 10^{-3}, \qquad \gamma_{xy} = -1.774 \times 10^{-4}$$

Note, to assist the reader, the values of the terms in the matrices are

$$A = \begin{bmatrix} 4.65 \times 10^5 & 2.47 \times 10^5 & 0 \\ 2.47 \times 10^5 & 2.82 \times 10^5 & 0 \\ 0 & 0 & 2.81 \times 10^5 \end{bmatrix} \text{N/mm} \quad B = \begin{bmatrix} 0 & 0 & 0 \\ 0 & 0 & 0 \\ 0 & 0 & 0 \end{bmatrix}$$

$$a = \begin{bmatrix} 4.03 \times 10^{-6} & -3.54 \times 10^{-6} & 0 \\ -3.54 \times 10^{-6} & 6.65 \times 10^{-6} & 0 \\ 0 & 0 & 3.54 \times 10^{-6} \end{bmatrix} \text{mm/N}$$

(3.15) Taking the x-direction as the hoop direction

$$\sigma_x = \frac{pr}{t} = \frac{3 \times 600}{3} = 600 \text{ MN/m}^2$$

$$\sigma_y = \frac{pr}{2t} = \frac{3 \times 600}{6} = 300 \text{ MN/m}^2$$

From the data given

$$\text{\emph{Compliance Matrix}} \qquad\qquad \text{\emph{Stiffness Matrix}}$$

Compliance Matrix *Stiffness Matrix*

$$S = \begin{bmatrix} \dfrac{1}{E_1} & -\dfrac{v_{12}}{E_1} & 0 \\ -\dfrac{v_{12}}{E_1} & \dfrac{1}{E_2} & 0 \\ 0 & 0 & \dfrac{1}{G_{12}} \end{bmatrix} \qquad Q = S^{-1}$$

Overall Stiffness Matrix *Overall Compliance Matrix*

$$\overline{Q}(\theta) = T_\sigma^{-1} \cdot Q \cdot T_\varepsilon \qquad\qquad \overline{S}(\theta) = \overline{Q}(\theta)^{-1}$$

$$\overline{Q}(60) = \begin{bmatrix} 9.82 \times 10^3 & 6.95 \times 10^3 & 2.71 \times 10^3 \\ 6.95 \times 10^3 & 2.21 \times 10^4 & 7.91 \times 10^3 \\ 2.71 \times 10^3 & 7.91 \times 10^3 & 7.50 \times 10^3 \end{bmatrix}$$

The Extension Stiffness matrix is then given by

$$A = 2 \cdot \overline{Q}(60) + 2 \cdot \overline{Q}(-60) + 6 \cdot \overline{Q}(0)$$

$$A = \begin{bmatrix} 2.35 \times 10^5 & 4.25 \times 10^4 & 0 \\ 4.25 \times 10^4 & 1.37 \times 10^5 & 0 \\ 0 & 0 & 4.80 \times 10^4 \end{bmatrix}$$

The compliance matrix is obtained by inverting [A]

$$a = A^{-1}$$

$$a = \begin{bmatrix} 4.49 \times 10^{-6} & -1.39 \times 10^{-6} & 0 \\ -1.39 \times 10^{-6} & 7.70 \times 10^{-6} & 0 \\ 0 & 0 & 2.08 \times 10^5 \end{bmatrix}$$

and the strains may be obtained as

$$\begin{bmatrix} \varepsilon_x \\ \varepsilon_y \\ \gamma_{xy} \end{bmatrix} = a \cdot \begin{bmatrix} \sigma_x \\ \sigma_y \\ \tau_{xy} \end{bmatrix} \cdot 10$$

$$\varepsilon_x = 0.023 \text{ or } 2.3\%$$

$$\varepsilon_y = 0.015 \text{ or } 1.5\%$$

$$\gamma_{xy} = 0$$

(3.16) (a) This is a symmetric laminate and the $\overline{Q}$ terms are as follows:

$$\overline{Q}_1 = \begin{bmatrix} 8.77 \times 10^4 & 2.98 \times 10^4 & 4.31 \times 10^4 \\ 2.98 \times 10^4 & 2.22 \times 10^4 & 1.35 \times 10^4 \\ 4.31 \times 10^4 & 1.35 \times 10^4 & 2.96 \times 10^4 \end{bmatrix}$$

$$\overline{Q}_2 = \begin{bmatrix} 8.77 \times 10^4 & 2.98 \times 10^4 & -4.31 \times 10^4 \\ 2.98 \times 10^4 & 2.22 \times 10^4 & -1.35 \times 10^4 \\ -4.31 \times 10^4 & -1.35 \times 10^4 & 2.96 \times 10^4 \end{bmatrix}$$

$$\overline{Q}_3 = \begin{bmatrix} 8.77 \times 10^4 & 2.98 \times 10^4 & -4.31 \times 10^4 \\ 2.98 \times 10^4 & 2.22 \times 10^4 & -1.35 \times 10^4 \\ -4.31 \times 10^4 & -1.35 \times 10^4 & 2.96 \times 10^4 \end{bmatrix}$$

$$\overline{Q}_4 = \begin{bmatrix} 8.77 \times 10^4 & 2.98 \times 10^4 & 4.31 \times 10^4 \\ 2.98 \times 10^4 & 2.22 \times 10^4 & 1.35 \times 10^4 \\ 4.31 \times 10^4 & 1.35 \times 10^4 & 2.96 \times 10^4 \end{bmatrix}$$

The A, B, D matrices are

$$A = \sum_{f=1}^{4} \overline{Q}_f \cdot (h_f - h_{f-1})$$

$$B = \frac{1}{2} \cdot \sum_{f=1}^{4} \overline{Q}_f \cdot [(h_f)^2 - (h_{f-1})^2] \qquad D = \frac{1}{3} \cdot \sum_{f=1}^{4} \overline{Q}_f [(h_f)^3 - (h_{f-1})^3]$$

So that with

$$h_0 = -0.2, \quad h_1 = -0.1, \quad h_2 = 0, \quad h_3 = 0.1, \quad h_4 = 0.2$$

The A, B, D matrices and hence the a^{-1}, d^{-1}, matrices may be determined as:

$$A = \begin{bmatrix} 3.51 \times 10^4 & 1.19 \times 10^4 & 0 \\ 1.19 \times 10^4 & 8.88 \times 10^3 & 0 \\ 0 & 0 & 1.18 \times 10^4 \end{bmatrix}$$

$$a = \begin{bmatrix} 5.25 \times 10^{-5} & -7.06 \times 10^{-5} & 0 \\ -7.06 \times 10^{-5} & 2.07 \times 10^{-4} & 0 \\ 0 & 0 & 8.42 \times 10^{-5} \end{bmatrix}$$

$$B = \begin{bmatrix} 0 & 0 & 0 \\ 0 & 0 & 0 \\ 0 & 0 & 0 \end{bmatrix} \qquad b = \begin{pmatrix} 0 & 0 & 0 \\ 0 & 0 & 0 \\ 0 & 0 & 0 \end{pmatrix} \qquad \beta = \begin{pmatrix} 0 & 0 & 0 \\ 0 & 0 & 0 \\ 0 & 0 & 0 \end{pmatrix}$$

$$D = \begin{pmatrix} 467.9 & 159.3 & 172.7 \\ 159.3 & 118.5 & 54.2 \\ 172.7 & 54.2 & 158.2 \end{pmatrix} \quad d = \begin{bmatrix} 5.58 \times 10^{-3} & -5.59 \times 10^{-3} & -4.18 \times 10^{-3} \\ -5.59 \times 10^{-3} & 0.016 & 7.62 \times 10^{-4} \\ -4.18 \times 10^{-3} & 7.62 \times 10^{-4} & 0.011 \end{bmatrix}$$

Then using

$$
\begin{bmatrix} \varepsilon_x \\ \varepsilon_y \\ \gamma_{xy} \\ \kappa_x \\ \kappa_y \\ \kappa_{xy} \end{bmatrix}
=
\begin{bmatrix}
a_{11} & a_{12} & a_{16} & b_{11} & b_{12} & b_{16} \\
a_{21} & a_{22} & a_{26} & b_{21} & b_{22} & b_{26} \\
a_{61} & a_{62} & a_{66} & b_{61} & b_{62} & b_{66} \\
\beta_{11} & \beta_{12} & \beta_{16} & d_{11} & d_{12} & d_{16} \\
\beta_{21} & \beta_{22} & \beta_{26} & d_{21} & d_{22} & d_{26} \\
\beta_{61} & \beta_{62} & \beta_{66} & d_{61} & d_{62} & d_{66}
\end{bmatrix}
\begin{bmatrix} N_x \\ N_y \\ N_{xy} \\ M_x \\ M_y \\ M_{xy} \end{bmatrix}
$$

The strains and curvatures are thus

$$\varepsilon_x = 2.101 \times 10^{-3}, \quad \varepsilon_y = -2.825 \times 10^{-3}, \quad \gamma_{xy} = 0,$$

$$\kappa_x = 0, \quad \kappa_y = 0, \quad \kappa_{xy} = 0$$

(b) This laminate is unsymmetrical. In the above analysis, $\overline{Q}_2$ and $\overline{Q}_3$ are reversed and this has the effect of making $[B] \neq 0$. The matrix values are.

$$
A = \begin{bmatrix}
3.51 \times 10^4 & 1.19 \times 10^4 & 0 \\
1.19 \times 10^4 & 8.88 \times 10^3 & 0 \\
0 & 0 & 1.18 \times 10^4
\end{bmatrix}
$$

$$
a = \begin{bmatrix}
5.75 \times 10^{-5} & -7.15 \times 10^{-5} & 0 \\
-7.15 \times 10^{-5} & 2.07 \times 10^{-4} & 0 \\
0 & 0 & 9.74 \times 10^{-5}
\end{bmatrix}
$$

$$
B = \begin{pmatrix}
0 & 0 & -863.8 \\
0 & 0 & -271 \\
-863.8 & -271 & 0
\end{pmatrix}
$$

$$
b = \begin{bmatrix}
0 & 0 & 1.91 \times 10^{-4} \\
0 & 0 & -3.49 \times 10^{-5} \\
1.91 \times 10^{-4} & -3.49 \times 10^{-5} & 0
\end{bmatrix}
$$

$$
\beta = \begin{bmatrix}
0 & 0 & 1.91 \times 10^4 \\
0 & 0 & -3.49 \times 10^{-5} \\
1.91 \times 10^{-4} & -3.49 \times 10^{-5} & 0
\end{bmatrix}
$$

$$
D = \begin{pmatrix}
467.9 & 159.3 & 0 \\
159.3 & 118.5 & 0 \\
0 & 0 & 1.58.2
\end{pmatrix}
$$

$$
d = \begin{bmatrix}
4.31 \times 10^{-3} & -5.36 \times 10^{-3} & 0 \\
-5.36 \times 10^{-3} & 0.016 & 0 \\
0 & 0 & 7.30 \times 10^{-3}
\end{bmatrix}
$$

Using the above equation relating strains and curvatures to N and M we get

$$\varepsilon_x = 2.3 \times 10^{-3}, \quad \varepsilon_y = -2.86 \times 10^{-3}, \quad \gamma_{xy} = 0$$

$$\kappa_x = 0, \quad \kappa_y = 0 \quad \text{and} \quad \kappa_{xy} = 7.67 \times 10^{-3} \ \text{mm}^{-1}$$

The difference in this case therefore is that we start to see some twisting of the laminate due to the non-symmetric nature of the lay-up.

(3.17) This is a non-symmetric laminate. The $\overline{Q}$ matrices are given in the text and the A, B and D matrices are determined from

$$A = \sum_{f=1}^{3} \overline{Q}_f \cdot (h_f - h_{f-1})$$

$$B = \frac{1}{2} \cdot \sum_{f=1}^{3} \overline{Q}_f \cdot [(h_f)^2 - (h_{f-1})^2] \qquad D = \frac{1}{3} \cdot \sum_{f=1}^{3} \overline{Q}_f \cdot [(h_f)^3 - (h_{f-1})^3]$$

$$h_0 = -0.15, \quad h_1 = -0.075, \quad h_2 = 0.075, \quad h_3 = 0.15$$

and a, b, β, d are determined from

$$\begin{bmatrix} a & b \\ \beta & d \end{bmatrix} = \begin{bmatrix} A & B \\ B & D \end{bmatrix}^{-1}$$

The strains and curvatures are then obtained from

$$\begin{bmatrix} \varepsilon_x \\ \varepsilon_y \\ \gamma_{xy} \\ \kappa_x \\ \kappa_y \\ \kappa_{xy} \end{bmatrix} = \begin{bmatrix} a_{11} & a_{12} & a_{16} & b_{11} & b_{12} & b_{16} \\ a_{21} & a_{22} & a_{26} & b_{21} & b_{22} & b_{26} \\ a_{61} & a_{62} & a_{66} & b_{61} & b_{62} & b_{66} \\ \beta_{11} & \beta_{12} & \beta_{16} & d_{11} & d_{12} & d_{16} \\ \beta_{21} & \beta_{22} & \beta_{26} & d_{21} & d_{22} & d_{26} \\ \beta_{61} & \beta_{62} & \beta_{66} & d_{61} & d_{62} & d_{66} \end{bmatrix} \begin{bmatrix} N_x \\ N_y \\ N_{xy} \\ M_x \\ M_y \\ M_{xy} \end{bmatrix} \quad \text{where } N_x = 30 \text{ N/mm}$$

from which

$$\varepsilon_x = 2.77 \times 10^{-3}, \quad \varepsilon_y = 2.55 \times 10^{-3}, \quad \gamma_{xy} = -1.026 \times 10^{-3}$$

$$\kappa_x = 0.015 \text{ mm}^{-1}, \quad \kappa_y = -0.021 \text{ mm}^{-1}, \quad \kappa_{xy} = 0.021 \text{ mm}^{-1}$$

The values in the relevant matrices are

$$A = \begin{bmatrix} 2.73 \times 10^4 & 8.06 \times 10^3 & 3.13 \times 10^3 \\ 8.06 \times 10^3 & 7.44 \times 10^3 & 1.48 \times 10^3 \\ 3.13 \times 10^3 & 1.48 \times 10^3 & 8 \times 10^3 \end{bmatrix}$$

$$a = \begin{bmatrix} 9.23 \times 10^{-5} & -8.51 \times 10^{-5} & -3.42 \times 10^{-5} \\ -8.51 \times 10^{-5} & 2.54 \times 10^{-4} & -3.94 \times 10^{-5} \\ -3.42 \times 10^{-5} & -3.94 \times 10^{-5} & 1.98 \times 10^{-4} \end{bmatrix}$$

$$B = \begin{pmatrix} -619.1 & 255.0 & -352.5 \\ 255.0 & 108.9 & -167.0 \\ -352.5 & -167.0 & 254.9 \end{pmatrix}$$

$$b = \begin{bmatrix} 5.07 \times 10^{-4} & -6.88 \times 10^4 & 6.84 \times 10^{-4} \\ -6.08 \times 10^{-4} & 3.85 \times 10^{-4} & 2.12 \times 10^{-4} \\ -3.23 \times 10^{-5} & 5.84 \times 10^{-4} & -1.31 \times 10^{-3} \end{bmatrix}$$

$$\beta = \begin{bmatrix} 5.07 \times 10^{-4} & -6.08 \times 10^{-4} & -3.23 \times 10^{-4} \\ -6.88 \times 10^{-4} & 3.85 \times 10^{-4} & 5.84 \times 10^{-4} \\ 6.84 \times 10^{-4} & 2.12 \times 10^{-4} & -1.31 \times 10^{-3} \end{bmatrix}$$

$$D = \begin{pmatrix} 236.0 & 47.7 & -29.3 \\ 47.7 & 50.3 & -13.9 \\ -29.3 & -13.9 & 47.2 \end{pmatrix}$$

$$d = \begin{bmatrix} 8.37 \times 10^{-3} & -8.00 \times 10^{-3} & 4.65 \times 10^{-3} \\ -8.00 \times 10^{-3} & 0.031 & -2.68 \times 10^{-3} \\ 4.65 \times 10^{-3} & -2.68 \times 10^{-3} & 0.036 \end{bmatrix}$$

(3.18) (a) Maximum Stress Criterion

The local stresses are given by

$$\begin{bmatrix} \sigma_1 \\ \sigma_2 \\ \tau_{12} \end{bmatrix} = T_\sigma \begin{bmatrix} \sigma_x \\ \sigma_y \\ \tau_{xy} \end{bmatrix}, \qquad \begin{bmatrix} \sigma_1 \\ \sigma_2 \\ \tau_{12} \end{bmatrix} = \begin{bmatrix} 61.6 \\ 38.4 \\ -29.4 \end{bmatrix} \text{ MN/m}^2$$

$$\frac{\hat{\sigma}_{1T}}{\sigma_1} = 20.1, \qquad \frac{\hat{\sigma}_{2T}}{\sigma_2} = 0.52, \qquad \frac{\hat{\tau}_{12}}{\tau_{12}} = 1.36$$

As the second term is less than 1, a tensile failure in the 2-direction would be expected.

(b) Maximum Strain Criterion

From the data given

$$\hat{\varepsilon}_{1T} = 2.49 \times 10^{-3}, \qquad \hat{\varepsilon}_{2T} = 3.777 \times 10^{3}, \qquad \hat{\varepsilon}_{1C} = 5.84 \times 10^{-5}$$

$$\hat{\varepsilon}_{2C} = 5.66 \times 10^{-3}, \qquad \hat{\gamma}_{12} = 7.14 \times 10^{-3}$$

The 1–2 strains are given by

$$\begin{bmatrix} \varepsilon_1 \\ \varepsilon_2 \\ \gamma_{12} \end{bmatrix} = S. \begin{bmatrix} \sigma_1 \\ \sigma_2 \\ \tau_{12} \end{bmatrix} \qquad \text{So,} \begin{bmatrix} \varepsilon_1 \\ \varepsilon_2 \\ \gamma_{12} \end{bmatrix} = \begin{bmatrix} 0.1 \\ 7.2 \\ -5.25 \end{bmatrix} 10^{-3}$$

$$\frac{\hat{\varepsilon}_{1T}}{\varepsilon_1} = 24.9, \qquad \frac{\hat{\varepsilon}_{2T}}{\varepsilon_2} = 0.5, \qquad \frac{\hat{\gamma}_{12}}{\gamma_{12}} = 1.36$$

Hence once again a failure is predicted in the 2-direction.

(c) Tsai–Hill Criterion

The Tsai–Hill equation gives

$$\left(\frac{\sigma_1}{\hat{\sigma}_{1T}} \right)^2 - \left(\frac{\sigma_1 \sigma_2}{\hat{\sigma}_{1T}} \right)^2 + \left(\frac{\sigma_2}{\hat{\sigma}_{2T}} \right)^2 + \left(\frac{\tau_{12}}{\hat{\tau}_{12}} \right) = 4.22$$

Therefore failure is predicted.

(3.19) (i) For an applied stress of $\sigma_x = 1$ MN/m^2, the stresses in the 1–2 direction are given by

$$\begin{bmatrix} \sigma_1 \\ \sigma_2 \\ \tau_{12} \end{bmatrix} = T_\sigma \begin{bmatrix} \sigma_x \\ \sigma_y \\ \tau_{xy} \end{bmatrix} \qquad \text{So} \begin{bmatrix} \sigma_1 \\ \sigma_2 \\ \tau_{12} \end{bmatrix} = \begin{bmatrix} 0.75 \\ 0.25 \\ -0.433 \end{bmatrix} \text{ MN/m}^2$$

$$\frac{\hat{\sigma}_{1T}}{\sigma_1} = 1867, \frac{\hat{\sigma}_{2T}}{\sigma_2} = 200, \frac{\hat{\tau}_{12}}{\tau_{12}} = 92.4$$

Hence an applied stress of $\sigma_x = 92$ MN/m^2 would initiate a shear failure before tensile or compressive failures occurred.

(ii) Maximum Strain Criterion

From the data given

$$\hat{\varepsilon}_{1T} = \frac{\hat{\sigma}_{1T}}{E_1} = 0.035, \, \hat{\varepsilon}_{2T} = \frac{\hat{\sigma}_{2T}}{E_2} = 5.56 \times 10^{-3}$$

$$\hat{\varepsilon}_{1C} = \frac{\hat{\sigma}_{1C}}{E_1} = 5.5 \times 10^{-3}, \qquad \hat{\varepsilon}_{2C} = \frac{\hat{\sigma}_{2C}}{E_2} = 0.011$$

$$\hat{\gamma}_{12} = \frac{\hat{\gamma}_{12}}{G_{12}} = 0.01$$

Also,

$$\begin{bmatrix} \varepsilon_1 \\ \varepsilon_2 \\ \gamma_{12} \end{bmatrix} = S \begin{bmatrix} \sigma_1 \\ \sigma_2 \\ \tau_{12} \end{bmatrix} \quad \text{So} \quad \begin{bmatrix} \varepsilon_1 \\ \varepsilon_2 \\ \gamma_{12} \end{bmatrix} = \begin{bmatrix} 1.68 \\ 2.19 \\ -10.8 \end{bmatrix} 10^{-5}$$

$$\frac{\hat{\varepsilon}_{1T}}{\varepsilon_1} = 2082, \qquad \frac{\hat{\varepsilon}_{2T}}{\varepsilon_2} = 253, \qquad \frac{\hat{\gamma}_{12}}{\gamma_{12}} = 92.4$$

Hence once again a stress of $\sigma_x = 92$ MN/m^2 is the limiting condition.

(iii) Tsai–Hill Criterion

Letting X be the multiplier for σ_x

$$\left(\frac{X \cdot \sigma_1}{\hat{\sigma}_{1T}} \right)^2 - \left(\frac{X^2 \cdot \sigma_1 \sigma_2}{\hat{\sigma}_{1T}} \right)^2 + \left(\frac{X\sigma_2}{\sigma_{2T}} \right)^2 + \left(\frac{X\tau_{12}}{\hat{\tau}_{12}} \right)^2 = 1$$

Solving for X gives $X = 83.8$.

Hence the limiting condition is $\sigma_x = 83$ MN/m^2.

(3.20) Let $N_x = 1.2$ N/mm (ie $\sigma_x = 1$ N/mm^2).

By calculating the matrix $[a]$ we get (for 30° ply).

$$\begin{bmatrix} \varepsilon_x \\ \varepsilon_y \\ \gamma_{xy} \end{bmatrix} = a \begin{bmatrix} N_x \\ N_y \\ N_{xy} \end{bmatrix} = \begin{bmatrix} 1.235 \\ -1.248 \\ 0 \end{bmatrix} \times 10^{-5}$$

Then

$$\begin{bmatrix} \varepsilon_1 \\ \varepsilon_2 \\ \gamma_{12} \end{bmatrix} = T_\varepsilon \begin{bmatrix} \varepsilon_x \\ \varepsilon_y \\ \gamma_{xy} \end{bmatrix} = \begin{bmatrix} 6.14 \\ -6.27 \\ -21.5 \end{bmatrix} \times 10^{-6}$$

$$\frac{\hat{\varepsilon}_{1T}}{\varepsilon_1} = 2379, \qquad \frac{\varepsilon_{2c}}{\varepsilon_2} = 3624, \qquad \frac{\hat{\gamma}_{12}}{\gamma_{12}} = 606$$

Hence the limiting condition is $\sigma_x = 606$ MN/m^2.

Also,

$$\begin{bmatrix} \sigma_x \\ \sigma_y \\ \tau_{xy} \end{bmatrix} = \overline{Q}_f \begin{bmatrix} \varepsilon_x \\ \varepsilon_y \\ \gamma_{xy} \end{bmatrix} \quad \text{and} \quad \begin{bmatrix} \sigma_1 \\ \sigma_2 \\ \tau_{12} \end{bmatrix} = T_\sigma \begin{bmatrix} \sigma_x \\ \sigma_y \\ \tau_{xy} \end{bmatrix}$$

which gives

$$\begin{bmatrix} \sigma_1 \\ \sigma_2 \\ \tau_{12} \end{bmatrix} = \begin{bmatrix} 0.787 \\ -0.039 \\ -0.148 \end{bmatrix} \text{ MN/m}^2$$

$$\frac{\hat{\sigma}_{IT}}{\sigma_1} = 2415, \qquad \frac{\hat{\sigma}_{2C}}{\sigma_2} = 5102, \qquad \frac{\hat{\tau}_{12}}{\tau_{12}} = 606$$

Hence the Maximum Stress Criterion also predicts $\sigma_x = 606$ MN/m^2 as the limiting condition.

For the Tsai–Hill Criterion, letting X = multiplier on stress σ_x, then solving

$$\left(\frac{X \cdot \sigma_1}{\hat{\sigma}_{IT}}\right)^2 - \left(\frac{X^2 \cdot \sigma_1 \sigma_2}{\hat{\sigma}_{IT}}\right)^2 + \left(\frac{X\sigma_2}{\sigma_{2T}}\right)^2 + \left(\frac{X\tau_{12}}{\hat{\tau}_{12}}\right)^2 = 1$$

gives $X = 583$.

This predicts a stress of $\sigma_x = 583$ MN/m^2.

If this is repeated for all the plies then we get

Table 1

Maximum permissible values of σ_x, to cause failure in each ply

Ply	Max. stress or Max. strain	Tsai–Hill
$\theta = 0°$	$\sigma_x = 1201$ MN/m^2	$\sigma_x = 1067$
$\theta = -30°$	$\sigma_x = 606$ MN/m^2	$\sigma_x = 583$
$\theta = 30°$	$\sigma_x = 606$ MN/m^2	$\sigma_x = 583$

The limiting condition is thus $\sigma_x = 583$ MN/m^2 as predicted by Tsai–Hill.

(3.21) $$K_c = \sigma(\pi a_c)^{1/2} \left(\frac{W}{\pi a} \tan\left(\frac{\pi a}{W}\right)\right)^{1/2}$$

$$= \frac{P}{Wb}(\pi a_c)^{1/2} \left(\frac{W}{\pi a} \tan\left(\frac{\pi a}{W}\right)\right)^{1/2}$$

$$43 = \frac{P}{30 \times 5}\left(\pi \times 5 \times 10^{-3}\right)^{1/2}(1.05)$$

$$P = 49 \text{ kN}$$

(3.22) CSM: $$K_c = \sigma(\pi a_c)^{1/2} \rightarrow a_c = \left(\frac{13.5}{80}\right)^2 \frac{1}{\pi} = 9.06 \times 10^{-3} \text{ m}$$

$$N = \frac{2((9.06 \times 10^{-3})^{-5.35} - (1000 \times 10^{-6})^{-5.35})}{3.3 \times 10^{-18}(80)^{12.7}\pi^{6.35}(-10.7)}$$

$$= 3 \times 10^5 \text{ cycles}$$

Woven Roving: $a_c = 34.9 \times 10^{-3}$, so in a similar way

$N = 1.14 \times 10^6$ cycles

(4.1) $Q_D = 1/2\pi^2 D^2 N H \sin\phi \cos\phi$

$= 1/2\pi^2 (50 \times 10^{-3})^2 (100/60)(2.4 \times 10^{-3})\sin 17.7 \cos 17.7$

$= 1.43 \times 10^{-5} \ \text{m}^3/\text{s}$

$$Q_P = \frac{\pi D H^3 \sin^2\phi}{12\eta}\left(\frac{P}{L}\right)$$

$$= \frac{\pi\left(50 \times 10^{-3}\right)\left(2.4 \times 10^{-3}\right)^3 \sin^2 17.7}{12(200)}\left(\frac{20 \times 10^6}{1}\right)$$

$= 1.67 \times 10^{-6} \ \text{m}^3/\text{s}$

Total flow $= Q_D - Q_P = 1.26 \times 10^{-5} \ \text{m}^3/\text{s}$

The shear rate in the metering zone will be given by

$$\dot\gamma = \frac{V_d}{H} = \frac{\pi D N \cos\phi}{H} = \frac{\pi\left(50 \times 10^{-3}\right)(100/60)\cos 17.7}{2.4 \times 10^{-3}}$$

$\dot\gamma = 104 \ \text{s}^{-1}$

(4.2) For the die, $Q = \dfrac{\pi P R^4}{8L\eta} = KP$

where $K = \dfrac{\pi R^4}{8L\eta} = \dfrac{\pi(1.5)^4 10^{-9}}{8 \times 40 \times 200 \times 10^{-6}} = 2.485 \times 10^{-7}\left(\dfrac{\text{m}^s/\text{s}}{\text{MN/m}^2}\right)$

for the extruder, $P_{max} = \dfrac{6\pi\eta LDN}{H^2 \tan\phi}$

$$P_{max} = \frac{6\pi \times 200 \times 10^{-6} \times 1000 \times 50 \times 100}{60 \times (2.4)^2 \times 0.3191} = 171 \ \text{MN/m}^2$$

$Q_{max} = 1/2\pi^2 D^2 N H \sin\phi \cos\phi = 1.436 \times 10^{-5} \ \text{m}^3/\text{s}$

From graph, operating point is

$P = 43 \ \text{MN/m}^2, Q = 1.075 \ \text{m}^3/\text{s}$

New operating point is

$P = 85 \ \text{MN/m}^2, Q = 1.075 \ \text{m}^3/\text{s}$

(4.3) Extruder: $Q_{max} = 1/2\pi^2 D^2 N H \sin\phi \cos\phi$

$$= 1/2\pi^2 (25)^2 \left(\frac{100}{60}\right)(2)\sin 17°42' \cos 17°42'$$

$= 2.98 \times 10^{-6} \ \text{m}^3/\text{s}$

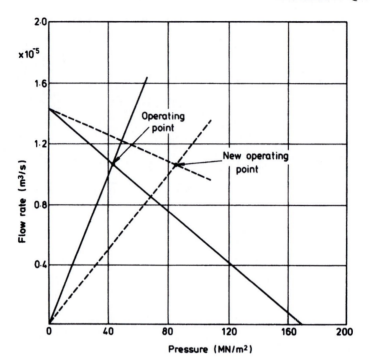

$$P_{max} = \frac{6\pi DLN\eta}{H^2 \tan \phi}$$

$$= \frac{6\pi(25)(500)\left(\dfrac{100}{60}\right)(400 \times 10^{-6})}{2^2 \times \tan 17°42'}$$

$$= 123 \text{ MN/m}^2$$

Die:
$$Q = 2\left\{\frac{\pi R^4}{8\eta L_d}\right\} P$$

$$= 2\left\{\frac{\pi(0.75)^4}{8 \times 400 \times 10^{-6} \times 10}\right\} P$$

Hence the characteristics shown may be drawn.
The operating point is at the intersection of the two characteristics.

$$\text{Output, } Q = 2.13 \times 10^{-6} \text{ m}^3/\text{s}$$

(4.4)
$$Q = 1/2\pi D^2 NH \sin\phi \cos\phi - \pi DH^3 \sin^2\phi \left(\frac{P}{12\eta L}\right)$$

but
$$Q = \frac{K_1 P}{\eta}$$

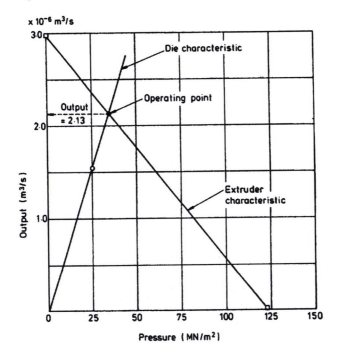

So $$Q = 1/2\pi D^2 NH \sin\phi \cos\phi - \frac{\pi D H^3 \sin^2\phi}{12\eta} \cdot \frac{Q\eta}{K_1 L}$$

$$Q = \frac{AH \sin\phi \cos\phi}{1 + BH^3 \sin^2\phi}$$

where $$A = 1/2\pi^2 D^2 N \text{ and } B = \frac{\pi D}{12LK_1}$$

So

$$\frac{dQ}{d\phi} = \frac{(1 + BH^3 \sin^2\phi)AH(2\cos^2\phi - 1) - AH\sin\phi\cos\phi(2H^3 B \sin\phi\cos\phi)}{(1 + BH^3 \sin^2\phi)} = 0$$

$$(1 + BH^3 \sin^2\phi)(2\cos^2\phi - 1) = 2H^3 B \sin^2\phi\cos^2\phi \tag{1}$$

Also $$\frac{dQ}{dH} = \frac{(1 + BH^3 \sin^2\phi)A\sin\phi\cos\phi - AH\sin\phi\cos\phi(3BH^2 \sin^2\phi)}{(1 + BH^3 \sin^2\phi)} = 0$$

So $$(1 + BH^3 \sin^2\phi) = 3BH^3 \sin^2\phi \tag{2}$$

Using (2) in (1)

$$(3BH^3 \sin^2\phi)(2\cos^2\phi - 1) = 2H^3 B \sin^2\phi\cos^2\phi$$

thus $$(3BH^3 \sin^2\phi)(\cos 2\phi) = 2H^3 B \sin^2\phi\cos^2\phi$$

$$2\cos 2\phi = 2\cos^2\phi \rightarrow 3(1 - 2\sin^2\phi) = 2(1 - \sin^2\phi)$$

which can be solved to give $\phi = 30°$

Also
$$[1 + BH^3 \sin^2 \phi] = 3BH^3 \sin^2 \phi$$

$$H^3 = \frac{1}{2B \sin^2 \phi} = \frac{2}{B}$$

$$H = \left[\frac{24K_1L}{\pi D}\right]^{1/3}$$

(4.5)
$$F = \pi R^2 P_0 \left(\frac{m}{m+1}\right) = \pi(0.25)^2(50 \times 10^6)\left(\frac{0.6}{2.6}\right)$$

$$= 2.26 \text{ MN}$$

(4.6)

$AB = 100$ mm

$$\text{Flow ratio} = \frac{CBA}{3} = \frac{150}{3} = 50$$

$$\text{Clamp force} = \pi R^2 P_0 \left(\frac{m}{m+2}\right)$$

$$= \pi(110)^2 140 \left(\frac{0.5}{2.5}\right)$$

$$= 1.06 \text{ MN}$$

(4.7)

(i)
$$\frac{c.s.a.}{s.a.} = \frac{1/4 \pi D^2}{\pi D} = 0.25D$$

(ii)

$DC = D \tan 10°$, $BC = D/\cos 10°$

$$c.s.a. = D^2 + D^2 \tan 10° = D^2(1 + \tan 10°)$$

$$s.a. = 2D + 2D/\cos 10 + 2D \tan 10°$$

$$\frac{c.s.a.}{s.a.} = \frac{D^2(1 + \tan 10°)}{2D(1 + (\cos 10°)^{-1} + \tan 10°)}$$

$$= 0.268 D$$

(iii)

$FG = (D/4) \tan 30°$

$$c.s.a. = D^2(1.5)/4 \tan 30°$$

$$s.a. = 6(D/2) = 3D$$

$$\frac{c.s.a.}{s.a.} = \frac{D}{8 \tan 30} = 0.216 D$$

(iv)

$$c.s.a. = D^2/2$$

$$s.a. = 2D + D = 3D$$

$$\frac{c.s.a.}{s.a.} = \frac{D^2}{6D} = 0.167\,D$$

(4.8)
$$V_d = 2\pi RN = \frac{2\pi \times 0.15 \times 5}{60} = 0.078 \text{ m/s}$$

$$Q = 2000 \text{ kg/hour} = 0.396 \times 10^{-3} \text{ m}^3/\text{s}$$

but
$$Q = HWV_d \text{ So } H = \frac{0.396 \times 10^{-3}}{1 \times 0.078} = 5.04 \text{ mm}$$

$$x = \sqrt{(5.04 - 4.5)150} = 9 \text{ mm}$$

Now
$$w = 1.786 \text{ (ie } x/H)$$

So
$$P_{max} = \frac{3 \times 1.5 \times 10^4 \times 0.078}{4.5 \times 10^{-3}}(3.57 - 0.64(1.786 + 5.77(0.33)))$$

$$= 0.93 \text{ MN/m}^2$$

(4.9) For
$$H_0 = 0.8, x = \sqrt{(H - H_0)R} = 10.9 \text{ mm}, w = 5.45$$

$$P_{max} = 24.3 \text{ MN/m}^2 \text{ (same equation as above)}$$

For
$$H_0 = 1.2 \text{ mm}, P_{max} = 4.9 \text{ MN/m}^2$$

For
$$H_0 = 1.6 \text{ mm}, P_{max} = 0.86 \text{ MN/m}^2$$

For
$$H_0 = 1.9 \text{ mm}, P_{max} = 0.073 \text{ MN/m}^2$$

(4.10) Initial volume of sheet $= \left(\dfrac{\pi D^2}{4}\right)4 = \pi D^2$

Since the dome formed is a free surface it may be assumed to have a constant thickness, t

$$\text{Final volume} = \left(\frac{\pi D^2}{2}\right)t$$

$$\frac{\pi D^2}{2}t = \pi D^2 \rightarrow t = 2 \text{ mm}$$

(4.11)

$$\alpha = \tan^{-1}\left(\frac{60}{10}\right) = 80.54°$$

$$L = 35 \text{ um}, t_0 = 3 \text{ mm}$$

$$H = 30 \tan \beta = 180$$

$$t_{AA} = t_0\left(\frac{1 + \cos \alpha}{2}\right)\left(\frac{H - L}{H}\right)^{\sec \alpha - 1}$$

$$= 0.58 \text{ mm}$$

When sheet just touches the base

$$S_c = (D(1 + \cos \alpha)/ \sin \alpha) - R = 40.82$$

So $L_c = 40.27$ mm and thus $t_B = 0.48$ mm

(4.12) (a) $B_{SH} = 1.8$ So Parison thickness $= 1.8 \times 2 = 3.6$ mm

$B_{ST} = \sqrt{1.8} = 1.34,$ Parison diameter $= 38 \times 1.34 = 50.98$ mm

(b) $h = B_{ST}^3 \, h_d \left(\dfrac{D_d}{D_m} \right) = (1.34)^3 \times 2 \times \left(\dfrac{38}{70} \right) = 2.62$

(c) $P = \dfrac{2\sigma h}{D_m} = \dfrac{2 \times 10 \times 2.62}{70} = 0.75$ MN/m^2

(4.13) $\dfrac{O_{MD}}{O_{TD}} = \dfrac{\text{die gap}}{\text{film thickness} \times (\text{blow-up})^2} = 2$

So $\text{blow-up} = \sqrt{\dfrac{1}{0.1 \times 2}} = 2.24$

$D_{\text{bubble}} = 2.24 \times 99 = 221.76$ mm

$\text{Lay flat width} = \dfrac{\pi D_b}{2} = 348.3$ mm

(4.14)
$$\Delta P = \frac{6\eta VL(2h - H)}{H^3}$$

$$= \frac{6 \times 100 \times 0.5 \times 6 \times 10^{-2}(0.6 - 0.5)10^{-3}}{(0.5)^3 \times 10^{-9}}$$

$$= 14.4 \text{ MN/m}^2$$

(4.15)
$$\left(\frac{T_0 - T}{T_0 - T_i}\right) = \exp\left(\frac{-h\beta t}{\rho C_p}\right) = \exp\left(\frac{-h\alpha t}{Hk}\right)$$

where $\alpha = k/\rho C_p$ = thermal diffusivity and H = wall thickness = $1/\beta$

$$\frac{300 - 250}{300 - 23} = \exp\left(\frac{-28.4 \times 8.6 \times 10^{-5}t}{3 \times 10^{-3} \times 230}\right)$$

$$t = 484.4 \text{ s} = 8.07 \text{ minutes}$$

(4.16)
$$\text{volume} = \frac{\text{mass}}{\rho} = \frac{150 \times 10^{-3}}{1200} \text{ m}^3$$

So,
$$\pi R^2 H = \frac{150 \times 10^{-3} \times 10^9}{1200}$$

$$H = \frac{150 \times 10^6}{\pi \times 1200 \times (150)^2} = 1.77 \text{ mm}$$

Also
$$t = \frac{3\eta V^2}{8\pi F H^4} = \frac{3 \times 10^4 \times (150 \times 10^6)^2}{8\pi \times 10^6 \times 100 \times 10^3 \times 1.77^4 \times (1200)^2} = 19 \text{ s}$$

(5.1)
$$\tau = \frac{3T}{2\pi R^3} = \frac{3}{2}\frac{WR}{\pi R^3} = \frac{3W}{2\pi R^2} = 7.49 \text{ W}$$

Also
$$\dot{\gamma} = \frac{\dot{\theta}}{\alpha} = \frac{\theta}{40 \times 5} = \frac{\theta}{200}$$

W(g)	$\tau(\text{N/m}^2)$	θ	$\dot{\gamma}(\text{s}^{-1})$	$\eta = \tau/\dot{\gamma}$
50	374.5	0.57	2.85×10^{-3}	1.31×10^5
100	749	1.25	6.25×10^{-3}	1.2×10^5
200	1498	2.56	1.28×10^{-2}	1.17×10^5
500	3745	7.36	3.68×10^{-2}	1.02×10^5
1000	7490	17.0	8.5×10^{-2}	8.8×10^4
2000	14980	42.0	0.21	7.13×10^4

A graph of η vs. τ gives

$$\tau = 1 \times 10^4 \text{ N/m}^2 \text{ and } \eta = 8 \times 10^4 \text{ Ns/m}^2$$

(5.2) For a Power Law fluid

$$\dot{\gamma} \propto (\tau)^{1/n}$$

So,

$$Q \propto (\Delta P)^{1/n}$$

or

$$\log Q \propto 1/n \log \Delta P$$

So the slope of a graph of $(\log Q)$ against $(\log \Delta P)$ will be a straight line of slope $1/n$ if the Power Law is obeyed. From the attached graph $1/n = 1.268$ so $n = 0.7886$

Consider the point, $Q = 0.8$ kg/min and $\Delta P = 5.2$ MN/m².

$$\tau = \frac{H(\Delta P)}{2L} = \frac{2 \times 10^{-3} \times 5.2 \times 10^6}{2 \times 50 \times 10^{-3}} = 1.04 \times 10^5 \text{ N/m}^2$$

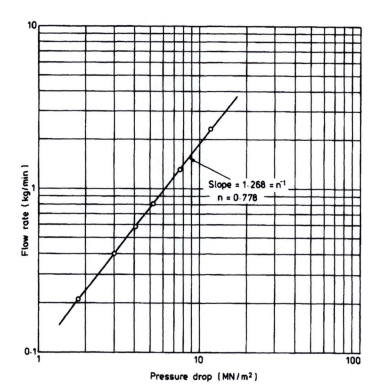

Slope = 1·268 = n^{-1}
n = 0·778

Flow rate (kg/min)

Pressure drop (MN / m²)

Now $\dot{\gamma} = \left(\dfrac{2n+1}{3n}\right)\dfrac{6Q}{TH^2} = \left(\dfrac{2(0.7886)+1}{3(0.7886)}\right)\left(\dfrac{6 \times 0.8 \times 10^6}{60 \times 0.94 \times 6 \times 2^2}\right) = 3863 \text{ s}^{-1}$

but

$$\tau = \tau_0 \dot{\gamma}^n$$

So

$$\tau_0 = \frac{1.04 \times 10^5}{(3863)^{0.7886}} = 154.3$$

(5.3) Cone and Plate

True shear rate, $\dot{\gamma} = \dfrac{\dot{\theta}}{\alpha} = \dfrac{0.062 \times 180}{3\pi} = 1.18 \text{ s}^{-1}$

Shear Stress, $\tau = \dfrac{3T}{2\pi R^3} = \dfrac{3 \times 1.18}{2\pi \times (37.5 \times 10^{-3})^3} = 10.7 \text{ kN/m}^2$

Ram Extruder

$Q = 70 \times 10^9 \begin{cases} \dot{\gamma} = \dfrac{4Q}{\pi R^3} = \dfrac{4 \times 70 \times 10^{-9}}{\pi \times 10^{-9}} = 89.1 \text{ s}^{-1} \\[2mm] \tau = \dfrac{PR}{2L} = \dfrac{3.9 \times 10^6 \times 1}{2 \times 32} = 60.9 \text{ kN/m}^2 \end{cases}$

$Q = 200 \times 10^{-9} \begin{cases} \dot{\gamma} = \dfrac{4 \times 200 \times 10^{-9}}{\pi \times 10^{-9}} = 254.6 \text{ s}^{-1} \\[2mm] \tau = \dfrac{5.7 \times 10^6}{2 \times 32} = 89.1 \text{ kN/m}^2 \end{cases}$

but $\tau = \tau_0 \dot{\gamma}^n$

$$\left(\frac{60.9}{89.1}\right) = \left(\frac{89.1}{254.6}\right)^n$$

So $n = 0.362$ and $\tau_0 = 11.98$

Now the cone and plate gives true shear rate whereas the ram extruder uses apparent shear rate. The Non-Newtonian correction factor is

$$\left(\frac{4n}{3n+1}\right) = \left(\frac{4 \times 0.362}{3(0.362)+1}\right) = 0.69$$

Therefore the true shear rate on the cone and plate is equivalent to a shear rate of $0.69(1.18) = 0.817$ on the ram extruder

At $\qquad \dot{\gamma} = 0.817 \quad \tau = 11.98(0.817)^{0.362} = 11.13 \text{ kN/m}^2$

$$\% \text{ difference} = \frac{11.13 - 10.7}{10.7} = 4.06\%$$

(5.4) If the material is assumed Newtonian and the shear rates are equated then

$$\frac{6Q_2}{TH^2} = \frac{4Q_1}{\pi R^3}$$

$$Q_2 = \frac{4TH^2 Q_1}{6\pi R^3} \qquad (1)$$

If the material is Non-Newtonian then the true shear rates should be equated

$$\left(\frac{2n+1}{3n}\right)\frac{6Q_2}{TH^2} = \left(\frac{3n+1}{4n}\right)\frac{4Q_1}{\pi R^3}$$

$$Q_2 = \left(\frac{3n+1}{4n}\right)\left(\frac{3}{2n+1}\right)\frac{4TH^2 Q_1}{6\pi R^3} \qquad (2)$$

So for $n = 0.37$ correction factor $= 0.9094$

Error $= 1 - 0.9094 = 0.0906 \rightarrow 9.06\%$

Now $\dot{\gamma} \propto Q$ and $\tau \propto \dot{\gamma}^n \propto Q^n$

and $P \propto \tau$ then $P \propto Q^n$

So Correction Factor $= \left(\left(\dfrac{3n+1}{4n} \right) \left(\dfrac{3n}{2n+1} \right) \right)^n$

for $n = 0.37$ Factor $= 0.9655$

Error $= 1 - 0.9655 = 0.0345 \rightarrow 3.45\%$

(5.5) $\dot{\gamma} = \dfrac{4Q}{\pi R^3} = \dfrac{4 \times 2.12 \times 10^{-6}}{\pi (3 \times 10^{-3})^3} = 100 \ s^{-1}$

From flow curves $\eta = 1.3 \times 10^3 \ Ns/m^2, G = 3 \times 10^4 \ N/m^2$

So Natural time $= \dfrac{\eta}{G} = \dfrac{1.3 \times 10^3}{3 \times 10^4} = 4.3 \times 10^{-2} \ s$

Dwell time $= \dfrac{V}{Q} = \dfrac{\pi (3 \times 10^{-3})^2 \times 37 \times 10^{-3}}{2.12 \times 10^{-6}} = 0.5 \ s$

So flow is mostly plastic.

(5.6) $\dot{\gamma} = \dfrac{6Q}{TH^2} = \dfrac{6 \times 0.75 \times 10^{-9}}{(5 \times 10^{-3})(1 \times 10^{-3})^2} = 0.9 \ s$

From flow curves $\eta = 2 \times 10^4 \ Ns/m^2, G = 1 \times 10^4 \ N/m^2$

So Natural time $= \dfrac{\eta}{G} = 2 \ seconds$

Dwell time $= 1$ second so flow is predominantly elastic.

(5.7) $Q = AV = \dfrac{\pi}{4} \times 25^2 \times 10^{-6} \times 50 \times 10^{-3} = 24.54 \times 10^{-6} \ m^3/s$

$\dot{\gamma} = \dfrac{6Q}{TH^2} = \dfrac{6 \times 24.54 \times 10^{-6}}{\pi \times 25 \times 10^{-3} \times 1 \times 10^{-6}} = 1953 \ s^{-1}$

From flow curves at $\dot{\gamma} = 1953 \ s^{-1}, \tau = 3.2 \times 10^5 \ N/m^2$.

$\tau = \dfrac{PH}{2L} \rightarrow P = \dfrac{3.2 \times 10^5 \times 2 \times 30}{1} = 19.2 \ MN/m^2$

(5.8) Flow rate = 3 m/min

$$\text{Area} = 2\{(15 + 15) + \pi(8.3)\} = 112 \times 10^{-6} \text{ m}^2$$

$$Q = \frac{3 \times 112 \times 10^{-6}}{60} = 5.6 \times 10^{-6} \text{m}^3/\text{s}$$

$$\dot{\gamma} = \frac{6Q}{TH^2} = \frac{6 \times 5.6 \times 10^{-6}}{(30 + 26) \times 2^2 \times 10^{-9}}$$

$$= 150 \text{ s}^{-1}$$

from flow curves, at $\dot{\gamma} = 150 \text{ s}^{-1}$, $\tau = 1.5 \times 10^5 \text{ N/m}^2$

$$\Delta P = \frac{2L\tau}{H} = \frac{2 \times 10 \times 1.5 \times 10^5}{2} = 1.5 \text{ MN/m}^2$$

$$\text{Power} = PQ$$

$$\text{Power} = 1.5 \times 10^6 \times 5.6 \times 10^{-6}$$

$$= 8.4 \text{ Nm/s } (= 8.4 \text{ W})$$

The temperature rise ΔT is then given by

$$\Delta T = \frac{\Delta P}{\rho C_p} = \frac{1.5 \times 10^6}{3.3 \times 10^6} = 0.45°C$$

(5.9) Flow rate = 3 m/min

$$\text{Area} = (30 \times 2) + 2(13 \times 2) = 112 \times 10^{-6} \text{ m}^2$$

$$Q = \frac{3 \times 112 \times 10^{-6}}{60} = 5.6 = 10^{-6} \text{ m}^3/\text{s}$$

$$\dot{\gamma} = \frac{6Q}{TH^2} = \frac{6 \times 5.6 \times 10^{-6}}{(30 + 26) \times 2^2 \times 10^{-9}} = 150 \text{ s}^{-1}$$

from flow curves, at $\dot{\gamma} = 150 \text{ s}^{-1}$, $\tau = 1.5 \times 10^5 \text{ N/m}^2$

$$\Delta P = \frac{2L\tau}{H} = \frac{2 \times 10 \times 1.5 \times 10^5}{2} = 1.5 \text{ MN/m}^2$$

$$\text{Power} = PQ$$

$$= 1.5 \times 10^6 \times 5.6 \times 10^{-6}$$

$$= 8.4 \text{ Nm/s} = 8.4 \text{ W}$$

From flow curves, at $\dot{\gamma} = 150$

$$G = 3.05 \times 10^4 \text{ N/m}^2$$

So, $$\gamma_R = \tau/G = 4.92$$

From graphs for swell ratio $B_{ST} = 1.41$, $B_{SH} = 1.99$

Thus approximate dimensions of extrudate would be 42.3 mm deep × 21.2 mm wide × 3.98 mm thick.

(5.10)

$$\dot{\gamma} = \frac{6Q}{TH^2}$$

$$= \frac{6 \times 46 \times 10^{-3}}{(\pi \times 30)(1.5 \times 10^{-3})^2}$$

$$= 1.3 \times 10^3 \text{ s}^{-1}$$

So from the flow curves, at $\dot{\gamma} = 1300 \text{ s}^{-1}$

$$\tau = 2.6 \times 10^5 \text{ Nm}^{-2}$$

$$G = 5 \times 10^4 \text{ Nm}^{-2}$$

$$\gamma_R = \frac{\tau}{G} = 5.2$$

From swell ratio curves

at $\gamma_R = 5.2$, $B_{ST} = 1.41$, $B_{SH} = 2$

$$h = B_{ST}^2 hd \left(\frac{D_d}{D_m} \right)$$

$$h = (1.41)^2 (1.5) \left(\frac{30}{60} \right) = 1.5 \text{ mm}$$

(5.11) From the flow curves, $\eta_0 = 3 \times 10^4$ and $n = 0.33$

$$L = \left(\frac{n+1}{2n+1} \right)^{n/(n+1)} \left(\frac{P}{\eta_0} \right)^{1/(n+1)} \left(\frac{H}{2} \right)^{n/(n+1)}$$

$$L = \left(\frac{1.33}{1.66} \right)^{0.33/1.33} \left(\frac{100 \times 10^6}{3 \times 10^4} \right)^{1/1.33} \left(\frac{1}{2} \right)^{0.33/1.33} = 355 \text{ mm}$$

(5.12) $$L = \left(\frac{P}{\eta_0} \right)^{1/(n+1)} \left(\frac{H}{2} \right) 6t_f \left(\frac{n}{4n+1} \right) \left(\frac{n}{3n+1} \right) \left(\frac{n(n+1)}{(2n+1)^2} \right)^{n/(n+1)}$$

$$\Delta T = \frac{T_3 - T_2}{T_1 - T_2} = \frac{128 - 60}{170 - 60} = 0.62 \rightarrow F_0 = 0.295 = \frac{\alpha t_f}{x^2}$$

So $\quad t_f = \dfrac{0.295 \times 0.5^2 \times 10^{-6}}{10^{-7}} = 0.74$ seconds

$$L = \left(\frac{100 \times 10^6}{3 \times 10^4}\right)^{\frac{1}{1.33}} \left(\frac{1}{2}\right) \left(\frac{6 \times 0.74 \times 0.33^3 \times 1.33}{2.32 \times 1.99 \times 1.66^2}\right)^{\frac{0.33}{1.33}} = 80.6 \text{ mm}$$

For a Newtonian fluid, $n = 1$

$$L = 0.13 \left(\frac{P t_f}{\eta}\right)^{1/2} H = 0.13 \left(\frac{100 \times 10^6 \times 0.74}{1.2 \times 10^2}\right)^{1/2} = 101.9 \text{ mm}$$

(5.13) $$\Delta T = \frac{120 - 50}{230 - 50} = 0.39$$

So, $$F_0 = 0.48 = \frac{\alpha t}{x^2} \text{ where } x = 1 \text{ mm}$$

$$t = \frac{0.48 \times 1 \times 10^{-6}}{1 \times 10^{-7}} = 4.8 \text{ seconds}$$

(5.14) $$\Delta T = \frac{T_3 - T_2}{T_1 - T_2} = \frac{90 - 30}{230 - 30} = 0.3$$

$$F_0 = 0.58 = \frac{\alpha t}{x^2}$$

$$t = \frac{0.58 \times (2 \times 10^{-3})^2}{1 \times 10^{-7}} = 23.2 \text{ s}$$

for the runner, ΔT also equals 0.3 so $F_0 = 0.29$. To cause freeze-off in 23.2 seconds then

$$0.29 = \frac{1 \times 10^{-7} \times 23.2}{R^2}$$

$R = 2.83$ mm $\qquad$ runner diameter $= 5.66$ mm

(5.15) Since the power law equations are going to be used rather than the flow curves, it is necessary to use the true shear rates.

ie $\quad \dot{\gamma} = \left(\dfrac{3n + 1}{4n}\right) \dfrac{4Q}{\pi R^3} = \left(\dfrac{3(0.35) + 1}{4(0.35)}\right) \left(\dfrac{4 \times 5 \times 10^{-5}}{\pi \times 1.5^3 \times 10^{-9}}\right)$

$\qquad = 2.76 \times 10^4 \text{ s}^{-1}$

Now $\tau = \tau_0 \dot{\gamma}^n = 40 \times 10^3 \left(2.76 \times 10^4\right)^{0.35} = 1.43 \text{ MN/m}^2$

Now $\tau = \dfrac{PR}{2L}$ $\quad$ So $\quad P = \dfrac{2L\tau}{R} = \dfrac{2 \times 25 \times 1.43}{1.5} = 47.8 \text{ MN/m}^2$

(5.16) Volume flow rate,

$$Q = \frac{\pi R^2 H}{t} = \frac{\pi \left(60 \times 10^{-3}\right)^2 \left(3 \times 10^{-3}\right)}{1} = 3.39 \times 10^{-5} \text{ m}^3/\text{s}$$

From Fig. 5.3, $\eta_0 = 3 \times 10^4$ Ns/m^2 (ie η at $\dot{\gamma} = 1$ s^{-1}) and $n = 0.33$.
So from illustrative example (5.3)

$$P = Q^n \left(\frac{2n+1}{2n\pi} \right)^n \frac{\eta_0(2)^{n+1}R^{1-n}}{(H)^{2n+1}(1-n)}$$

$$= 1.05 \text{ MN/m}^2$$

(5.17) $$\Delta T = \frac{114-50}{170-50} = 0.531 \rightarrow F_0 = 0.35 = \frac{\alpha t}{x}$$

$$x = \sqrt{\frac{1 \times 10^{-7} \times 3}{0.35}} = 0.926 \text{ mm}$$

gate depth $= 2x = 1.85$ mm

Now $$\dot{\gamma} = \frac{6Q}{TH^2} \rightarrow T = \frac{6Q}{\gamma H^2} = \frac{6 \times 4 \times 10^{-6} \times 10^9}{1000 \times 3 \times 1(1.85)^3} = 1.26 \text{ mm}$$

From Fig. 5.3 at $\dot{\gamma} = 1000$ s^{-1}, $\tau = 2.9 \times 10^5$ N/m^2

$$\tau = \frac{PH}{2L} \text{ so } P = \frac{2L\tau}{H} = \frac{2 \times 0.6 \times 2.9 \times 10^5}{1.85} = 0.19 \text{ MN/m}^2$$

(5.18)

$$H_0 = H_1 + 2L\tan\alpha$$
$$H_0 = 0.5 + 30\tan 3° = 2.07 \text{ mm}$$
$$T = \pi D = 40\pi = 126 \text{ mm}$$

$$\dot{\gamma}_1 = \frac{6Q}{TH^2} = \frac{6 \times 10 \times 10^{-6}}{126 \times 10^{-3}(0.5 \times 10^{-3})^2} = 1910 \text{ s}^{-1}$$

$$\dot{\gamma}_3 = 1910\left(\frac{0.5}{2.07}\right)^2 = 111 \text{ s}^{-1}, \quad \dot{\varepsilon}_3 = \frac{1}{3}(111)\tan 3° = 2 \text{ s}^{-1} \rightarrow \sigma_3 = 0.45 \text{ MN/m}^2$$

$$\dot{\varepsilon}_2 = \frac{1}{3}\dot{\gamma}_2 \tan\alpha = 1/3(1910)\tan 3° = 33 \text{ s}^{-1} \rightarrow \sigma_2 = 7.5 \text{ MN/m}^2$$

For the tapered section,

$$P_0 = \frac{4\dot{\gamma}(\eta\lambda)^{1/2}}{(3n+1)} = \frac{4 \times 111(1150 \times 2.25 \times 10^5)^{1/2}}{1.99}$$

$$= 3.6 \text{ MN/m}^2$$

$$P_s = \frac{\tau_1}{2n \tan \alpha} \left(1 - \left(\frac{H_1}{H_0}\right)^{2n}\right) = \frac{3.6 \times 10^5}{0.66 \tan 3°} \left(1 - \left(\frac{0.5}{2.07}\right)^{0.66}\right)$$

$$= 6.3 \text{ MN/m}^2$$

$$P_E = 1/2\sigma_1 \left(1 - \left(\frac{H_1}{H_0}\right)^2\right) = 1/2 \times 7.5 \left(1 - \left(\frac{0.5}{2.07}\right)^2\right)$$

$$= 3.53 \text{ MN/m}^2$$

Parallel Section,

$$P_s = \frac{2L\tau}{H} = \frac{2 \times 6 \times 10^{-3} \times 3.6 \times 10^5}{0.5 \times 10^{-3}} = 8.6 \text{ MN/m}^2$$

Total Pressure loss = 22.1 MN/m^2

At exit, $\dot{\gamma} = 1910$ So $\tau = 3.6 \times 10^5$ N/m^2, $G = 5 \times 10^4$ N/m^2

So $\gamma_R = \frac{36}{.5} = 7.2 \rightarrow B_H = 2.43, B_T = 1.56$

So wall thickness = 1.21 mm, diameter = 62.4 mm

(5.19) Assuming no swelling at the die exit

$$\dot{\gamma} = \frac{6V}{H} = \frac{6 \times 15 \times 10^{-3}}{0.75 \times 10^{-3}} = 120 \text{ s}^{-1}$$

From the flow curves, $\tau = 1.3 \times 10^5$ N/m^2 $G = 3 \times 10^4$ N/m^2
So $\gamma_R = \tau/G = 4.4$ and so $B_H = 1.85, B_T = 1.36$

$$\therefore \quad \text{New } H = \frac{0.75 \times 10^{-3}}{1.85} = 0.41 \times 10^{-3} \text{ m}$$

$$\text{New } V = 15 \times 10^{-3} \times 1.85 \times 1.36 = 37.7 \times 10^{-3} \text{ m/s}$$

So, $$\dot{\gamma} = \frac{6 \times 37.7 \times 10^{-3}}{0.41 \times 10^{-3}} = 552 \text{ s}^{-1}$$

This gives $\tau = 2.2 \times 10^5$ N/m^2, $G = 4 \times 10^4$ N/m^2

So, $\gamma_R = 5.5$ and $B_H = 2.08, B_T = 1.455$

$$\text{New } H = \frac{0.75 \times 1^{-3}}{2.08} = 0.36 \times 10^{-3} \text{ m}$$

$$\text{New } V = 15 \times 2.08 \times 1.455 \times 10^{-3} = 45.39 \times 10^{-3} \text{ m/s}$$

So, $$\dot{\gamma} = \frac{6 \times 45.39}{0.36} = 756.6 \text{ s}^{-1}$$

And so on until the iteration converges, at which point

$$H = 0.405 \times 10^{-3} \text{ m}, \quad D = 29.2 \text{ mm}$$

(5.20) $$\dot{\gamma} = \frac{6Q}{TH_1} = \frac{6 \times 100 \times 10^{-6}}{\pi\left(250 \times 10^{-3}\right)\left(1 \times 10^{-3}\right)^2} = 764 \text{ s}^{-1}$$

so from flow curves $\tau = 2.4 \times 10^5 \text{ N/m}^2$

$$P_{s_1} = \frac{2L\tau}{H} = \frac{2 \times 5 \times 2.4 \times 10^5}{1} = 2.4 \text{ MN/m}^2$$

$$H_3 = H_2 + 2L\tan\alpha \quad (\text{where } \alpha = 1/2 \text{ angle})$$

$$H_3 = 1 + 2(12)\tan(1)^\circ = 1.42 \text{ mm}$$

At $$B_2, \dot{\gamma} = 764 \text{ s}^{-1}, \dot{\varepsilon} = 1/3\dot{\gamma}\tan\alpha$$

$$= 1/3(764)\tan(1) = 4.45 \text{ s}^{-1}$$

So

$$P_{s_2} = \frac{\tau_2}{2n\tan\alpha}\left[1 - \left(\frac{H_2}{H_3}\right)^{2n}\right]$$

$$= \frac{2.4 \times 10^5}{2(0.33)\tan(1)}\left[1 - \left(\frac{1}{1.42}\right)^{0.66}\right]$$

$$= 4.3 \text{ MN/m}^2$$

$$P_{E_2} = \frac{\sigma_2}{2}\left[1 - \left(\frac{H_2}{H_3}\right)^2\right]$$

$$= \frac{1 \times 10^6}{2}\left[1 - \left(\frac{1}{1.42}\right)^2\right]$$

$$= 0.25 \text{ MN/m}^2$$

Total Pressure loss $= 6.95 \text{ MN/m}^2$

At die exit, $$\dot{\gamma} = \frac{6Q}{TH^2} = \frac{6 \times 100 \times 10^{-6}}{\pi(250 \times 10^{-3}) \times (1 \times 10^{-3})^2}$$

$$= 764 \text{ s}^{-1}$$

From flow curves, $\tau = 2.4 \times 10^5 \text{ N/m}^2$

$$G = 4.1 \times 10^4 \text{ N/m}^2$$

$$\gamma_R = \frac{\tau}{G} = \frac{2.4 \times 10^5}{4.1 \times 10^4} = 5.85$$

From swelling ratio curves

$$B_{SH} = 2.19 \quad \text{and} \quad B_{ST} = 1.48$$

So   Bubble thickness $= 1 \times 2.19 = 2.19$ mm

Bubble diameter $= 250 \times 1.48 = 370$ mm

(5.21)

$$\dot\gamma_1 = \frac{6Q}{TH^2} = \frac{6 \times 4 \times 10^{-6}}{10^{-2}(2 \times 10^{-3})^2}$$

$$= 600 \text{ s}^{-1}$$

$$\dot\gamma_4 = 6667 \text{ s}^{-1}$$

$$\dot\varepsilon_2 = 1/3\dot\gamma_2 \tan\alpha$$

$$= 1/3(600)\tan 5°$$

$$= 17.5 \text{ s}^{-1}$$

$$\dot\varepsilon_3 = 194 \text{ s}^{-1}$$

$$P_0 = \frac{4\dot\gamma(\eta\lambda)^{1/2}}{(3n+1)} = \frac{4 \times 600(4 \times 10^2 \times 2.25 \times 10^5)^{1/2}}{1.99} = 11.4 \text{ MN/m}^2$$

$$P_{s_1} = \frac{2L\tau}{H} = \frac{2 \times 10^{-2} \times 2.4 \times 10^5}{2 \times 10^{-3}} = 2.4 \text{ MN/m}^2$$

$$P_{s_2} = \frac{\tau_3}{2n\tan\alpha}\left(1 - \left(\frac{H_3}{H_2}\right)^{2n}\right) = \frac{5.7 \times 10^5}{0.66\tan 5}\left(1 - \left(\frac{0.6}{2}\right)^{0.66}\right) = 5.4 \text{ MN/m}^2$$

$$P_E = 1/2\sigma_3\left(1 - \left(\frac{H_3}{H_2}\right)^2\right) = 1/2(43.6 \times 10^6)\left(1 - \left(\frac{0.6}{2}\right)^2\right) = 19.8 \text{ MN/m}^2$$

$$P_{s_3} = \frac{2L\tau}{H} = \frac{2 \times 10 \times 10^{-3} \times 5.7 \times 10^5}{0.6 \times 10^{-3}} = 19 \text{ MN/m}^2$$

Total pressure loss = 58 MN/m^2

(5.22) (a) As shown in the text, the flow rate of a power law fluid in a circular section is given by

$$Q = \left(\frac{n+1}{3n+1}\right)\pi R^2 \left(\frac{n}{n+1}\right)\left(\frac{P}{2\ell\eta_0}\right)^{1/n} R^{(n+1)/n} \qquad (1)$$

Now for a volume flow rate, Q, the volume in any increment of time, dt, is Qdt and this is equal to $\pi R^2 d\ell$

$$Qdt = \pi R^2 d\ell$$

So from (1)

$$\pi R^2 d\ell = \left(\frac{n}{3n+1}\right)\pi R^2 \left(\frac{P}{2\ell\eta_0}\right)^{1/n} R^{(n+1)/n} dt$$

$$\ell^{1/n} d\ell = \left(\frac{n}{3n+1}\right)\left(\frac{P}{2\eta_0}\right)^{1/n} R^{(n+1)/n} dt$$

$$\frac{n\ell}{(n+1)} = \left(\frac{n}{3n+1}\right)\left(\frac{P}{2\eta_0}\right)^{1/n} R^{(n+1)/n} t$$

So
$$\ell = \left(\frac{n+1}{3n+1}\right)^{n/(n+1)} \left(\frac{P}{2\eta_0}\right)^{1/(n+1)} R t^{n/(n+1)}$$

(b) If the cross-section is freezing-off then the radius of the channel will be changing with time and is given by

$$r = R\left(1 - \left(\frac{t}{t_f}\right)^{1/3}\right)$$

So since
$$\frac{d\ell}{dt} = \left(\frac{n}{3n+1}\right)\left(\frac{P}{2\ell\eta_0}\right)^{1/n} r^{(n+1)/n}$$

ie
$$\frac{d\ell}{dt} = \left(\frac{n}{3n+1}\right)\left(\frac{P}{2\ell\eta_0}\right)^{1/n}\left(R\left(1 - \left(\frac{t}{t_f}\right)^{1/3}\right)\right)^{(n+1)/n}$$

So
$$\ell^{1/n}d\ell = \left(\frac{n}{3n+1}\right)\left(\frac{P}{2\ell\eta_0}\right)^{1/n} R^{(n+1)/n}\left(\left(1 - \left(\frac{t}{t_f}\right)^{1/3}\right)\right)^{(n+1)/n} dt$$

$$\frac{n\ell^{(n+1)/n}}{(n+1)} = \left(\frac{n}{3n+1}\right)\left(\frac{P}{2\eta_0}\right)^{1/n} R^{(n+1)/n}\left(6t_f\left(\frac{n}{4n+1}\right)\left(\frac{n}{3n+1}\right)\left(\frac{n}{2n+1}\right)\right)$$

$$\ell = \left(\frac{P}{2\eta_0}\right)^{1/(n+1)} R\left(6t_f\left(\frac{n}{4n+1}\right)\left(\frac{n(n+1)}{(3n+1)^2}\right)\left(\frac{n}{2n+1}\right)\right)^{n/(n+1)}$$

(5.23) From the equations in the previous example

$$\frac{\ell_f}{\ell} = \left(6\left(\frac{n}{4n+1}\right)\left(\frac{n}{3n+1}\right)\left(\frac{n}{2n+1}\right)\right)^{n/(n+1)}$$

For a Newtonian fluid, $n = 1$ so $\ell_f/\ell = 0.316$

For a Non-Newtonian fluid, $n = 0.3 \ell_f/\ell = 0.424$

(5.24) (a) As shown in the previous examples, if the pressure is held constant

$$\ell = \left(\frac{n+1}{3n+1}\right)^{n/(n+1)} \left(\frac{P}{2\eta_0}\right)^{1/(n+1)} R t^{n/n+1)}$$

but
$$Q = \left(\frac{n+1}{3n+1}\right)\pi R^2 \left(\frac{n}{n+1}\right)\left(\frac{P}{2l\eta_0}\right)^{1/n} R^{(n+1)/n}$$

So substituting for ℓ gives

$$Q = \left(\frac{n+1}{3n+1}\right)^{n/(n+1)}\pi R^3 \left(\frac{n}{n+1}\right)\left(\frac{P}{2\eta_0 t}\right)^{1/(n+1)}$$

$$= \left(\frac{1.3}{1.9}\right)^{0.3/1.3}\pi(5 \times 10^{-3})^3 \left(\frac{0.3}{1.3}\right)\left(\frac{140 \times 10^6}{2 \times 10^4 t}\right)^{1/1.3} = \frac{7.53 \times 10^{-5}}{t^{0.769}}$$

(b) As shown in the text for the flow of a power law fluid in a capillary

$$Q = \left(\frac{n+1}{3n+1}\right)\left(\frac{\frac{dR}{d\ell}}{2\eta_0}\right)^{1/n} R^{(n+1)/n} \pi R^2 \left(\frac{n}{n+1}\right)$$

$$2\eta_0 Q^n = \left(\frac{n}{3n+1}\right)^n \left(\frac{dP}{d\ell}\right) \pi^n R^{3n+1}$$

but

$$Qdt = \pi R^2 d\ell \quad \text{So} \quad d\ell = (Qdt)/\pi R^2$$

$$\therefore \quad 2\eta_0 Q^n = \left(\frac{\pi n}{3n+1}\right)^n \frac{dP}{Qdt} \cdot \pi R^2 \cdot R^{3n+1}$$

$$2\eta_0 Q^n dt = \left(\frac{\pi n}{3n+1}\right)^n \pi R^{3n+3} \cdot dP$$

which may be integrated to give

$$P = \frac{2\eta_0 Q^{n+1}}{\pi R^{3n+3}}\left(\frac{3n+1}{\pi n}\right)^n t = \frac{2 \times 10^4 (7 \times 10^{-5})^{1.3}}{\pi (5 \times 10^{-3})^{3.9}}\left(\frac{1.9}{0.3\pi}\right)^{0.3} t$$

$$= (29.36t) \text{ MN/m}^2$$

The variations of Q and P in each case are shown in the attached graph.

(5.25) Volume of cavity $= 50 \times 10 \times 3 \times 10^{-9}$ m³

So time to fill this $= 1.5$ seconds

Gate: volume $= 2 \times 4 \times 0.6 \times 10^{-9}$ m³

So time to pass through $= 4.8 \times 10^{-3}$ seconds

$$\dot{\gamma} = \frac{6Q}{TH^2} = \frac{6 \times 10^{-6} \times 10^9}{4 \times 2^2} = 375 \text{ s}^{-1}$$

From flow curves, $\tau = 2.1 \times 10^5$ N/m²

$$\therefore \qquad P = \frac{2L\tau}{H} = \frac{2 \times 0.6 \times 2.1 \times 10^5}{2} = 0.13 \text{ MN/m}^2$$

Runner: volume $= \pi(4)^2 \times 20 \times 10^{-9}$ m³

time to fill this $= 1$ second

$$\dot{\gamma} = \frac{4Q}{\pi R^3} = \frac{4 \times 1 \times 10^{-6} \times 10^9}{\pi(4)^3} = 20 \text{ s}^{-1}$$

So $\qquad \tau = 7.6 \times 10^4$ N/m² $\rightarrow P = \dfrac{2 \times 20 \times 7.6 \times 10^4}{5} = 0.6 \text{ MN/m}^2$

Sprue, etc: time to pass through $= 1$ second

$\qquad\qquad\qquad$ pressure loss $= 0.6$ MN/m²

Total time to fill $= 1 + 1 + 1.5 = 3.5$ seconds

Now, as shown in previous example, for a constant flow rate situation the pressure build-up in the machine is given by

$$P = \frac{2\eta_0 Q^{n+1}}{\pi R^{3n+3}} \left[\frac{3n+1}{\pi n} \right]^n t$$

For P_E, $n = 0.33$ and $\eta_0 = 3 \times 10^4$ (from flow curves)

$$P = \frac{2 \times 3 \times 10^4 (1 \times 10^{-6})1.33}{\pi(4 \times 10^{-3})^{3.99}} \left[\frac{1.99}{0.33\pi} \right]^{0.33} t = 1.34 \times 10^6 t$$

Therefore after 3.5 seconds $P = 4.7$ MN/m².

But Pressures losses $= (0.6 + 0.6 + 0.13) = 1.33$ MN/m²

So packing pressure $= 3.37$ MN/m²

(5.26) The temperature gradient, $\Delta T = \dfrac{T_t - T_m}{T_i - T_m}$

$$\Delta T = \frac{70 - 20}{190 - 20} = 0.29$$

From Fig. 5.23, $\qquad\qquad F_0 = 0.29 = \dfrac{\alpha t}{x^2}$

$$t = \frac{0.29 \times (1.5 \times 10^{-3})^2}{1 \times 10^{-7}} = 6 \text{ s}$$

Since the haul-off speed is 0.4 m/s, the water bath would need to be at least (0.4×6) m long ie 2.4 m.

(5.27) Using the equation derived in the text:

$$\frac{L}{H} = \frac{1}{2} \left(\frac{P}{\eta_0}\right)^{\frac{1}{n+1}} \left[6t_f \left(\frac{n}{4n+1}\right) \left(\frac{n}{3n+1}\right) \left(\frac{n(n+1)^2}{(2n+1)}\right)\right]^{\frac{n}{n+1}}$$

where the freeze-off time may be calculated from

$$t_f = \frac{4x^2}{\pi^2 \alpha} \ln\left[\frac{4}{\pi}\left(\frac{T_m - T_M}{T_f - T_M}\right)\right]$$

Using the data given, the flow ratios will be

Material	Flow Ratio (L/H)
LDPE	240
polypropylene	220
polystyrene	165
PVC	60
POM	121
acrylic	86
polycarbonate	32
nylon 66	30
ABS	123

Note the η_0 and n data from Table 5.2 has been used even though the temperatures quoted are not always the same as the moulding temperature.

(5.28)

$$h = B_{ST}^3 h_d \left(\frac{D_d}{D_m}\right)$$

So

$$B_{ST} = \sqrt[3]{\frac{2.5 \times 100}{2 \times 40}} = 1.46$$

From Fig. 5.11 at $B_{ST} = 1.46$, $\gamma_R = 5.5 = \tau/G$

From the flow curves the value of shear stress to give $\tau/G = 5.5$ is $\tau = 2.25 \times 10^5$ N/m^2 (also $G = 4.1 \times 10^4$ N/m^2). From the flow curves at $\tau = 2.25 \times 10^5$ N/m^2,

$$\dot{\gamma} = 620 = \frac{6Q}{TH^2} \rightarrow Q = 52 \times 10^{-6} \text{ m}^3/\text{s} = 147 \text{ kg/hour}$$

$$P = \frac{2\sigma h}{D_m} = \frac{2 \times 2.5 \times 4 \times 10^6}{100} = 0.2 \times 10^6 \text{ N/m}^2$$

So the suggested pressure would cause melt fracture.

(5.29) Stress $= \rho L = 730 \times 9.81 \times 0.4 = 2.86 \times 10^4$ N/m^2

From flow curves $\lambda = 2.25 \times 10^5$ and $E = 4.6 \times 10^4$

So characteristic time $= \lambda/E = 4.95$

Also $\varepsilon = \rho L \left(\dfrac{1}{2E} + \dfrac{t}{\lambda} \right) = 730 \times 9.81 \times 0.4 \left[\dfrac{1}{2 \times 4.6 \times 10^4} + \dfrac{1}{2.25 \times 10^5} \right] = 0.044$

So $\qquad\qquad\qquad \delta L = 0.044 \times 0.4 = 0.017 \text{ m} = 17 \text{ mm}$

(5.30) For the overall change of temperature and pressure

$$\eta/\eta_R = 10^{A\Delta T + B\Delta P}$$

For acrylic, $\qquad\qquad \eta/\eta_R = 10^{[(28.32\times-40)+(9.54\times-50)]10^{-3}} = 4.53$

Similarly for the others $\eta/\eta_R = 0.95, 1.415, 2.03$ and 1.28.

Also for $\qquad\qquad\qquad \eta/\eta_R = 1, A\Delta T = -B\Delta P$

For acrylic $\qquad\qquad \dfrac{\Delta P}{\Delta T} = \dfrac{28.32}{9.54} = 2.97 \text{ MN/m}^2 {}^\circ\text{C}$

Similarly for the others, $(\Delta P)/(\Delta T) = 1.17, 1.87, 3.07$ and 1.93

Index